INFOR

Wiley Essentials in
OPERATIONAL RESEARCH AND MANAGEMENT SCIENCE

INFORMS Analytics Body of Knowledge

Edited by James J. Cochran

WILEY

This edition first published 2019

Registered Office
John Wiley & Sons, Inc., 111 River Street, Hoboken, NJ 07030, USA

Editorial Office
111 River Street, Hoboken, NJ 07030, USA

For details of our global editorial offices, customer services, and more information about Wiley products visit us at www.wiley.com.

Wiley also publishes its books in a variety of electronic formats and by print-on-demand. Some content that appears in standard print versions of this book may not be available in other formats.

Library of Congress Cataloging-in-Publication Data

ISBN: 9781119483212

Set in 10/12 pt WarnockPro-Regular by Thomson Digital, Noida, India

V10004499_091418

Contents

Preface

A body of knowledge (BOK) is a comprehensive compilation of the core concepts and skills with which a professional in a specific discipline should be familiar. BOKs are generally produced and maintained by members of an academic society or professional association, and a BOK serves as the means by which the academic society or professional association communicates its vision, both internally and externally.

The broad objective of this BOK, entitled *INFORMS Analytics Body of Knowledge* (*ABOK*), is to provide those interested in the development and application of the tools of analytics with an understanding of what analytics is and how analytics can be used to solve complex problems, make better decisions, and formulate more effective strategies. *ABOK* is produced by the Institute for Operations Research and the Management Sciences INFORMS[1] and represents the perspectives of some of the organization's most respected members on a wide variety of analytics-related topics.

We use INFORMS' definition of analytics–*the scientific process of transforming data into insight for making better decisions*–as the foundation for this book. But each chapter also reflects the unique insights and experiences of the chapter's author(s). This is intentional; analytics is a nascent, diverse, and complex discipline (or perhaps a collection of disciplines) that is defined somewhat differently by various practitioners and organizations. The various perspectives within this book will provide the reader with a better understanding of this dynamic field.

This book is a valuable resource for professionals in business and industry who are looking for ways to fully and effectively integrate analytics into their organizations' problem-solving, decision-making, and strategic planning.

1 INFORMS (www.informs.org) is the largest international association of operations research (O.R.) and analytics professionals and students. INFORMS provides unique networking and learning opportunities for individual professionals, and organizations of all types and sizes, to better understand and use O.R. and analytics tools and methods to transform strategic visions and achieve better outcomes.

Instructors who are developing or revising/modernizing analytics courses and programs will also find *ABOK*'s chapters illuminating.

ABOK's chapters are written by colleagues recognized for their expertise in various areas of analytics (Philip T. Keenan, Jonathan H. Owen, and Kathryn Schumacher of General Motors; Karl Kempf of Intel Corporation; Thomas H. Davenport of Babson College; Brian Downs of Accenture LLC; Mary E. Helander of the IBM T.J. Watson Research Center; Gerald G. Brown and Samuel H. Huddleston of the Naval Postgraduate School; Arnie Greenland of the University of Maryland; Ramesh Sharda of Oklahoma State University and Pankush Kalgotra of Clark University). We have solicited input from colleagues in industry, government, and academia, and each chapter has been peer-reviewed by respected colleagues in analytics in order to ensure that *ABOK* will be a useful practical resource.

An appendix on writing and teaching analytics with cases is also included in this book. This appendix is included to support the development of courses in analytics and foster the case approach in analytics courses. It is also intended to encourage colleagues in business and industry to work with academicians to develop and publish analytics cases for use in analytics courses as a means to improve student understanding and appreciation of the importance and relevance of this discipline. Although *ABOK* is not intended to be a comprehensive source for preparation for INFORMS Certified Analytics Professional (CAP®) and Associate Certified Analytics Professional (aCAP™) examinations, its contents will be very helpful to those preparing for these examinations.

Each chapter and the appendix also feature relevant portions of interviews with other well-respected practitioners and instructors of analytics. These interviews were conducted by the INFORMS' *Analytics Body of Knowledge Committee* and provided by Eric Stephens of the Vanderbilt University Medical Center; Alan Taber of Lockheed Martin Missiles and Fire Control; Jeff Camm of Wake Forest University; Katya Scheinberg of Lehigh University; Harrison Schramm of the United States Navy (retired); Greta Roberts of Talent Analytics; Susan Martonosi of Harvey Mudd College; Russell Walker of Northwestern University's Kellogg School of Management; Robert Clark of RTI International; Cole Smith of Clemson University; and Matt Drake of Duquesne University.

Major undertakings, such as a body of knowledge, can only succeed if all members of a large and talented team work toward a common objective, and *ABOK* is certainly no exception to this rule. Several colleagues from industry and academia provided detailed reviews of the chapters. Tasha Inniss of INFORMS; Cole Smith of Clemson University; Manoj Chari of SAS; J. Antonio Carbajal of Turner Broadcasting System, Inc.; Ashley Cowall of Booz Allen Hamilton; Graciela Chadwick of Chick-fil-A; Nick Wzientek of Rocky Mountain Resources; Linda Schumacher of ABB, Inc.; Alan Taber of Lockheed Martin Missiles and Fire Control; Susan Martonosi of Harvey Mudd College; Sean

MacDermant of International Paper; and Matt Drake of Duquesne University each generously reviewed chapters and provided valuable input.

INFORMS' *Analytics Body of Knowledge* Committee, which is chaired by Terry Harrison (Penn State University) and includes Michael Rappa (North Carolina State University), Jim Williams (FICO), Alan Briggs (Elder Research), Eric Stephens (Vanderbilt University Medical Center), Alan Taber (Lockheed Martin Missiles and Fire Control), Jeanne Harris (Columbia University), and Layne Morrison (IBM), has provided valuable input. Lisa Greene and Bob Clark of RTI, International were instrumental in executing the interviews and advised on several issues.

Other members of INFORMS who provided advice and feedback include Donald Baillie (Anzac Finance Solutions), James Taylor (Decision Management Solutions), Irv Lustig (Princeton Consultants), Harrison Schramm (retired Naval officer), Thomas Reid (Booz Allen Hamilton), Charley Tichenor (Marymount University), Selene Crosby (Tesoro Companies, Inc.), Jack Levis (UPS), Anne Robinson (Verizon Wireless), Mike Gorman (University of Dayton), Glenn Wegryn (independent consultant), Ira Lustig (Princeton Consultants), and Brenda Dietrich (Cornell University). Several members of INFORMS' staff, including Jeff Cohen, Bill Griffin, Tasha Inniss, Jan paul Miller, Melissa Moore, and Louise Wehrle, have made vital contributions to *ABOK*. Danielle LaCourciere, Mindy Okura-Marszycki, Lauren Olesky, Kathleen Pagliaro, and Andrew Prince of John Wiley & Sons, Inc. have also made critical contributions.

I am very excited about what *ABOK* can do for the analytics community, and I am confident you will share my enthusiasm once you have read *ABOK*. This is a living resource that will be updated and revised in the future to ensure it remains current, timely, and cutting-edge, and I encourage you to contact me with suggestions for how to improve it.

James J. Cochran, PhD

Associate Dean for Research, Professor of Applied Statistics, and the Rogers-Spivey Faculty Fellow
Culverhouse College of Business
The University of Alabama

List of Contributors

Gerald G. Brown
Operations Research Department
Naval Postgraduate School
Monterey, CA
USA

James J. Cochran
Culverhouse College of Business
The University of Alabama
Tuscaloosa, AL
USA

Thomas H. Davenport
Technology, Operations, and Information Management
Babson College
Wellesley, MA
USA

Brian T. Downs
Accenture Digital
Data Science Center of Excellence
Dallas, TX
USA

Arnie Greenland
Robert H. Smith School of Business
University of Maryland
College Park, MD
USA

Mary E. Helander
Data Science Department
IBM T. J. Watson Research Center
Yorktown Heights, NY
USA

Samuel H. Huddleston
Operations Research Department
Naval Postgraduate School
Monterey, CA
USA

Pankush Kalgotra
Graduate School of Management
Clark University
Worcester, MA
USA

Philip T. Keenan
General Motors
Global Research & Development
Warren, MI
USA

Karl G. Kempf
Decision Engineering
Intel Corporation
Chandler, AZ
USA

Jonathan H. Owen
General Motors
Global Research & Development
Warren, MI
USA

Kathryn Schumacher
General Motors
Global Research & Development
Warren, MI
USA

Ramesh Sharda
Spears School of Business
Oklahoma State University
Stillwater, OK
USA

1

Introduction to Analytics

Philip T. Keenan, Jonathan H. Owen, and Kathryn Schumacher

General Motors, Global Research & Development, Warren, MI, USA

1.1 Introduction

We all want to make a difference. We all want our work to enrich the world. As analytics professionals, we are fortunate–this is our time! We live in a world of pervasive data and ubiquitous, powerful computation. This convergence has inspired new applications and accelerated the development of novel analytic techniques and tools, while breathing new life into decades-old approaches that were previously too data- or computation-intensive to be of practical value. The potential for analytics to have an impact has been a call to action for organizations of all types and sizes. Companies are creating new C-level positions and departments to grow analytic capability. A torrent of new start-ups have formed to sell analytics products and services. Even governments have created new high-profile offices to leverage analytics. These changes have driven a surge in demand for analytics professionals, and universities are creating departments, curricula, and new program offerings to fill the gap.

But what exactly do we mean when we say "analytics"? The term is widely used, but has vastly different meanings to different people and communities. A number of well-established disciplines, including statistics, operations research, economics, computer science, industrial engineering, and mathematics, have some claim to "analytics" and interpret it to have specialized meaning within their domains. The popular usage of the term is often comingled with other widely used but equally overloaded terms such as "big data," "data science," "machine learning," "artificial intelligence," and "cognitive computing." As a result, this seemingly innocuous term has led to much confusion over the last decade as people using the same language often talk right past each other. In the authors' own experience, frustration at all levels of an organization is inevitable when well-intentioned and intelligent people believe they have a shared

INFORMS Analytics Body of Knowledge, First Edition. Edited by James J. Cochran.

understanding–on a new project initiative, for example–only to discover weeks or months later that there was a fundamental misunderstanding of what work was to be performed or insights delivered.

In a 2016 article intended to reduce some of this confusion, Robert Rose identified three main usages of the term "analytics" [1]:

1) As a synonym for metrics or summary statistics
2) As a synonym for "data science" (another overloaded term)
3) As a very general term to represent a quantitative approach to organizational decision-making

Our use of the term is closest to the last of these; we consider analytics broadly as a process by which a team of people helps an organization make *better decisions* (the objective) through the *analysis of data* (the activity). This chapter gives a brief, high-level introduction to the subject. We first describe a conceptual framework for analytics, and define three primary categories of analytics (descriptive, predictive, and prescriptive). We then discuss considerations for applying analytics within an organization, and briefly discuss the ethical implications of using analytics. Subsequent chapters dive more deeply into each component of the process of applying analytics, including developing a request for a new project, building a cross-functional team, collecting data, analyzing data with a wide variety of mathematical and statistical methods, and communicating results back to the client.

Interview With Alan Taber

Alan Taber, System Engineer with Lockheed Martin Missiles and Fire Control, defines analytics in the following way:

Analytics is both a mindset and a process. The mindset is that instead of simply reacting to what you perceive your environment to be that you gather data understanding the limits and bounds of that data. You feed it into a model. It can be a very detailed model or a simple model about how situations evolve over time if you do take options A or B or C, or some combination thereof, and then you test that hypothesis. You have the continual feedback loop to say if what you're doing makes sense and also keep an eye on your surroundings because what may have made sense a year ago or a month ago may no longer make sense. That's the mindset, to always be paying attention rather than running on autopilot.

The process is to make sure you understand the root problem, figure out if you can frame that as a problem that's amenable to being solved with data, figure out your data sources, and don't limit yourself to the data you have on hand and know how to collect. If you need a different data set, go get it. Once you have your data and can run your test, do that. Over and under and around all that, you're working with your stakeholders so that when you deploy people are familiar enough with what you're doing that they're willing to try it out rather than saying, "I don't understand the model and therefore I'm busy, I don't have time to learn, I'm not interested." If you are overwhelming people with information but not helping them actually solve the problems that they perceive they have, you simply will not get very far. You will have wasted all your time. So that's the mindset and that's the process.

This is an excerpt from one of a series of interviews with analytics professionals and educators commissioned by the *INFORMS Analytics Body of Knowledge* Committee.

1.2 Conceptual Framework

As shown in Figure 1.1, the generic analytics process can be viewed as a continuous cycle where the analysis of data produces insights that inform better decision-making. We use this simple figure to highlight two fundamentally different approaches to analytics: *data-centric* and *decision-centric.*

1.2.1 Data-Centric Analytics

The philosophy behind *data-centric* analysis is to "let the data speak freely." Working under this philosophy generally involves pulling together as much relevant data as possible, analyzing that data to identify patterns that lead to insight, and serving up those insights to a decision-maker who (hopefully) will make better informed decisions. As shown in Figure 1.2, this follows the natural (clockwise) flow of the analytics process.

Not surprisingly, the data-centric approach has gained popularity with the surge in "big data." Many of the analytic methodologies employed in this arena–including data mining and classification, machine learning, and artificial intelligence–increase in effectiveness with the volume of data available for analysis. Advocates believe that we are in a new "machine age" that is changing the landscape of business and the world [2–4]. Some argue that the data-centric "big data" paradigm is really about eliminating sampling error; they claim that we are

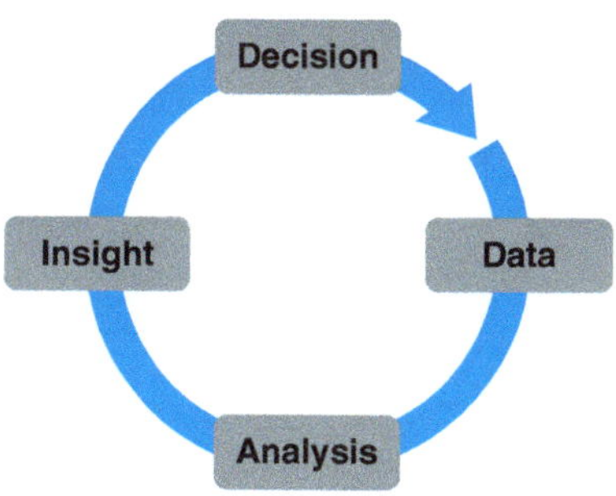

Figure 1.1 Simplified visual representation of the analytics process.

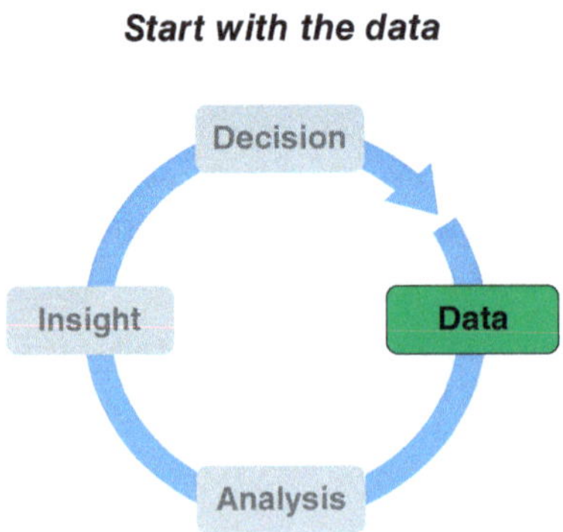

Figure 1.2 The *data-centric* approach starts with the data to surface insights.

no longer reliant on small samples since we have storage capacity to hold and computing power to process vast amounts of data [5]. Others have observed that the promised insights have not always materialized, and that the challenge is "to solve new problems and gain new answers–without making the same old statistical mistakes on a grander scale" [6].

1.2.2 Decision-Centric Analytics

Decision-centric analytics begins with an understanding of the *decision* that needs to be made and what *insights* would lead to better expected outcomes. Decision-centric models typically encapsulate subject matter expertise (SME) and codify domain knowledge in order to relate decision variables to the target objective. Data requirements are determined by the chosen analytical model; ideally these data already exist in a convenient form, but often they must be extracted from disparate sources or collected through new instrumentation or market research. As summarized in Figure 1.3, this approach starts with the final outcome–the decision–and works backward (counterclockwise) at each step to define and develop needed analysis and data resources.

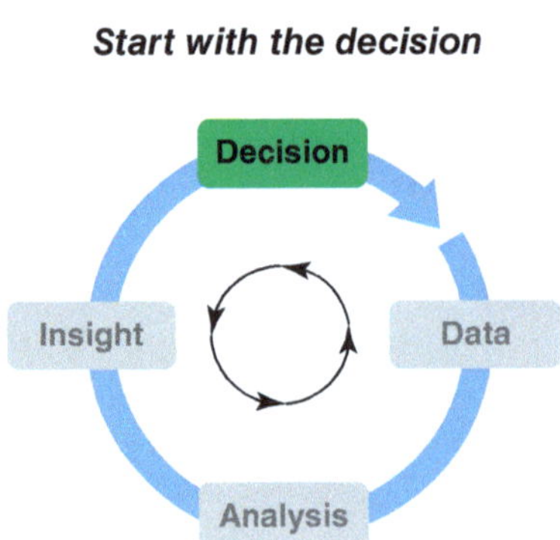

Figure 1.3 The *decision-centric* approach starts with the problem and works backward.

Decisions are often defined as an "irrevocable allocation of resources" [7]. Improving decision-making requires an understanding of the desired outcome (the objective), alternative actions (decision variables), and boundary conditions (constraints), but also the richer context of possible future conditions (scenarios). It also requires that we answer several softer questions: Who is making the decision? What is her or his scope of control and influence? What information is already available to the decision-maker(s) and where are the gaps? In a decision-centric approach, many of these questions are considered as part of upfront framing activities that look ahead toward operational implementation.

1.2.3 Combining Data- and Decision-Centric Approaches

Analytic practitioners and professional communities are often predisposed to either data-centric or decision-centric approaches. In the authors' view, this is attributable to different pedagogical perspectives and experiences. Given the centrality of computing and information technologies for handling large amounts of data, it is not surprising that many organizational IT functions are naturally aligned with a data-centric view. Business operations and the analytic teams that support them often have a natural affinity for decision-centric approaches that leverage their deep understanding of key problems and models that support improvements. Table 1.1 summarizes salient features of the two approaches.

Important opportunity arises from combining elements of the two approaches. There is undeniable potential to leverage increasingly pervasive data and computational power associated with data-centric analysis, but contextual knowledge and subject matter expertise provide needed guardrails so that the resulting insights are meaningful.

Acknowledging the natural tendencies of individuals or analytics organizations toward data- or decision-centric approaches may help practitioners to identify growth opportunities. For example, traditionally decision-centric organizations may benefit by expanding the amount of data used in their analyses, including unstructured data sources. Typically, data-centric groups may improve the fit and predictive power of their models by incorporating domain-specific expertise.

Evidence of the benefit of utilizing a combined approach is seen in recent movements to incorporate "thick data" into marketing analytics (see Refs [8,9], for example). Combining thick data, such as ethnographic studies or focus group responses (see Figure 1.5), with big data, such as transaction data, enables a more complete understanding of customers' preferences and behaviors. Decision-centric framing, domain knowledge, and deep subject matter expertise collectively provide scaffolding that helps big data insights take shape.

Table 1.1 Comparison of data-centric and decision-centric approaches.

	Data-centric analysis **(Data science, computer science)**	**Decision-centric analysis** **(Decision science, operations research)**
Mantra	"Start with the data"	"Start with the decision"
Philosophy	Leverage large amounts of data. Let the data "speak freely" by identifying patterns and revealing implicit (hidden) factor relationships	Leverage domain knowledge and subject matter expertise to model explicit variable relationships
Data	More is better, especially for "big data" applications (e.g., speech or image recognition)	Custom collection of curated data sets
Computing	High-performance computing is often price of entry. Potential need for specialized processors (e.g., GPUs, TPUs) for acceptable execution speeds, especially in contexts requiring real-time analysis	Desktop or server-based computing is typical. Trade-offs between potential benefits of leveraging high-performance computing versus added overhead in development and maintenance
Pros	• Increasingly automatable • Potential to extract weak signals from large, unstructured data sets	• Causal focus • Strategic value beyond historical observations
Cons	• Risk of conflating correlation with causation • Analysis inferences are limited by history • Noisy data with confounded effects	• Human subject matter expertise required • Cost of data acquisition can be high
Key disciplines	• Computer science • Data science • Machine learning and unstructured data mining • Artificial intelligence (AI), deep learning	• Management and decision sciences • Operations research • Mathematics • Classical statistics
Example applications	• Image classification • Speech recognition • Autonomous vehicle scene recognition	• Supply chain optimization • Scenario planning • New business model development

1.3 Categories of Analytics

A well-known and useful classification scheme for analytics was proposed by Lustig et al., at IBM [10]. Based on their experience with a variety of companies across a diverse set of industries, they defined three broad categories of analytics:

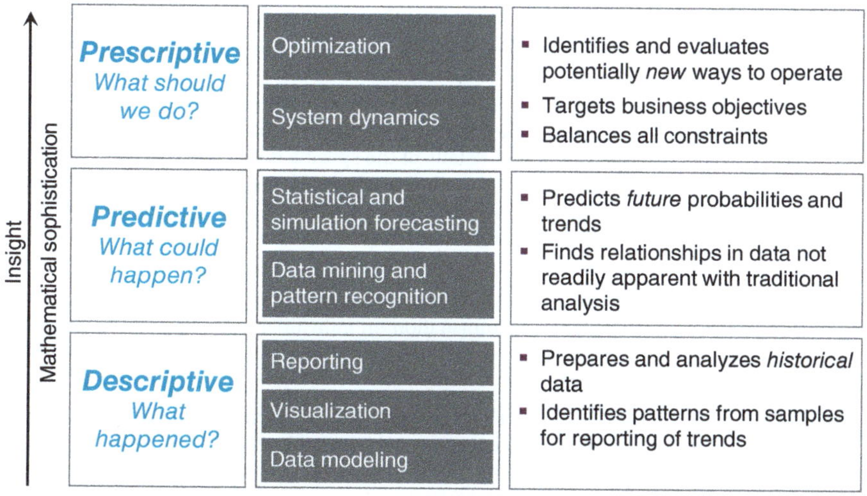

Figure 1.4 Categories of analytics.

descriptive, *predictive*, and *prescriptive*. As summarized in Figure 1.4, there is a natural progression in the level of insight provided–and potential value–as an organization moves from descriptive to predictive and ultimately to prescriptive analytics. Typically there is also a progression in the mathematical sophistication of the analysis techniques, as well as the organizational maturity required to absorb and act on resulting insights.

1.3.1 Descriptive Analytics

The purpose of descriptive analytics is to reveal and summarize facts about what has happened in the past or, in the case of real-time analysis, what is happening in the present. This is done by examining and synthesizing data collected from a variety of sources. Raw data are captured and recorded in source systems, eventually to be cleaned, retrieved, and normalized such that entities and relationships can be meaningfully understood. The audience for descriptive analytics is broad, potentially reaching all functions and levels of an organization. Descriptive analytics are at the heart of most business intelligence (BI) systems.

Data Modeling

Many organizations have access to vast quantities of data. Useful descriptive analytics generally involves processing the raw facts into higher level abstractions. Data scientists think in terms of *entities* and *relationships*. For example, a customer database might contain entities like "Household" and "Product," linked by relationships like "Purchased," with data elements

Table 1.2 Potential sources of data.

Source	Examples
Transaction data	Data associated with a transactional event. Example: a purchase transaction with details of the specific item purchased, where and when it was purchased, the price paid and any discounts applied, how the customer paid (e.g., cash, credit card, finance), and other contextually relevant data (e.g., inventory of other items for sale at the same time and location)
Customer data	Data associated with customers. Examples: detailed demographic or psychographic information on individuals and households, history of interactions (past purchases, Web site visits, customer service requests)
Sensor data	Data collected through electronic or mechanical instrumentation. Examples: web browser cookies tracking customer activity, electronic sensors monitoring weather conditions, airplane flight data recorder information
Public data	Open-source data from individuals, organizations, and governments. Example: aggregated census data
Unstructured	Data without known structure. Examples: text and images from social media, call center recordings, qualitative data from focus groups or ethnographic studies
Curated data	Data collected for a specific purpose with downstream analysis in mind. Examples: consumer surveys, designed market research experiments

including the demographics of the households and the price, cost and features of the products.

Sources of data can be highly varied (see Table 1.2 for examples), as can the size and information density of any given data set (see Figure 1.5). There is also high variability in the expense and effort required to collect different types of data. On one end of the spectrum, ethnographic studies require social scientists to spend many hours shopping with or interviewing individual customers, and thus the data are very carefully curated and very expensive to collect. On the other end of the spectrum, "data exhaust" is logged nearly for free, including data generated from smartphones and online activity [11]. Data exhaust is collected without a specific intended purpose and can be especially messy, so substantial cleanup effort is usually necessary before this type of data are usable.

Developing a data model that captures the structure and relationships among the different data elements is a fundamental task. Generic data models are often constructed to efficiently store ingested data, without specific analytic use cases in mind. Although such data models can be useful for general-purpose reporting and data exploration, purpose-built data models are typically needed for efficient

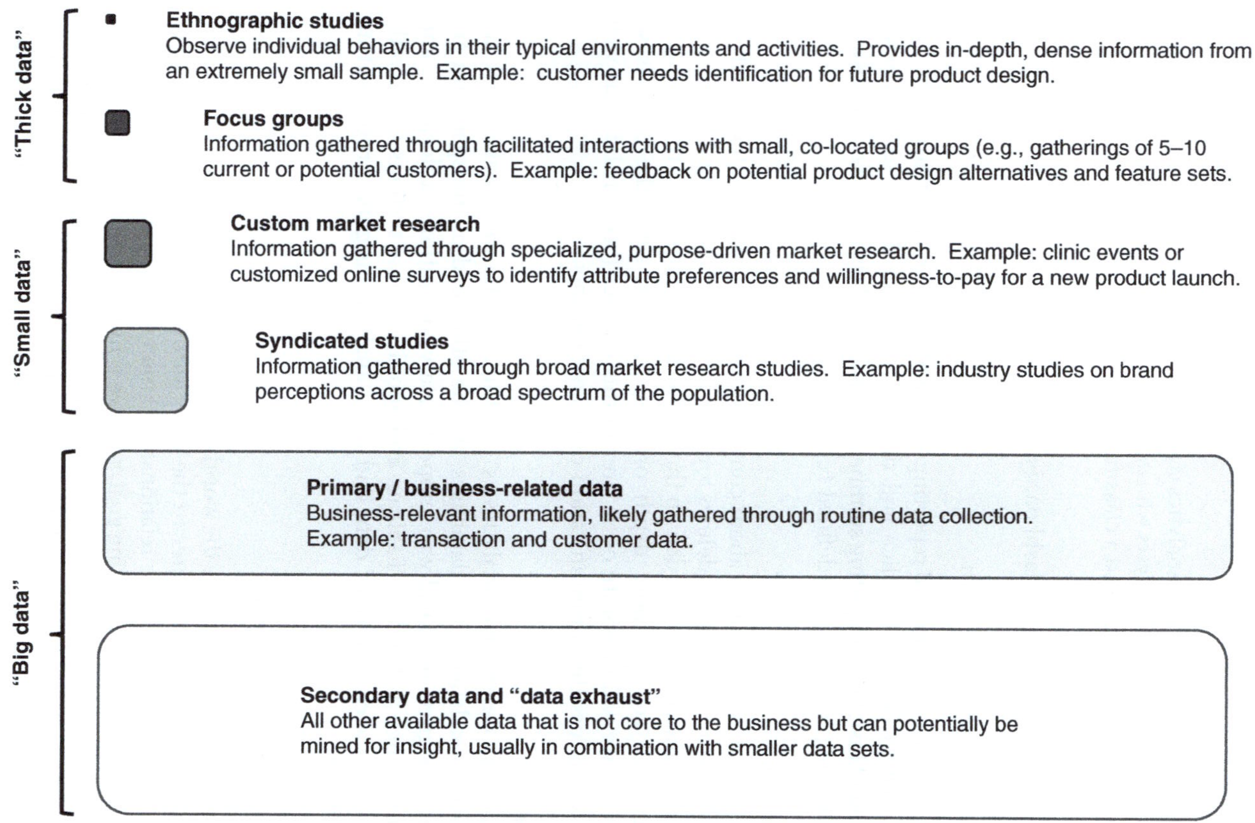

Figure 1.5 Illustration of variability in the size and information density of different data sets.

analysis. Depending on the size of the organization and the speed with which new data arrives, substantial IT support may be required to run systems that capture and record data, clean it, and store it in a warehouse or lake for eventual retrieval and analysis.

Reporting

The real value of descriptive analytics comes from putting access to this plethora of data into the hands of analysts who can use it to rapidly answer questions. To this end, Lustig et al. proposed a classification of descriptive analytics into three areas [10]:

1) Standard reporting and dashboards
2) Ad-hoc reporting
3) Analysis/query/drill-down

In our experience, standard reporting and dashboards are useful to a point, but users need to be able to "slice and dice" the data on the fly to gain more meaningful insights, computing summary statistics and visualizing comparisons without being limited to predefined reports.

Visualization

Descriptive analytics is often about communication, not math. Authors such as Tufte [12] provide useful guidelines for describing and visualizing data in ways that reduce the cognitive burden on those who must interpret the results. Later chapters will go into more depth on this; however, since the topic is so important, we will elaborate on it later in this chapter (Section 1.4.2) as well when we discuss the communication of project insights.

Software

Software for descriptive analytics is plentiful. At the most basic level are ordinary spreadsheets and databases. At the other end of the spectrum are systems designed specifically to support data visualization, exploration, and reporting–such as Cognos, Tableau, and Spotfire. These systems can greatly increase the accessibility of data and basic analytic insights throughout an organization.

1.3.2 Predictive Analytics

Descriptive analytics describe the world as it is (or as it recently was). In contrast, *predictive analytics* seek to forecast the likely future state of the world through a deeper understanding of the relationships among data inputs and outcomes. This is a much more demanding goal, so there is much more that can go wrong. Inexperienced analysts and leaders often imagine that once you have a good descriptive model, you can use it to make good forecasts. Not true! Statisticians have long understood that correlation does not imply causation. As a result, teams that wish to forecast the future need to use more sophisticated modeling approaches and follow more rigorous validation procedures if they want to have confidence that their forecasts make sense.

As a very simple example of the difference between descriptive and predictive analytics, consider television programs that cover the stock market. Every day, talking heads explain why the stock market behaved the way it did the previous day. But can any of them accurately forecast what the market will do tomorrow? Not a one. If they could, they would be billionaires living on a beach, not reading off a teleprompter in a TV studio. Hindsight may be 20–20, but foresight certainly is not.

Data Mining and Pattern Recognition

The starting point for predictive analytics is often mining data to identify meaningful relationships and patterns. As we work with increasingly large and diverse data sets, there is a growing opportunity to identify hidden relationships that relate disparate data. For example, clustering analysis might be used to segment customer populations into groups that go beyond simple demographic or psychographic characteristics. Or we might apply various machine learning techniques to identify objects and trajectories for autonomous vehicle scene recognition and navigation.

The set of available data mining techniques is highly varied, and practitioners need to be adept at selecting appropriate methods based on an understanding of the pros and cons of each within a given application context. Many methods are based on classical statistical models, often to classify populations into distinct groups (e.g., classification and regression trees) or to estimate the impact of a set of descriptor variables on a metric of interest (regression). Machine learning and artificial intelligence techniques can arguably answer a broader set of questions (e.g., image recognition), but trade the transparent simplicity of classical models for a harder-to-explain "black box" capable of representing more complex relationships. Regardless of the methodology, analysts must be alert to the danger of false positives. Given enough computer time and input data, one can *always* find some sort of "statistically significant" effect that is actually pure noise.[1]

Predictive Modeling, Simulation, and Forecasting

Predicting the future requires a model. Simply collecting and reporting data, or identifying interesting patterns about the past and present is not sufficient.

One of the simplest models assumes that the future will behave like the past; for obvious reasons, this is often referred to as a naive model. For an established company, sales next month will likely be similar to sales last month. However, leaders who request analytics projects generally want deeper insights than that!

---

1 The reader is encouraged to see https://imgs.xkcd.com/comics/significant.png for a lighthearted cartoon illustrating the dangers of false significance.

The next simplest model is trend extrapolation. If sales were 100 units in January, 110 in February, and 120 in March, it seems plausible to predict that they will be 130 in April and 140 in May. Projecting simple trends can be useful, but it is not always appropriate. Suppose you are selling tax preparation software; this forecast would be inaccurate, as sales in May will instead be close to zero, since most customers will have filed their taxes with the IRS by April 15. In this context, a more advanced model that "seasonally adjusts" the data would be appropriate.

More sophisticated models often include other explanatory variables in addition to time. For example, when trying to predict the number of vehicles the US automotive industry will sell next year, it is often helpful to consider macroeconomic data such as the unemployment rate, interest rates, and inflation. The automotive industry is cyclical–sales fall during recessions and rise during periods of economic expansion. Predicting the timing of the next recession can be almost as challenging as predicting the future course of the stock market. As a result, predictive models generally need to report ranges, or uncertainty bounds, rather than simple point forecasts. Unfortunately, many clients have difficulty consuming range estimates and prefer to pretend that point forecasts suffice. This is one of the many challenges the analytics practitioner faces when trying to communicate results in a form accessible to decision-makers.

Deciding what variables to include in a model can also be challenging. Leave out an important causal factor and the model's predictions may be seriously wrong. Including extraneous factors can also cause difficulties. For instance, classical regression models can fail if several input variables are closely correlated, an issue known as multicollinearity.

Analysts often attempt to assess the goodness of fit of their proposed model. For example, when fitting a regression model, most software packages report the "R-squared" metric, a measure of how closely the model matches the data. Analysts often construct a variety of models (perhaps using different subsets of variables in each) and pick the one with the highest R-squared. Unfortunately, this technique of "chasing R-squared" is not, in fact, a good approach–it can easily lead to overfitting, which in turn can lead to poor performance when predicting future values.

To avoid this pitfall, analysts can instead divide the data into a "training sample" used for fitting the model, and a "validation sample" used for assessing and comparing models after they have been fitted. Executed properly, this methodology can dramatically reduce the risk of overfitting, so it should be standard operating policy for all analysts whenever sufficient data are available.

Leveraging Expertise

There are a great many methodologies available for building predictive models. Frequentist statistical models have been used for over a century. Bayesian statistical models became widely used starting around 1995, when faster computers and algorithms made them computationally practical. Machine learning methods have become popular in recent decades, made possible by faster computers and larger data sets. Statistical and machine learning methods work well for analyzing a vast array of situations, but they tend to rely on the computer to *discover patterns* in the historical data and assume these patterns will repeat in the future. However, sometimes the future is different from the past. For example, when launching a new product, historical sales data are not available. How then to predict future sales?

Potential solutions have been developed for such cases, but they are substantially more complicated and time consuming (i.e., expensive) than methods that make use of existing data. For example, when launching a new product, one such approach is to perform primary market research to test how potential customers react to the new product.

In some situations, a practitioner has abundant knowledge of the structure of the real world, and incorporating that knowledge into the model building process can be extremely valuable. Simulation models are particularly useful in such situations. Simulation is based on the understanding of how some entities–individuals, components, or other actors–behave in isolation, and how their interactions lead to consequences under different scenarios. Simulation techniques can be classified based on what interacts and how the interactions occur. Table 1.3 summarizes key differences between three common types of simulation models: discrete event, agent-based, and system dynamics.

Table 1.3 Comparative summary of three common simulation models.

Discrete event simulation	Models a system using a central global mechanism, often a network, within which entities interact according to centrally specified rules at discrete points in time (events). Interactions are defined by standardized structures such as queues. Example: call center and discrete manufacturing operations analysis
Agent-based simulation	Models a system using autonomous agents (representing both individuals and collective groups), each with their own rules for behavior. Interactions are determined by domain-specific rules potentially based on the state of the agents involved and the overall state of the system. The overall system behavior emerges from the interactions of the agents. Example: flight simulation for a flock of birds
System dynamics	Models a system using stocks and flows. Interactions are defined by feedback loops and control policies. System dynamics is to agent-based simulation as thermodynamics is to molecular simulation, in that it aims to reduce the computational and cognitive burden through aggregation. Example: Bass diffusion model of the impact of advertising

Simulation models require a lot of effort to calibrate to observed history. However, because they model the underlying "physics" (e.g., microeconomics) of the situation, they can incorporate additional data from subject matter experts or market research. Simulation models can be used to evaluate "what-if" scenarios, a capability that is very useful to decision-makers, and is not possible with basic forecasting models.

1.3.3 Prescriptive Analytics

Prescriptive analytics seek to go further than forecasting a future state, to make actionable recommendations about what the decision-maker should do to achieve a particular objective, such as maximize profit. With descriptive and predictive analytics, the analytics team shoulders most of the burden of interpreting the results and developing recommendations for action. With prescriptive analytics, the computer helps with that process by evaluating a large number of potential alternative courses of action and reporting the best ones. The team still needs to apply a level of business judgment in interpreting the answers, since all models are incomplete descriptions of reality. Nonetheless, this sort of analytics has the greatest potential to help decision-makers realize tangible benefits through better decision-making.

However, automating the process of generating actionable recommendations requires a higher standard for defining causal relationships. Consider the following hypothetical example. Suppose you develop a time series model that attempts to forecast US automotive sales using imports of cheese from Mexico as the explanatory variable. You may find that the model fits the data well (it is descriptive). You may well also find that the prediction it makes (more cheese imports correlates with more vehicle sales) also turns out to be accurate year after year into the future (it is predictive). Nevertheless, if you were to then make the prescriptive recommendation that auto manufacturers should lobby Congress to reduce tariffs on Mexican cheese in order to stimulate car sales in the United States, you would be making a very foolish error. The relationship is spurious. There is no causal connection, so reducing tariffs would have no actual effect on vehicle sales. Instead, both cheese sales and vehicle sales are correlated with overall gross domestic product (GDP): when people have more money to spend, they use it for cheese and for cars; when they have less, they defer both kinds of purchases.

The lesson of the tale is clear: you need to first understand how the real-world business situation works, and model it appropriately. One huge risk of "big data" is that analysts will simply throw a huge quantity of data at a machine learning system with no thought about what kinds of relationships are plausible. In some settings this is not an issue (think "people who shopped for X also shopped for Y" recommendation engines). But in other settings, recommending nonsensible actions may destroy credibility.

No one knows the future. What we can hope to achieve with prescriptive analytics is simply to help decision-makers make the best decision possible, given the best data available at the time.

Prescriptive analytics typically require a combination of simulation and optimization. You begin by determining what quantity you wish to maximize–for example, the net present value of operating your business. Next, you list the decision levers available to you, such as investments in advertising, new product development, or price cuts for existing products. Next, you build and calibrate a model that is robust under a wide variety of ways of pulling the levers. This may require something like a system dynamics model, since it may need to capture scenarios in which the future does not look like a simple trend extrapolation of the past. Finally, you embed the simulator inside an optimization loop that evaluates a large number of different ways of setting the decision levers and tells you which one maximizes your objective, for example, is most profitable. The optimizer frequently needs to deal with various sorts of constraints, for instance, some decision levers are discrete, others are continuous, and some economic variables, like price and sales volume, cannot be negative.

Prescriptive models must also consider how entities outside of your control (e.g., competitors) will behave or react to your decisions. These may be "random," as in Monte Carlo simulation, or "strategic," as in Game Theory. Real life generally includes both.

For a real-life example, consider "Modeling General Motors and the North American Automobile Market" [13]. The client was the then-President of GM North America. The goal was to maximize future profitability. The team developed a system dynamics simulation model combining internal activities such as engineering, manufacturing, and marketing with external factors such as the competition for consumer purchases in the new and used vehicle marketplaces. Eight groups of automotive manufacturers competed for a decade across 18 vehicle segments, making monthly segment-by-segment decisions about price, volume, and investment in future products. The model included Monte Carlo simulation of random effects, such as how attractive future competitor vehicles turned out to be once they entered the marketplace, and when the next recession would occur. This was then embedded inside an optimization loop that evaluated alternative strategies. Instead of point forecasts, it generated probability distributions on future profitability, as illustrated in Figure 1.6. Ultimately it was able to show that despite future uncertainty, following a particular proposed strategy (B) would produce a probability density shifted to the right (i.e., toward higher profits) as compared to following an initial strategy (A). This supported a *prescriptive* recommendation to enact strategy B.

Just as with descriptive and predictive models, prescriptive models require substantial amounts of business judgment and work best when the team iterates between analyzing scenarios and discussing them with subject matter experts. No computer model is perfect. The data may contain valuable information, but

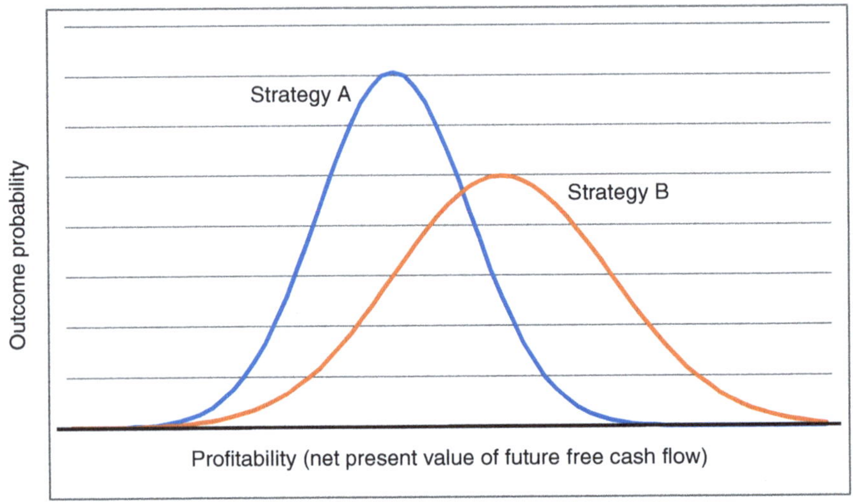

Figure 1.6 Output from an example prescriptive analysis of alternative policies [13].

inevitably you will get better results if you also incorporate subject matter expertise. At a minimum, this expertise is necessary for qualitatively interpreting the results, and when possible can also be quantitatively incorporated into the model itself.

1.4 Analytics Within Organizations

Suppose you have decided you want to do analytics within your organization. How do you get started?

Until recently, in many large organizations this involved a lot of pushing. Analytically minded employees would see an opportunity, perhaps even build a prototype analysis tool for a particular business challenge, show it to management, and then often watch it die a quiet death at the hands of leaders who did not understand the potential benefits of analytics, or who felt threatened by the thought of being replaced by a computer program.

In the last decade, however, things have changed dramatically. Analytics has become a senior management buzzword and a prominent topic of articles in publications like *Harvard Business Review* and the *McKinsey Quarterly*. These days, it is no longer a question of you, an individual employee, wanting to get more involved. Now the question is: "Your organization has decided it needs to do more analytics. How does it get started?"

The answer is of course unique to each organization, but we will make some general comments, first about the life cycle of an individual analytics project, and then about the alternative ways an organization can implement such projects.

1.4.1 Projects

Analytics projects work best when you have three key ingredients: (1) quantitative analytics professionals who are well-versed in the data and appropriate analytic techniques, collaborating closely with (2) subject matter experts who understand the problem domain, and (3) leadership sponsors in the core business who understand the value of better data-driven decisions and will champion implementation in the organization.

A new analytics project typically begins with a conversation between executives, one with operating responsibility for a difficult business decision and the other with experience doing analytics projects. If they are able to communicate effectively, they will be able to jointly write a framing document: a statement of the problem to be solved that also describes the scope, outputs to be delivered, and a high-level description of the kinds of input data and analytical frameworks that will likely be helpful in creating the desired outputs. The framing document should also include a list of stakeholders whose engagement will be needed to see their project through to implementation.

Next comes a stage we call "invent and pilot." This is a highly iterative process. The stakeholders assemble a cross-functional team combining analytical experts with business experts. The team gets up to speed on the business problem, obtains samples of available data, tries a variety of methods for analyzing it, discusses the results of each, and eventually settles on an approach that is feasible to execute within the time and resource constraints of the project while also delivering results that make actual business sense to the end clients.

Next comes "productionization." In a small organization, this could be as simple as providing the client with a spreadsheet. In a large organization, this may be a much longer and more expensive process involving the internal IT organization. Typically IT support is necessary to automate the data feed into the analytical environment, and to provide data security for both the inputs and the results of the analysis. Ideally, IT also provides services such as data cleaning, although often this is beyond their scope and falls to the analytics team instead. This can be a huge undertaking, since a great many real-world data sets have missing values, incorrect values, and are inconsistent with other data sets that are needed for the same project.

IT may also choose to develop some sort of delivery platform, such as a custom app or Web site, in order to simplify the user experience for end client users and to help maintain control of the data for security purposes.

Finally, IT deploys the solution to the client. Typically the analysis team continues to play a major role for the first year or so, conducting ongoing analysis and presenting it to leadership, as well as training people in the client organization to use the system. Often a change management process is required, since the new analytics based method of making decisions may involve a very different process than the one people in the organization are familiar with. It is

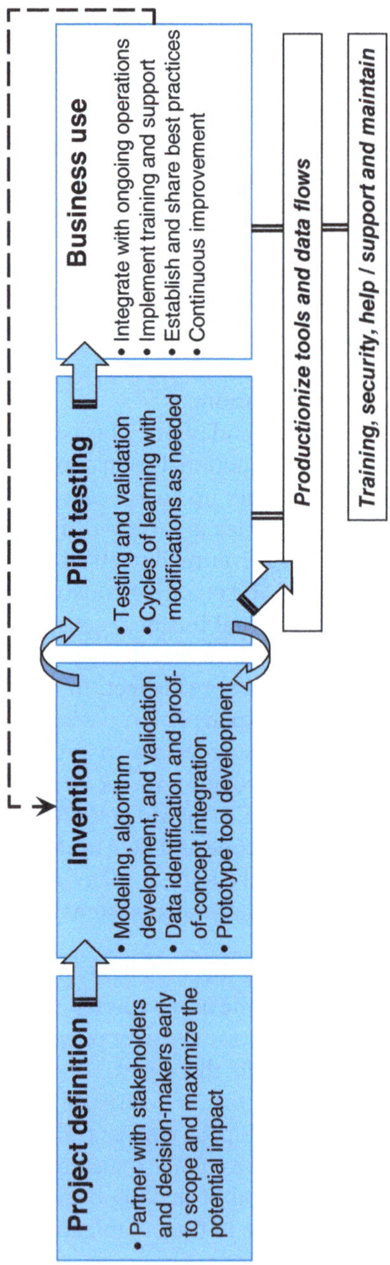

Figure 1.7 Life cycle of analytics projects.

best if some members of the client organization were participants in the cross-functional analytics team from the beginning, but at a minimum, some members of the client team must be trained as "superusers"–people who can load data, run the model, and present and interpret results, all without requiring much support from the analytics experts who built the system initially.

Additional activities (e.g., training, security, help/support) are often needed to sustain an analytics capability over time and support ongoing business use. As users become more sophisticated with experience and grow in their ability to leverage insights, new questions arise that require model enhancements. The complete life cycle of a typical analytics project is summarized in Figure 1.7.

Interview With Eric Stephens

When asked to identify the key skills needed to obtain the problem definition/ problem statement, Eric Stephens, Manager of Population Health Analytics at the Vanderbilt University Medical Center, responded as follows:

These aren't necessarily going to be in any particular order, but first and foremost I think is communication. This means the ability to listen, as well as to speak and write. In fact, listening is probably even more critical in this context than it may be in others because the ability to listen–and to comprehend and understand the situation–is extremely critical to framing the problem properly.

Although it is typically not something an analytics practitioner can influence, the culture of the organization can have a significant effect on the ability to properly define the problem. In my previous organization, there were many cases where I worked very closely with the president. He would frequently call and ask, "I need this data for this time period" or "I need to see this and this," and that's all the information he would consider. This is problematic because there may be parameters, circumstances, or other attributes that aren't stated that could significantly impact the output or the result. I would always have to push back on him a little bit to say, "OK, can we step back just a moment and can you give me a little bit more information about the problem you're trying to solve? What is it you're trying to accomplish? What's the overall objective?" Toward the end of my tenure there things got a little better, but I remember when I tried to initiate this type of conversation early on, it was usually met with something like, "it doesn't really matter," "you don't need to know," "it's not important right now," or "I don't have time to go into it." My effort was to try to communicate with him in order to better understand from his perspective what he was trying to accomplish. In situations like this, it's incumbent upon the analytics professional to convey that he or she is simply trying to provide the executive with the most appropriate solution for their problem.

The communication element is important in terms of being able to really listen and understand what the situation is; this includes the ability to empathize with the other person. From an analytic standpoint, this means being able to understand what the other person's overall situation is. For example, they may be under a lot of pressure from the president of the organization. Let's say that they're a VP or someone who reports to the senior executive team. Their sales may be significantly down, and they're trying to understand why so that they can either reorganize their product selection, or hire new salespeople, or whatever the case may be. That person may be thinking such things as "what could this mean in terms of my employment?" or "what impact would this decision have on the overall organization?" Being able to put yourself in another person's shoes really gives a lot of perspective into what the overall problem is and how it could potentially be addressed with an analytic solution.

Another important skill is the ability to think at the level of the person who is presenting the problem. It goes along with empathy, but it's really more concrete. In other words, if you are dealing with an executive, then the ability to think from the executive's perspective in terms of the business implications of the decision is important. It's not just a problem that you throw some data at and you build some models and that's it. It is important to be able to think at a higher level: to comprehend and understand the business as executives do. Certainly, it doesn't mean that every analytics professional needs to have an MBA in business strategy, but the more accomplished or the more adept the analytics professional is at thinking at that level, the more it opens up or exposes additional potential analytic solutions that may not necessarily have come to mind.

All else being equal, being able to communicate with empathy can make all the difference in how successful an analytics professional is in addressing business problems. Consider a situation in which you've got Analyst A, who is not able to think or converse at an executive level. They're mired in the statistical minutia or spend most of their day thinking in computer language rather than in the language of business. This person may be incredibly skilled at developing technical solutions, but has difficulty communicating with those in the business who are requesting their assistance. Contrast that with Analyst B, who is also very adept at building models and at programming whatever tool necessary to do the work that they need to do, but at the same time can switch perspectives so that they can converse with the business owner or executive at their level. Oftentimes, what I see are analytics professionals who can't bridge that gap, resulting in communication breakdowns at best, and a lack of trust at worst. When this occurs, the executive or businessperson asking the question may feel like the analyst lacks the understanding necessary to be able to deliver effectively. This is definitely not a recipe for analytics success.

This is an excerpt from one of a series of interviews with analytics professionals and educators commissioned by the *INFORMS Analytics Body of Knowledge* Committee.

1.4.2 Communicating Analytics

The best model in the world is of no value if the team is unable to persuade the decision-maker to act on the recommendation, so clear and transparent communication of recommendations and their rationale is essential. Writing good presentations takes effort. That effort is extremely important, even though it is completely irrelevant to the underlying mathematics. Analysts and executives frequently have very different perspectives and cognitive styles. Analysts are comfortable with mathematical formulae and inherently interested in computation, whereas executives are more focused on people, products, relationships, and results that impact business outcomes.

Junior analysts are prone to presentation pitfalls such as pasting a data table directly into a presentation (complete with six significant digits) and giving the slide a generic topic title like "Future Profit." Executives look at the mass of numbers and wonder why the analyst is so naive as to believe they can actually distinguish between 10.5678 and 10.5679. Wondering if the analyst is equally naive about other, less obvious issues, the entire analysis is now suspect.

Unfortunately, even experienced analysts can get so caught up in the mathematically interesting details of their work that they neglect to take the time to properly frame their communication. A good presentation uses "sentence titles," so that a reader who only reads the titles and does not look further into the slide can still follow the gist of the story. Good slides make their point clearly while also looking visually balanced and simple. This requires careful thought. Who is the audience? What is my goal for this meeting? What do I need to tell them to accomplish that goal? Business presentations are not mystery novels: they should lead with the answer and provide supporting details only in backup, for reference in the event they are needed. The analyst has to think about the important themes and illustrate them carefully. This usually means selecting a few key metrics and showing a relevant comparison, such as "benefit if you follow our recommendation versus benefit under the *status quo* plan."

1.4.3 Organizational Capability

The sketch of the life cycle of an analytics project in Figure 1.7 highlights some important issues. One is that the analytics experts who build the initial prototype solution tend to be scarce commodities. Whether the organization maintains its own internal pool of analytics talent or hires external consultants for each project, either way these people are expensive and difficult to recruit and retain. That is why it is essential to train a group of "superusers" who can support and maintain the project after the initial stage, so that the analytics specialists can be reassigned to new projects.

This scarcity leaves organizations with two key questions: how to prioritize analytic initiatives, and whether to use internal or external talent.

Prioritizing opportunities should be based on the impact to the organization as a whole. For businesses, this generally means improving the net present value of future free cash flows, or in simpler words, prioritizing the opportunities with the biggest potential bang for the buck.

There are two main ways this can occur: push and pull. In the "push" version, someone–either a central analytics organization or a central planning function–attempts to model the key drivers of business performance and the available levers for influencing those drivers. Applying a sensitivity analysis to this model results in a prioritized list of opportunities for intervention that have the highest potential to improve profitability. The leader of the organization must then "push" this agenda by socializing it with the leaders of the prioritized functions, who may or may not be receptive to the idea that some outsider thinks they can run the area more efficiently or more profitably. However, depending on the culture of the company, some of these leaders will be intrigued by the possibility of improvement, and will champion the initial projects. If those succeed, other leaders will generally become interested as well.

Over the past decade, many organizations have switched from push to pull, as analytics has become more visible in the C-suite. In the "pull" version, the central analytics organization prioritizes requests as they come in from leaders around the business. This version generally works much better than pushing, because the leaders themselves initiate the project and are pulling for it to happen. Someone still needs to set priorities, however, so it is still valuable to model key performance drivers and have a means for estimating the potential impact of each new project. Generally speaking, a project with a billion dollar potential impact requires only modestly more analytics resources than a project with a million dollar impact, so prioritizing based on the estimated size of the impact can be very helpful.

The prioritization decision is closely linked with the question of using internal or external talent. There are pros and cons to both approaches. External consultants can get up to speed quickly, draw upon a deep experience base within their firm, and already have a base of talented analytics professionals available. However, they are expensive. Moreover, "consulting makes the consultant smarter"–unfortunately, the client rarely gets as much of that benefit. Far too often, consultant-based projects turn out to be difficult to productionize without essentially paying the consulting company forever, because only the consultants really understand the analytics process at a deep level. Moreover, despite internal firewalls within consulting companies that keep specific details of competing clients strategies private, once a consulting firm develops a methodology for solving a particular business problem with one client, they are likely to want to leverage that investment by applying the more "generic" elements of that methodology with other clients. Initially, those new clients may indeed be in different industries, but over time the knowledge often diffuses more broadly, with the risk of eventually benefiting competitors of the original client.

As a result, companies that view analytics as a competitive advantage generally prefer to hire their own permanent analytics staff. This strategy too has downsides however, since it may be difficult to attract and retain sufficiently qualified people. Moreover, sometimes internal groups become insular, cut off from the advances in other industries, whereas consultants in a large firm may benefit from seeing many applications across a variety of industries.

If an organization does hire its own analytics staff, where should they fit in the organizational structure? Some companies centralize them under a Chief Analytics Officer, others spread them among a variety of client organizations, and some use a mix of both approaches. Sometimes analytics is viewed as part of the IT function, other times it is separate. Not surprisingly, it is difficult to make one-size-fits-all recommendations–the right answer depends on the size and shape and culture of the organization. For example, if the IT function's role and culture is primarily to manage infrastructure costs, they will probably not be a good fit for an analytics organization, which by nature is more like a small start-up or internal consulting company. In such cases, a centralized analytics group in conjunction with centers of expertise within client functions may be a good approach.

1.5 Ethical Implications

As analytics become increasingly pervasive, the ethical implications of collecting data and partially or fully automating decision-making become increasingly important. Analytics methods have the potential to provide tremendous value to individual companies and organizations, and to broader society. However, widespread collection of data raises privacy and security concerns. Additionally, broad adoption of algorithms to make decisions may have negative unintended consequences. Analytics professionals should be aware of these potential pitfalls and take actions to ensure that models are deployed in a responsible way.

In many countries, particularly in Europe, laws limit the kind of personal data companies are allowed to collect, store, and share, or provide consumers with the right to have their data erased. Even countries that allow collection of personal information often have laws mandating public notification if the data are inadvertently released or maliciously accessed by hackers. As a result, all organizations that analyze data must now stay informed about the potential legal implications of their data sets and take appropriate security measures to comply with applicable laws.

New technology for collecting data will raise new questions around "who owns data?" For example, who should have access to or be able to sell your Internet search history, your Fitbit health record, or your autonomous vehicle's sensor data? Similarly, as organizations lean more heavily on automated analytics, who will bear responsibility for errors, when for instance a driverless car is involved in a crash? There are many open questions that need to be resolved before the potential advantages of these technologies can be fully realized. In the coming decades, conversations regarding these topics will likely continue and will involve policy makers, lawyers, academics, politicians, and analytics professionals. Analytics professionals have a responsibility to honestly represent the capabilities and limitations of these technologies in these discussions, and to work toward solutions that serve the public good.

Algorithms have the potential to make decisions in ways that are more transparent and objective than a human decision-maker. For example, decades ago loan officers explicitly considered applicants' race when deciding whether to approve their loan applications. Modern credit scoring algorithms explicitly do not consider race as a factor. While not perfect, these algorithms are less discriminatory. However, the predictions or recommendations that come out of a model can be perceived as being completely objective, when in reality they are subject to biases in the data or in the modeling decisions. For example, data collected from smartphone apps are not representative of the whole population, as avid smartphone users skew young, affluent, and urban. Distribution of public services based on smartphone data may potentially exclude individuals who are invisible in the digital data set [14]. Similarly, racial biases in crime data can lead to racial biases in crime predictions, such as those used in predictive policing models [15].

Widespread deployment of certain algorithms can also create self-reinforcing feedback loops. For example, in most states in the United States, auto insurance premiums are substantially higher for people with poor credit [16]. These people then face much higher expenses, increasing the likelihood that their credit remains poor. Cycles like these are not a new phenomenon, but as price discrimination algorithms become more prevalent and more precise, the implications of the cycles become more profound.

Most countries have anti-discrimination laws that forbid discrimination based on factors like race, religion, gender, nationality, disability, and so on. Well-intended modelers who explicitly omit variables representing these categories may still inadvertently discriminate by including variables that correlate with these categories. For example, recidivism models are used to predict the likelihood that criminal defendants will reoffend. While these models do not explicitly use race, they use variables that correlate with race, such as education level and employment history, and thus defendants of different races may receive different risk scores [17].

The potential hazards of using analytics vary widely with the specific application. Sentencing decisions, for example, are fundamentally more morally fraught than decisions regarding which ad to serve on a Web site. Nonetheless, all analytics professionals should be aware of these issues, and should consider the societal consequences of their work. Diakopoulos and Friedler [18] proposed the following five principles that can guide accountability in the application of analytics:

1) *Responsibility:* Someone should have the authority and resources to deal with adverse consequences. Fully automated decision-making does not require a human in the loop, but a human should be involved to monitor the system and be able to make changes if needed.
2) *Explainability:* Decisions should be explainable to people affected by those decisions. Explaining the outcomes of machine learning models is especially difficult, but efforts are underway to develop interpretable machine learning methods, such as the DARPA Explainable Artificial Intelligence program [19]. In some applications, like speech recognition, explainability may be less important. But when used in contexts that have serious consequences for people's lives, such as determining who should receive a loan or be released from prison, clear and accessible explanations are essential.
3) *Accuracy:* Sources of error and uncertainty should be identified, logged, and benchmarked. Any model can make inaccurate predictions or misleading recommendations if it is given flawed data.
4) *Auditability:* Just as third parties are often used to identify security vulnerabilities, auditing could be used to identify potential ethical implications. The third party could exist within the same company, to protect proprietary information, but would have a different perspective from the original algorithm designer and could creatively search for potential unintended consequences.
5) *Fairness:* Biases can be "baked in" to existing data, and automated decisions can amplify structural discrimination. Analysts should be aware of this risk, and evaluate for potential discriminatory effects.

Recognizing the increasing risk of unintended consequences in the growing field of analytics, some organizations and professional societies in the area have taken the step of establishing explicit ethics guidelines to heighten awareness and stress the importance of responsible behavior (see Figure 1.8, for example).

1.6 The Changing World of Analytics

The analytics landscape has changed rapidly in recent years, and the pace of change continues to accelerate. Dramatic reductions in the cost to store and transmit data combined with the "Internet of Things" have resulted in much larger and more readily available data sets. Additionally, use of analytics is becoming more widespread, as many influential people and organizations publicize the benefits. Universities are responding to the shortage of trained analysts by developing undergraduate and graduate programs of various levels of rigor, and conferences related to "Big Data and Analytics" abound. At some

SOCIETY

Whereas operations research and analytics can have a deep impact on society, with applications ranging from medical decisions to national defense, business strategy, public policy, and many other contexts, we aspire to be:

- Accountable for our professional actions and the impact of our work.
- Forthcoming about our assumptions, interests, sponsors, motivations, limitations, and potential conflicts of interest.
- Honest in reporting our results, even when they fail to yield the desired outcome.
- Objective in our assessments of facts, irrespective of our opinions or beliefs.
- Respectful of the viewpoints and the values of others.
- Responsible for undertaking research and projects that provide positive benefits by advancing our scientific understanding, contributing to organizational improvements, and supporting social good.

OUR ORGANIZATION

Whereas our work influences the success and standing of our organizations (universities, businesses, government, and nonprofit agencies) as well as our constituencies (students, clients, customers, and suppliers), we aspire to be:

- Accurate in our assertions, reports, and presentations.
- Alert to possible unintended or negative consequences that our results and recommendations may have on others.
- Informed of advances and developments in the fields relevant to our work.
- Questioning of whether there are more effective and efficient ways to reach a goal.
- Realistic in our claims of achievable results, and in acknowledging when the best course of action may be to terminate a project.
- Rigorous by adhering to proper professional practices in the development and reporting of our work.

THE PROFESSION

Whereas we are part of the profession of operations research and analytics and have an obligation to help advance the profession and to uphold high standards on behalf of our colleagues and future generations, we aspire to be:

- Cooperative by sharing best practices, information, and ideas with colleagues, young professionals, and students.
- Impartial in our praise or criticism of others and their accomplishments, setting aside personal interests.
- Inclusive of all colleagues, and rejecting discrimination and harassment in any form.
- Tolerant of well-conducted research and well-reasoned results, which may differ from our own findings or opinions.
- Truthful in providing attribution when our work draws from the ideas of others.
- Vigilant by speaking out against actions that are damaging to the profession.

Figure 1.8 Guidelines on ethics from analytics professional society INFORMS [20].

point the hype will diminish, but because the results are real, analytics will not go away. Indeed, we expect organizations will continue to rely even more on analytics-based decision support in the future, as the benefits become increasingly well understood.

Increased volume of data has motivated the rise of parallel and distributed computing systems and the development of new algorithms for efficiently storing and retrieving data in these systems. Although particular vendors and platforms may rise or fall in popularity, the general theme is clear: problems that involve more data than comfortably fits on a single computer can be distributed over many computers in a way that makes answering certain common types of questions very efficient.

Certain kinds of data, such as real-time transaction data, or web browsing data, can be particularly massive, and the future storage requirements will likely grow astronomically. Distributed information systems that store this type of data are particularly well suited to descriptive analytics. It is straightforward to divide a giant database across multiple machines and let each one report back on the subset of data elements that match a given query. This can make report-generating systems run much faster.

Predictive and prescriptive analytics generally require more sophisticated mathematical models that are more difficult to fit into a distributed computing paradigm. This has led to the development of new algorithms for old methods that are better suited to distributed environments, as well as to entirely new methods. For example, "deep learning" methods are a form of machine learning that rely heavily on access to vast quantities of data. Traditional statistical techniques designed for small data situations rely on structure imposed by the analyst. Deep learning is attractive because (in theory) it allows the computer to find structure in the data without the human analyst having to first teach it a great deal. In practice, this depends on having a sufficiently large and rich data set available with a sufficiently high signal within the noise. Deep learning shows particular promise in situations such as voice and image recognition, where defining a structure is especially challenging, and where vast quantities of data are indeed readily available.

Traditional statistical methods are still the preferred choice in many other settings, where there is known structure and the amount of available data are more limited. For example, market research data are integral in many common business decisions, providing considerable value despite being "small" data.

We expect that both "big" and "small" data situations will remain important. Big data often follows the data-centric framing and bottom-up collection process, whereas small data generally start from the top-down decision-centric view. We predict that the most significant advances in applied analytics will come from combining the best of both worlds–leveraging the deep subject matter expertise required for small data applications to make the most of big data opportunities.

Some people are working on approaches to try to automate the analytics process further. At the moment, it is a very labor-intensive process requiring people with significant levels of education and experience. Will it be possible for computers to automate much of that? Perhaps someday, but at least for the foreseeable future, it seems to us that subject matter experts will continue to play a key role in many analytics projects. The real world is infinitely complex. Explaining the world to a computer is not easy. Cleaning data and interpreting results are complex cognitive tasks not easily replaced by current forms of "artificial intelligence." Applying existing mathematical methods to a problem once the data are clean can be reasonably straightforward, but that is not the time-consuming, rate-limiting step in the analytics process, so automating it will not really solve the problem of scarce talent. Creating new mathematical methods suited to emerging new decision questions will long remain solely the province of human experts.

1.7 Conclusion

This chapter broadly defined analytics, using conceptual frameworks (data-centric and decision-centric) and high-level classifications (descriptive, predictive, and prescriptive). We introduced considerations for implementing analytics in organizations, and potential ethical implications. The following chapters will describe in more depth how analytics can be successfully implemented, including how to get started, data and organizational requirements, solution methodologies, and management considerations.

Analytics offers exciting and vast possibilities. The analytics landscape is rapidly evolving, and new methods, data sources, and computing resources create new opportunities. Businesses have opportunities to improve profit by growing revenue or reducing cost. Governments and nonprofits have opportunities to use resources more efficiently and deliver better services. For society more broadly, there are opportunities to improve health outcomes, reduce environmental impact, improve quality of life, and increase transparency and fairness. However, capturing these potential gains is not easy. Effectively implementing analytics requires the right data, the right tools, the right people, and the right systems.

References

1 Rose R (2016) Defining analytics: a conceptual framework. OR/MS Today, 34–38.
2 Rosenbush S, Totty M (2013) How big data is changing the whole equation for business. The Wall Street Journal, March 10.

3 Brynjolfsson E, McAfee A (2014) *The Second Machine Age: Work, Progress, and Prosperity in a Time of Brilliant Technologies* (W. W. Norton & Company).
4 McAfee A Brynjolfsson E (2017) *Machine, Platform, Crowd: Harnessing Our Digital Future* (W. W. Norton & Company).
5 Siebel T (2017) IOT: AI at enterprise scale. Comments made at MIT Sloan CIO symposium.
6 Hartford T (2014) Big data: are we making a big mistake? *Financial Times,* March 29.
7 Howard RA (1968) The foundations of decision analysis. *IEEE Trans. Syst. Sci. Cybern.* SCC-4 (3): 211–219.
8 Madsbjerg C, Rasmussen MB (2014) The power of 'thick' data: businesses need to know how a product or service fits into the emotional lives of their customers. The Wall Street Journal, March 21.
9 Rasmussen MB, Hansen AW (2015) Big data is only half the data marketers need. *Harvard Business Review,* digital article, November 16.
10 Lustig I, Dietrich B, Johnson C, Dziekan C (2010) The analytics journey, *Analytics Magazine,* 11–18.
11 Noyes K (2016) 5 things you need to know about data exhaust. *PC World,* digital article, May 13. Available at http://www.pcworld.com/article/3069507/5-things-you-need-to-know-about-data-exhaust.html.
12 Tufte ER (1986) *The Visual Display of Quantitative Information* (Graphics Press, Cheshire, CT).
13 Keenan P, Paich M (2004) Modeling General Motors and the North American Automobile Market. The 22nd International Conference of the System Dynamics Society, Oxford, England.
14 Crawford K (2013) The hidden biases in big data. *Harvard Business Review Blog,* April 1.
15 Eckhouse L (2017) Big data may be reinforcing racial bias in the criminal justice system. *Washington Post,* February 10.
16 Consumer Reports (2015) The truth about car insurance, July 30. Available at http://www.consumerreports.org/cro/car-insurance/auto-insurance-special-report/index.htm.
17 Barry-Jester AM, Casselman B, Goldstein D (2015) Should prison sentences be based on crimes that haven't been committed yet? *FiveThirtyEight,* August 4.
18 Diakopoulos N, Friedler S (2016) How to hold algorithms accountable. *MIT Technology Review,* November 17.
19 Gunning D Defense Advanced Research Projects Agency Explainable Artificial Intelligence (XAI). Available at https://www.darpa.mil/program/explainable-artificial-intelligence (accessed July 18, 2017).
20 INFORMS Ethics Guidelines. Available at https://www.informs.org/content/download/357082/3750804/file/Ethics_Guidelines.pdf (accessed July 19, 2017).

2

Getting Started with Analytics

Karl G. Kempf

Decision Engineering, Intel Corporation, Chandler, AZ, USA

> *"The secret of getting ahead is getting started. The secret of getting started is breaking your complex overwhelming task into smaller manageable tasks, and then starting on the first one."*
>
> Mark Twain [1]

2.1 Introduction

In 1965, Gordon Moore made a prediction that computing would dramatically increase in power and decrease in relative cost at an exponential pace over time [2]. True to this speculation, computing power measured in millions of instructions per second (MIPS) per dollar has grown by a factor of 10 every 4–5 years [3]. Advances in computing have driven or enabled similar results in memory, storage, and networking that have in turn enabled the World Wide Web and a sequence of revolutions in analytics.

Davenport identifies Analytics 1.0, 2.0, and 3.0 [4] and since advances in computing and associated technologies show no signs of slowing, we can expect Analytics 4.0, 5.0, and so on into the future. Prior to 2010, Analytics 1.0 was characterized by small, structured, internally generated data sets. Analytics were confined to reporting and what we would now think of as descriptive analytics. Results took weeks if not months to produce and so organizations could not think of analytics as a competitive advantage. The rise of Analytics 2.0 has dramatically changed each part of that characterization. Data are assembled from a variety of internal and external sources, in a variety of formats, sometimes including real-time streams. Analytics has slowly started to include predictive and prescriptive techniques. Speed in collecting and analyzing data has become paramount. Organizations think of and use analytics as a competitive weapon.

INFORMS Analytics Body of Knowledge, First Edition. Edited by James J. Cochran.

Analytics 3.0 is continuing the trend to larger and more varied data sets as well as faster and more powerful analytics including machine learning. Analytics soon will support internal decisions by being embedded into operational processes and will enhance data-based products and services for customers. Predictive and prescriptive analytics are becoming more commonplace and indispensable to organization strategy.

In the construction of this chapter, we assume the reader has no analytics experience or some experience in Analytics 1.0, and in either case is interested in getting started with Analytics 2.0. The goal is utilization of a methodology for examining large data sets to expose as yet undiscovered patterns and correlations, market trends and customer preferences, and other information to help businesses make more informed decisions. Applications enable analytics professionals to analyze growing volumes of data that are often untapped by conventional business intelligence programs. This requires the ability to collect, integrate, manage, and leverage relevant data sources to help identify actionable improvement opportunities. It includes technologies for data manipulation and governance, application of analytics, and communication of results as well as organizational components. Properly executed analytics can point the way to multiple business benefits, including new revenue opportunities, more effective marketing, better customer service, higher impact research and development activities, faster product design-to-market cycles, improved operational efficiency in manufacturing, and faster and more reliable supply chains, all providing competitive advantages in the marketplace [5,6].

2.2 Five Manageable Tasks

In this chapter we introduce and explain the five manageable tasks required to succeed at the seemingly complex overwhelming task of getting started in analytics [7]. These include (i) choosing the business problem on which to focus, (ii) assembling the team, (iii) acquiring and preparing the data, (iv) selecting and applying the analytic tools, and (v) executing to produce an actionable result. Each task is treated in detail in later chapters. The tasks described here for getting started with analytics are ubiquitous. Whether the business problem selected is Descriptive (what events happened and when) or Diagnostic (why events happened in the past) or Predictive (what is likely to happen in the future), the same basic tasks must be completed. Whether the data used are structured, semistructured, or unstructured, internally available or acquired from outside, stored over time or streaming in real time, or a combination of all of these types and sources, the basic tasks must be executed faithfully. One of the tasks is building the team by selecting contributors from across the company and perhaps including personnel from consulting firms or universities. Another task involves selecting from the plethora of analytics tools commercially available today. Armed with a focus on the problem, a strong team, and appropriate data and tools, the last task is to supply an explainable and actionable result that can be communicated across the business as a successful example of applying analytics.

The importance of the five tasks claimed here as foundational and ubiquitous to all analytics is substantiated by a "Big Data" survey conducted by Capgemini Consulting [8]. The survey covered 226 respondents from across Europe, North America, and the Asia-Pacific (APAC) region and spanned multiple industries including retail, manufacturing, financial services, energy and utilities, and pharmaceuticals. When senior executives were asked to identify the major roadblocks to analytics, they pointed specifically to the five tasks described here:

Task 1: 39% answered "absence of a clear business case for funding and implementation."

Task 2: 35% replied "ineffective coordination of big data and analytics teams across the organization" and 27% "lack of sponsorship from top management."

Task 3: 46% indicated "scattered data lying in silos across various teams," 27% "ineffective governance for big data and analytics," and 15% "data security and privacy concerns."

Task 4: 25% mentioned "lack of big data and analytic skills," 22% "lack of clarity on big data and analytics tools and technologies," and 18% "cost of tools and infrastructure for big data and analytics."

Task 5: 12% pointed to "resistance to change within the organization."

2.2.1 Task 1: Selecting the Target Problem (see also Chapter 1)

The first task for successfully getting started with analytics is framing the target problem. Carefully determine what you are trying to achieve with your analysis before beginning the initial project. Start with a focus, not a mandate. Otherwise you will try to find out everything and risk finding out nothing. Identify a number of internal business or operational problems to solve or processes to improve. Focus on challenges and clarify key questions and concerns related to the goal. In the early stages of working on Task 1, you may pose a variety of questions expecting to get each considered but not necessarily resolved. This will lead in later stages to the description of the specific problem you need and want to address.

This description must include the value proposition. How will the outcome be measured–more efficient use of a scarce resource, a speed improvement of an important process, higher revenue in a historical market? Will there be a performance impact or an organizational transformation? In considering target problems, scoring and ranking by potential value is a useful exercise. Additional filtering can be accomplished by considering how your result will drive decisions and actions. This will help ensure that there is demand for the answer produced by your initial project. Selecting a good target problem and then supplying insight that does not drive action will not have an impact. Without impact, your project will (and should) be labeled as a failure.

Right sizing is another selection criteria. On the one hand, the problem should be sufficiently large to provide you an opportunity to produce a solution that the organization will recognize as a substantial contribution. On the other hand, the problem should be small enough so that the necessary data can be acquired, managed, and manipulated by a team getting started with analytics. A small but high-impact problem should be the target.

2.2.2 Task 2: Assemble the Team (see also Chapter 3)

Interview With Greta Roberts

Greta Roberts, cofounder and CEO of Talent Analytics, shares her thoughts on the optimal analytics team size and the different challenges that teams of different sizes face?

Optimal size to me is relative. If you have a small team and they have less work and they need to do everything in the analytics workflow, then a team of one could be optimal. What we've seen in the questions we ask our customers and in some of the research work we do is that when we ask people, "How big is your analytics team?" there are still a lot of people who do just have one person out there because they're probably not yet part of this global Analytics Center of Excellence inside of their company.

We also see three to five people working perhaps in retail, and three to five people working on the corporate analytics team, and maybe one person working in HR, and so on. So to my knowledge, you have these little disparate analytics teams. What we've seen in some of our research is that it still is kind of like that–one to five people, except for organizations that are starting to build an analytics Center of Excellence. Is that optimal? I don't know. I think that's what's happening today. I expect that it will start to grow, and maybe there will be more Centers of Excellence that might have a larger presence. I suspect companies will still have analytics professionals embedded in lines of business like marketing and sales and workforce and other lines of business. It is completely correlated to the amount of work needed to be done. There are some people who can do the entire cycle of analysis and programming and algorithms and deploying models and making presentations. There are some that are optimized doing a slice of that workflow.

This is an excerpt from one of a series of interviews with analytics professionals and educators commissioned by the *INFORMS Analytics Body of Knowledge* Committee.

The goal of the second task for successfully getting started with analytics is assembling the team. This means establishing a core group that understands the importance of using data to drive business decisions. It involves finding folks with the right mix of skill sets who want to collaborate to develop a new capability for the company.

Collaboration is key. Too often some particular group initiates an effort while claiming that forming a cross-functional team will take too long. Unfortunately, it is seldom if ever the case that a single function in a company has all the skill sets needed for success. When their effort subsequently fails, interest in analytics is damaged. The attempt to speed up adoption in fact slows down the adoption. Getting started with analytics not only affords the opportunity to break down silos, but also demands the sharing of skills and information across departments. As you begin the analytics effort, proceed with collaboration between executives, business experts, operations personnel, information technology specialists, data scientists, engineering, and whomever else can help. Resolving issues and uncovering insights from your data for your target problem requires a variety of close working relationships. There are a number of roles listed, including (but not limited to) the following in no particular order.

Executive Sponsor [9]

This team member can set the vision and tone for the work while ensuring the target problem aligns with organizational goals. The executive sponsor can remove roadblocks and address funding issues as they arise. Serving as the communication channel to other executives, the sponsor ultimately broadcasts the success of the project.

Project Manager

The project needs a committed leader to manage the schedule and the budget while delivering the results. As a "getting started" project in a new endeavor including multi-group collaboration, an experienced project manager will be required to oversee moving the team through uncharted territory. Bonus points if this manager is known and trusted by the executive sponsor.

Domain Expert

A domain expert from the business can help the team better understand the target problem, measure the results of the endeavor, and learn to speak the sponsor's language. It should be the case that this team member was heavily involved in selecting the target problem as well as forecasting the action(s) the solution is intended to drive.

IT Expert

This team member has the necessary systems knowledge to help gather and organize data and information. She/he might know where the appropriate data silos are located and who controls them. Ideally her/his skills (or those of associates of the IT expert) should stretch to such important topics as data quality, security, and governance.

Data Scientist [10,11]

The data scientist understands modeling and algorithms as well as how to explore data sets for new insights. Especially important is the practical knowledge of which of the myriad of tools are available to select from the analytics toolbox for application to the specific target problem and associated data. Of the roles to fill, this might be the most difficult. If that is the case, consultants, advanced degree interns, or other experts from outside the organization can accelerate delivery, reduce risk, and expedite learning.

Stakeholders

You have to be sure to identify all of the affected stakeholders. If you fail to do so, the ones that you neglect are likely to raise objections as the project proceeds. It may be that the domain expert is one of the business stakeholders. Alternatively, the stakeholders may be in operations or engineering or sales and marketing. It may not be necessary to include stakeholders as team members working on the project on a daily basis, but it is certainly necessary to establish consensus on the goals and keep stakeholders updated on status as the project progresses.

To get started with analytics, the team does not necessarily have to be large, but it does need to be creative and fast. There may be many unique obstacles–technical and political–to overcome along the way. There will certainly be many interested parties who will need to be pulled along, so crisp clear reporting will be necessary–especially to help the executive sponsor spread the word once success is achieved. Note that although a successful initial project will have a positive impact on the business, the main benefit may be the team learning captured and the organizational confidence gained.

2.2.3 Task 3: Prepare the Data (see also Chapter 4)

The data task involves a number of stages and usually these stages proceed in parallel with input from various sources. The initial challenge is to understand the characteristics of the data needed to address the target problem. In the best of cases for getting started in analytics, a search of internal data sources yields most if not all of the data needed. A successful search is followed by acquiring access to data in batch mode or as a real-time stream. In the worst of cases, a major investment in time, personnel, and budget is required to find and access data before it is clear whether the data can be used to successfully solve the target problem. This heavy lift increases the chance of failure and the team might consider rerunning Task 1. Access is followed by validation and cleansing or correction [12] sometimes including transformation (e.g., standardization of units or formats) in preparation for analysis.

Most discussions of data in the context of analytics include the five "V's": Value, Veracity, Volume, Velocity, and Variety. Value and Veracity are critical to data for finding a good solution to the selected problem. Volume, Velocity, and

Variety are crucial in picking the data management approach used. Prior to Analytics 1.0, as well as for many projects during 1.0, databases based on relational algebra were able to manage a few thousand documents or a few thousand transactions per second. For Analytics 2.0 and beyond with higher Volume, higher Velocity, and higher Variety, relational databases are inadequate. Hence, the advent of alternative approaches such as Key-Value Pair databases, Wide-Column stores, Graph databases, and many more. Relational databases work with structured data and scale vertically but not horizontally. Alternative approaches work with unstructured and semistructured data and scale out very well horizontally. Relational databases are over-kill, damaging scalability for data that can be effectively used as key-value pairs, and are under-kill, decreasing performance for data that need more context than just relations like graph structures.

Value means finding the right data to address the target problem. This may not be obvious in the early days of the effort even if the team members have some idea of types of data needed. Obviously, the more potentially relevant data that can be acquired the better since data that are ultimately found to be irrelevant can be ignored. From the opposite perspective, if some relevant data are missing, then some potential insights could be missed in the analysis.

Veracity draws attention to the question of quality. Inaccuracies in the data can quickly compromise all of the effort invested by the team. Insight from the analysis will be lacking and the resulting decisions may be poor. The team must realize that data quality is more important than data quantity. Focus on gathering as much data as possible without considering whether it is accurate will not yield the desired result. On balance, there is the very real question of acceptable accuracy. The team must also guard against setting accuracy standards so high that they are neither relevant nor cost-effective in a business context.

The question of Volume raises an important trade-off. On the one hand, relative to the potential value to the company of the specific project and the general learning to do successful analytics in the future, storage is cheap as is compute time for analysis. Hence, the more data the better. On the other hand, the team resources required to discover, access, cleanse, and transform the data are nontrivial. The objective of the "getting started" project is to solve the target problem, not collect all data that might ever be useful to the company.

How fast data are generated and the speed with which they need to be acquired and used–Velocity–are relative to the speed of business. Companies must be in a position to respond to what is going on around them. The time companies have to make their decisions is decreasing. This means that data have a shelf life and their usefulness decrease with the passage of time. The team must be sensitive to the velocity with which the data must be moved from acquisition to action to satisfy the focus problem [13].

Variety means that useful data will come from many sources, some supplying structured data, some unstructured, and some semistructured. Structured data are comprised of clearly defined data types whose pattern makes it easy to manipulate. Structured data often reside in relational databases in IT data centers. In addition, most companies have a large number of spreadsheets scattered throughout the organization with well-defined data organized in rows and columns. Examples of unstructured data include text, audio and video files, and a majority of social media data. The data can be human or machine generated, but in any case are not structured via predefined data models or schema that make them easy to manipulate. E-mail is an example of semi-structured since it is structured in sender, recipient, date, and time, but is unstructured in the contents of the message. Some (or all) of the data required may be acquired from outside the company either publicly accessible or available for purchase. Not all meaningful data will be collected electronically. Thus, the data may be in all different formats and structures, and may be of variable completeness and quality.

In addition to these technical problems, the team may encounter political problems in collecting data. In some circles, data are power and the owners of the data may not be anxious to share. Under some circumstances, even within the same company, Group X might not want Group Y to have access to its data and so might be reluctant to share. Generalizing these specific problems leads to the topic of data governance. Who is responsible for controlling access to the data? Who is responsible for maintaining the data? These and related questions require the team to develop and communicate a set of rules to deal with privacy and security from day 1 of the project [14,15].

Addressing these issues is the job of the IT expert with support from the other team members, especially the data scientist. A step-by-step data process must be strictly followed as part of governance:

- Identify and document each internal and external data source, including location, contents, owner, quality, format, and origin if possible. Negotiate access and conditions of use with special attention to confidentiality.
- Maintain a detailed record of all data captured, including what, when, and from whom. Archive that data in their original format for future reference especially if modifications are required for integration into the analysis set.
- Develop a consistent format suitable for use by analytics. If a change is needed to incorporate new data or to improve the analytics, change the existing data as well and carefully document all format changes. If necessary, it is better to add a new field rather than change the name or meaning of an existing field.
- Granularity is important. Granularity beyond that required to address the target question is more difficult to acquire and manage, but it can be aggregated to suit various needs and purposes. Data of less granularity may be easier to obtain, but will not be as useful since it can't be disaggregated. Granular (disaggregated) data can be aggregated, but aggregated data cannot be disaggregated.

During the execution of these steps, the overriding concern should be quality. New data need to be as complete as possible as well as correct and consistent. It should be time stamped for temporal alignment with existing data. Bad data such as outliers and missing values should be detected and either repaired or eliminated as soon as possible (note that these may be informative anomalies!). A substantial part of data quality control can be accomplished with a new data set before it is introduced into the existing data set. Quality control continues as new data are checked as much as possible for consistency and alignment with the existing data. Documentation of issues detected and remedies employed is a good practice.

2.2.4 Task 4: Selecting Analytics Tools (see also Chapters 5 and 6)

Any attempt at listing, describing, or comparing available open-source or commercial analytic tools faces at least two challenges. The first challenge is that this list would be very long indeed. Even a basic online search will identify scores of possibilities. The second challenge is that this list would very quickly become obsolete. Research advances the state of the art, new tools are introduced, current tools are updated, and old tools are discontinued in short order. These challenges are not addressed here but can be managed by regularly consulting the appropriate periodicals [16,17]. In this section we can only describe the characteristics of tools that should be taken into consideration when making a selection [18].

The technical factors for selecting the analytic tools to be used are (a) the target problem selected and the type of solution desired, (b) the data available with which to solve the problem, (c) the computational infrastructure available to the team, and (d) visualization requirements.

Analytical Specificity or Breadth

Examples of analyses requiring very specific tools include video analytics for extracting information from video footage and voice or speech analytics for examining audio recordings. Sentiment analysis (also known as opinion mining) might require combined video and audio understanding. Image analytics for working with photographs or medical images and text analytics for use with large quantities of unstructured text data are other examples that require tools tailored to the specific task. For a "getting started" project, it is more likely that the data being examined is numeric and the analytics is some form of statistics, simulation, or optimization. In this case, breadth of techniques is important. For example, a single software package supporting clustering, segmentation, decision trees, time series, classification, and regression is more useful than a tool with only a single technique since many approaches will need to be considered when identifying which is most appropriate. Of course, the broader the functionality, the higher the price and the more the sophistication presumed of the user.

Access to Data

The factors for selecting among the various tools must be based on your team's specific requirements for accessing and processing data volumes and varieties. Whether your data are in a columnar or in-memory or nonrelational database on a private or public cloud, the analysis tool must be able to efficiently access the data. The IT expert and the data scientist need to collaborate in selecting storage technology and analysis technology that are compatible.

Execution Performance

The need for performance is determined by size and complexity of the data set and the velocity with which the analysis needs to run to drive the actions desired. Overall execution performance is determined by the analysis technique employed as well as the computational, storage and networking infrastructure available. Assuming your initial project is a great success, assessing how the tool scales to larger problems and more capable infrastructure will save money and resources in the future.

Visualization Capability

Clear communication of the results of the initial project will be crucial to success. The visualization capabilities of the analysis tool or some other specific visualization adjunct tool are therefore of great interest to the team. Equally important is the utilization of the extraordinary pattern recognition skills of human beings. Visualization of the raw data and the intermediate results of analysis for inspection by team members and stakeholders is often important–and sometimes crucial–to solving the target problem.

Nontechnical factors are also involved in the tool selection process. These include (a) the skill set of the data scientist(s) on the team, (b) the pricing from the vendor, (c) the budget available to the team, and (d) collaboration requirements.

Data Scientist Skillset

Available analysis tools range from those that target novice users to those engineered for expert users. One perspective is that the selected tool should match the skill level of the data scientist(s) on the team. Unfortunately, if the analytics the tool supplies are not appropriate to solve the target problem, the team will fail to deliver the desired solution. Another perspective (the correct one) is that the selected tool should match the needs of the project. If that exceeds the skill level of the data scientist(s), some skill enhancement is warranted. While this might delay the project, effective and relatively inexpensive training is available from multiple sources, including classes at conferences, universities, and tool vendors. Alternatively, the team may add a scientist who holds the required skill set.

Vendor Pricing

There are almost as many pricing models for analytics tools as there are vendors of analytics tools. At one extreme is a free or open-source version of a tool with charges for support. At the other extreme are large vendors who offer a massive portfolio of analytics tools and ideally (for them) are interested in site-wide or enterprise-wide licenses for the whole portfolio. Somewhere between these extremes are small vendors who license one or a few tools. Not surprisingly, each of these arrangements can include consulting (at an additional cost). Another common pricing determinant is number of simultaneous users, or number of processor cores in the infrastructure, or size of memory serviced. The latter two relate to the infrastructure available.

Team Budget

There is a range of strategies here. If budget is a constraint, then the least expensive path to providing a high-quality solution to the test problem may be the only feasible option. Even if budget is not a constraint, this may be the wise choice. But if budget is available and the team is confident of success, speculation concerning future projects can enter into the purchasing decision. In addition, perceived risk can be a consideration. Some teams feel safer dealing with a large vendor that offers well-tested software of proven reliability and has an extensive user community. Others prefer a small vendor that offers the potential of a closer working relationship.

Sharing and Collaboration

Although perhaps not critical for a small team working on a "getting started" project, a criterion looking to the future of larger teams is tool capability for sharing. Imagine a large team spread over multiple sites and multiple time zones (perhaps including multiple continents). Advantage could be gained in terms of velocity and brain power by sharing models and collaborating regarding interpretation of the results across multiple data scientists and potentially including business users. Some scientists have actually found that software and programming languages provide a common lingua franca that helps bridge language gaps; they can often communicate more succinctly and clearly through the code they write and the models they build than they can through their spoken and written languages.

Addressing each of these issues is the job of the data scientist supported by the other team members, especially the IT expert. During the decision process, it should be kept in mind that vendors often supply "evaluation" versions of their offerings for potential customers to test. Access is strictly time limited, and capacity is frequently limited; sometimes these vendors provide "typical" data sets for demonstration to help speed a purchasing decision.

2.2.5 Task 5: Execute (see also Chapter 7)

With the focus problem defined and at least some initial data and an analytic tool in hand, the team can start iterating. Although "getting started" in analytics should be equivalent to building the minimum viable system to provide actionable results, if the first round of data collection and application of analytics supplies a solution, there are only two possible conclusions: either the team is incredibly lucky or the focus problem is far too easy. A more realistic execution trajectory involves iterating around data, analytics, and learning. The initial data set will likely be inadequate and the learning will point to gathering more data, restructuring the existing data, eliminating some data, and improving the data in some way. Probably the initial analytic approach will be lacking in some way and the learning will lead to modifying the approach or trying a different technique or combining techniques–refining the analytics in some way to move closer to solution. This is the art of the team. This is the moving through uncharted territory that was mentioned earlier. It requires persistence, creativity, and confidence. This is where the learning how to succeed at analytics takes place. The more quickly the iterations can take place the better.

This process can be aided by establishing a schedule for reporting to interested stakeholders. Showing visualizations of the data as well as the intermediate analytics results to experts in the domain could produce insights and suggestions that will be of great value to the team. In any case, this will serve to keep the stakeholders engaged. The team must remember, however, that getting started in analytics is, with some stakeholders, an "old dogs, new tricks" exercise and there will be skeptics and naysayers. Useful guidance can come from these folks too, if the team has the patience to listen to and fully comprehend what these stakeholders are missing or not understanding. The stakeholder with a negative perspective may be correct (even insightful) and have especially useful feedback! In any case, these stakeholders have self-identified as folks to invite to the celebration once the focus problem has been cracked.

Once the project is successful and an actionable result is generated, make sure it is applied and implemented correctly. As mentioned repeatedly, validating the exercise with a real measurable business result is the real goal. Once that result is demonstrated in practice, project success needs to be socialized. The executive sponsor should drive this top-down, while the domain expert can broadcast bottom-up. This will convince the executive team that the organization can succeed and benefit with analytics, and educate employees on the value of creating a data culture. Documentation of lessons learned makes the success easier to duplicate by other teams with other team members. In any case, do not rest on your laurels! Next problem please!

Interview With Harrison Schramm

Harrison Schramm, who recently retired after a 20-year career as a helicopter pilot and operations research analyst in the U.S. Navy, suggests ways an analytics professional can overcome the challenges of data politics to educate clients and get stakeholder buy-in.

You have to present yourself as a human being. Before another human being will trust you, you have to present yourself as someone with whom they have at least some (maybe even superficial) commonality. Finding a way to have a relationship on which the work you are doing together can build is key. Let me tell you a parallel story about that (I'm going back to storytelling). I recently had a great deal of dental work done. I thought my dentist was fantastic. Why did I think my dentist was fantastic? I did not go to another dentist and have that dentist check to see if the drilling and filling and whatever my dentist did was right, but she sure seemed like she knew what she was doing. She had a great bedside manner, and she was someone I could relate to. It is interesting to consider that the criteria on which I judged the quality of my dentist had absolutely nothing to do with the quality of the way she repaired my teeth–I actually don't know how well she did it! Of course, the quality of the work you can do is very important, but you might not get an opportunity to demonstrate your capabilities if your potential client isn't comfortable with you.

This is an excerpt from one of a series of interviews with analytics professionals and educators commissioned by the *INFORMS Analytics Body of Knowledge* Committee.

2.3 Real Examples

Unfortunately, few groups publish information about the projects they successfully executed, especially while getting started with analytics. Small but high-impact projects for a particular company are not often deemed worthy of broadcast. Reports do sometimes come out in trade magazines [19–21] that are then collected and used as data to support broader abstractions and theories [22]. A few are briefly described here to provide examples of successful small but impactful projects using data from various sources and analytics of many types all delivering beneficial actionable results.

Case 1: Sensor Data and High-Velocity Analytics to Save Operating Costs [23]

U.S. Xpress is the third largest privately owned trucking company in the United States. Fortunately, it had already partially addressed Task 3 before the BDA case described here. It had embarked on a general overhaul of its information management system since business questions took at least weeks and sometimes months to answer. The task was selected out of exasperation during an industry downturn. The topic was how to cut costs, and the request from one executive was to focus on truck idling times as a potential key to saving on fuel. Since every truck had some sort of communications device, the IT team was able to respond in less than 6 weeks with an application to supply data on how much time trucks were stationary with their engines running and using up fuel but not going anywhere. Simple analysis across data from 8000 tractors and 22,000 trailers showed that implementing policy changes for the drivers could indeed result in substantial results. This insight gained through the collection, management, and analysis of operations data saved U.S. Xpress roughly $6 million per year.

Case 2: Social Media and High-Velocity Analytics for Quick Response to Customers [24,25]

At the time of this example activity, Starbucks had already become proficient at mining social media like Facebook and Twitter as well as niche coffee forum discussion groups. An obvious task was to assess the reaction of customers to newly introduced products and the sooner the better. (Here is a paradigmatic example of the value of acquiring, analyzing, and acting on data in near-real time as opposed to stockpiling data over time for analysis and response later.) The specific question was whether customers would think a particular brew tasted too strong. Real-time efforts began as soon as the first short/tall/grande/venti/trenta was poured. By mid-morning, it was clear from social media that the taste was agreeable but the price was not. It was perceived to be too high. The price was reduced across the Starbucks network by early afternoon, and at the end of the first day there were no further negative comments. Note the punch line: not end of quarter or end of month, not even end of week, but rather *end of the first day.*

Case 3: Sensor Data and High-Velocity Analytics to Save Maintenance Costs [26]

Petróleos Mexicanos (also known as Pemex) is a producer, refiner, and distributor of crude oil, natural gas, and petroleum products. It is one of the largest petroleum companies in the world, and it relies on analytics to solve basic but important operational question. Oil refineries use water to heat fluids and cool equipment during the refining process. A major component of the water system is the cooling tower, and a typical refinery has many of them. Each of these towers has a number of large cooling fans that regulate the temperature of the water contained in the tower. Due to mechanical wear, axis misalignment, oil leaks, and other problems, the fan motors and gear boxes sometimes begin to vibrate. This shortens their productive life and risks unexpected shutdowns that cost time and money and have a negative impact on refinery operations. Maintenance crews sometime waste effort by addressing a fan that is not yet vibrating, and in other instances do not arrive at a fan until it is too late to address the issue. Pemex has found a way to avoid the vast majority of fan-related problems by mounting wireless vibration sensors and collecting and analyzing high volumes of data in real time, thereby reducing parts and labor expenses as well as dramatically reducing shutdown risk.

Case 4: Using Old Data and Analytics to Detect New Fraudulent Claims [27]
Infinity Property & Casualty Corporation, headquartered in Birmingham, Alabama, is a national provider of nonstandard car insurance for individuals who are unable to secure coverage through standard insurance companies due to a driving record with accidents, tickets, prior DUI, or vehicle type. Considering its business model, it is not surprising that fraud management is a particularly important part of Infinity's operations. Thinking through this task, Infinity speculated that automobile insurance claims could be scored in the same way as consumer credit applications. Furthermore, Infinity realized that it had archived years of adjusters' reports that could be analyzed and correlated to instances of fraud. It built an analytics-based system around that data to assign fraud probability "scores" when initial accident reports are filed. Based on the score, suspicious claims are sent to fraud investigators within a day or two for deeper analysis. This has resulted in a roughly 50× decrease in time taken to identify attempted fraudulent claims and increased their success rate in catching fraudulent claims, which has resulted in $12 M in recoveries.

Case 5: Using Old and New Data Plus Analytics to Decrease Crime [28]
PREDPOL Inc. is located in Santa Cruz, California, and was incorporated in 2012. Its business is predictive policing: using data and predictive analytics to supply law enforcement agencies with predictions for the places and times where and when crimes are most likely to occur. Depending on the granularity of the input data, PREDPOL can provide predictions with a resolution of 500 ft × 500 ft segments on a map of the patrol area. It uses no personal information, eliminating any concerns about personal liberties or profiling. The predictions are based on the observation that certain crime types tend to cluster in time and space and are based solely on crime type, crime location, and crime date/time. Historical data are obtained from the target police department's records to build the initial model. Fresh data are collected and used daily to create updated predictions for each patrol area and shift. In areas of Los Angeles where the predictions have been used to focus policing efforts, there has been a 20^{+}% reduction in violent crimes and a 30^{+}% decrease in burglaries.

Case 6: Collecting the Data and Applying the Analytics Is the Business [29]
Chicago start-up Food Genius (FG) is a foodservice data provider. They scrape open content from the Internet, specifically menus with prices posted online by independent and chain restaurants around the country. When possible, they break down menu items into ingredients. With this data and analysis, FG can tell, for example, whether restaurants are still luring their burger customers with added bacon or whether customers' tastes have switched over to avocado.

Furthermore, FG can determine the asking price delta of such a switch. This gives FG the ability to determine what combinations of ingredients, flavors, and buzzwords are being offered in attempts to make dishes more appealing and perhaps worth an increased price to diners. Customers of the analysis supplied by FG fall into two categories. One is the restaurants: independent restaurants that want to understand what the local competition is doing, and chains that want to see regional trends across the country. The other is the companies that supply raw ingredients to the restaurants that are being analyzed.

References

1 Mark Twain. Available at https://www.brainyquote.com/quotes/quotes/m/marktwain118964.html (accessed May 20, 2017).
2 50 years of Moore's law. Available at http://www.intel.com/content/www/us/en/silicon-innovations/moores-law-technology.html (accessed May 27, 2017).
3 Trends in the cost of computing (2015) Available at http://aiimpacts.org/trends-in-the-cost-of-computing/ (accessed April 30, 2017).
4 Davenport TH (2013) Analytics 3.0. *Harv. Bus. Rev.* 91(12): 64–72.
5 Davenport TH (2006) Competing on analytics. *Harv. Bus. Rev.* 84(1): 98–107.
6 Barton D, Court D (2012) Making advanced analytics work for you: a practical guide to capitalizing on big data. *Harv. Bus. Rev.* 90(10): 78–83.
7 Davenport TH (2015) 5 Essential principles for understanding analytics. *Harv. Bus. Rev.*, October 21, 2015.
8 Cracking the data conundrum: how successful companies make big data operational. Available at https://www.capgemini-consulting.com/cracking-the-data-conundrum (accessed April 28, 2017).
9 McShea C, Oakley D, Mazzei C (2016) The reason so many analytics efforts fall short. *Harv. Bus. Rev.*, August 29, 2016.
10 Davenport TH, Patil DJ (2012) Data scientist: the sexiest job of the 21st century. *Harv. Bus. Rev.* 90(10): 70–76.
11 Data Scientist Report. Available at https://visit.crowdflower.com/WC-2017-Data-Science-Report_LP.html (accessed May 29, 2017).
12 Xiong H, Pandey G, Steinbach M, Kumar V (2006) Enhancing data analysis with noise removal. *IEEE Trans. Knowl. Data Eng.* 18(3): 304–319.
13 Davenport TH, Barth P, Bean R (2012) How big data is different. *MIT Sloan Manag. Rev.* 54(1): 43.
14 Richards NM, King JH (2013) Three paradoxes of big data. *Stanf. Law Rev. Online* 66, 41–46.
15 Richards NM, King JH (2014) Big data ethics. *Wake Forest Law Rev.* 49, 393–432.
16 Analytics Magazine Available at http://analytics-magazine.org/ (accessed May 14, 2017).

17 OR/MS Today. Available at https://www.informs.org/ORMS-Today (accessed May 14, 2017).
18 Redlon M (2015) A guide to selecting an analytics vendor. *Harv. Bus. Rev.* October 23, 2015.
19 Laskowski N (2013) Ten big data case studies in a nutshell. Available at http://searchcio.techtarget.com/opinion/Ten-big-data-case-studies-in-a-nutshell (accessed May 10, 2017).
20 Laskowski N (2015) Ten analytics success stories in a nutshell. Available at http://searchcio.techtarget.com/opinion/Ten-analytics-success-stories-in-a-nutshell (accessed May 10, 2017).
21 Petersen R (2016) 37 big data case studies with big results. Available at https://www.businessesgrow.com/2016/12/06/big-data-case-studies/ (accessed May 10, 2017).
22 He W, Wang F, Akula V (2017) Managing extracted knowledge from big social media data for business decision making. *J. Knowl. Manag.* 21(2): 275–294.
23 Lemos R (2011) Big data: how a trucking firm drove out big errors. Available at http://www.cio.com/article/2410714/data-center/big-data--how-a-trucking-firm-drove-out-big-errors.html (accessed April 26, 2017).
24 Gallaugher J, Ransbotham S (2010) Social media and customer dialog management at Starbucks. *MIS Q. Exec.* 9(4): 197–212.
25 Chua A, Banerjee S (2013) Customer knowledge management via social media: the case of Starbucks. *J. Knowl. Manag.* 17(2): 237–249.
26 Espinosa R, Spinoso R (2012) Wireless monitoring asset protection. Available at http://www.controleng.com/channels/process-control/process-control-news/single-article/wireless-monitoring-asset-protection/b60ecfe04dcf887c656d14522d4fe728.html (accessed April 28, 2017).
27 Dibble W (2011) Infinity property and casualty builds a smarter system for Fraud. InformationWeek Insurance & Technology. Available at http://www.insurancetech.com/infinity-property-and-casualty-builds-a-smarter-system-for-fraud/a/d-id/1313275 (accessed May 17, 2017).
28 PREDPOL: The Predictive Policing Company (2017) How PredPol works: we provide guidance on where and when to patrol. Available at http://www.predpol.com/ (accessed May 24, 2017).
29 Rekdal A (2016) Table tech: these 22 Chicago companies are changing the food industry. Available at http://www.builtinchicago.org/2016/10/07/chicago-food-tech-companies (accessed May 14, 2017).

Further Reading: Papers

Ahlawat T, Rambola R (2016) Literature review on big data. *Int. J. Adv. Eng. Technol. Manag. Appl. Sci.* 3(5): 21–30.

Chen Y, Chen H, Gorkhali A, Lu Y, Ma Y, Li L (2016) Big data analytics and big data science: a survey. *J. Manag. Anal.* 3(1): 1–42.
Chong D, Shi H (2015) Big data analytics: a literature review. *J. Manag. Anal.* 2(3): 175–201.
Elgendy N, Elragal A (2014) Big data analytics: a literature review paper, Industrial Conference on Data Mining. Springer International Publishing, pp. 214–227.
Gandomi A, Haider M (2015) Beyond the hype: big data concepts, methods, and analytics. *Int. J. Inform. Manag.* 35(2): 137–144.
Power D (2014) Using 'big data' for analytics and decision support. *J. Decis. Syst.* 23(2): 222–228.
Soon K, Lee C, Boursier P (2016) A chronology of big data adoption. *Int. J. Appl. Bus. Eco. Res.* 14(1): 521–544.
Tsai C, Lai C, Chao H, Vasilakos A (2015) Big data analytics: a survey. *J. Big Data* 2(1): 21–53.

Further Reading: Books

Ankam V (2016) *Big Data Analytics* (Packt Publishing, Birmingham, UK). ISBN 978–1785884696 (Hadoop).
Bunnik A, Cawley A, Mulqueen M, Zwitter A, eds. (2016) *Big Data Challenges: Society, Security, Innovation and Ethics* (Palgrave Macmillan, London, UK). ISBN: 978–1349948840 (Security).
Davis K, Patterson D (2012) *Ethics of Big Data: Balancing Risk and Innovation* (O'Reilly Media, Sebastopol, CA). ISBN-13: 978–1449311797 (Ethics).
Foreman J (2013) *Data Smart: Using Data Science to Transform Information into Insight* (John Wiley & Sons, Inc., Indianapolis, IN,). ISBN 978–1118661468 (Techniques).
Maheshwari A (2014) *Data Analytics Made Accessible* (Amazon Digital Services) (Case Studies).
Provost F, Fawcett T (2013) *Data Science for Business* (O'Reilly Media, Sebastopol, CA). ISBN 978–1449361327 (Concepts).

3

The Analytics Team

Thomas H. Davenport

Technology, Operations, and Information Management, Babson College, Wellesley, MA, USA

3.1 Introduction

Analytics are created or initiated by people. People frame the question to be answered by analytics, select the data to be analyzed, propose and test hypotheses, and determine how well the hypotheses are supported in the data. Even in relatively automated machine learning environments, analysts or data scientists select the data and the tools, and kick off the process of finding a model that fits the data. The capabilities of human analysts are among the most important factors in determining the success of an analytics initiative.

In organizations of any size, it is impossible for one analyst to do all the necessary analytics work. Therefore, the topic of human analytical resources quickly becomes one of assembling and managing an analytical team. In addition, there may often be too many different skills required for high-quality analytical work for one person to possess them all. It is usually the case that some sort of division of labor and skills across a team is necessary.

This chapter, then, will focus on assembling and managing teams of analytical people to analyze data and assist the organization in making analytical and data-driven decisions. It addresses not only an organization's requirements for analytical capabilities but also the individual skills required to make analytics successful. It will also address some of the ways in which analytical teams can be organized within a company.

Although this chapter appears in a book published by INFORMS, it is not a commercial for that organization. Nevertheless, there should be little doubt that certification of analytical skills is a useful exercise to ensure that the necessary skills are present in an individual's repertoire. INFORMS has created one of the more effective certification programs in its CAP, or Certified Analytics Professional. The CAP certification requires work experience in analytics, but the Associate CAP (aCAP) does not. I won't discuss these further–there are many

INFORMS Analytics Body of Knowledge, First Edition. Edited by James J. Cochran.

materials available on the program's website (https://www.certifiedanalytics.org/)–but as a member of the Analytics Certification Board, I will testify to its quality and urge individuals and organizations to pursue this certification.

3.2 Skills Necessary for Analytics

The skills necessary to work with analytics have evolved considerably over the several decades that companies have been pursuing business analytics. I'll describe the evolution in this chapter, beginning with the basic skills that have been necessary since the 1960s or 1970s, when the use of analytics in businesses began to take off.

Quantitative skills–broadly speaking, the ability to extract, meaning from numbers–are the core requirement for any type of quantitative analyst. But tuning a regression equation or manipulating a spreadsheet is only the beginning. Effective analysts need to be proficient not only with data but also with people.

- *Quantitative and technical skills* are the foundation. All analytical people must be proficient in both general statistical analysis and the quantitative disciplines specific to their industry or business function: lift analysis in marketing, stochastic volatility analysis in finance, biometrics in pharmaceutical, and informatics in health care firms, for example. Some types of analysts–those involved in "business intelligence" or reporting work–may get by without substantial statistical knowledge, but this lack would probably limit their careers today. Analytical people must also know how to use the specific software associated with their type of analytical work, whether it be to build statistical models, generate visual analytics, define decision-making rules, conduct "what-if" analyses, or present a business dashboard.
- *Business knowledge and design skills* enable analysts to be more than simple backroom statisticians. They must be familiar with the business functions and processes to which analytics are being applied–marketing, finance, HR, new product development, and the like. They need enough general business background to work at the interfaces of business processes and problems. They also must have insight into the key opportunities and challenges facing the company, and know how analytics can be used to drive business value. One study of quantitative analysts suggested that they have more business acumen than their nonanalytical counterparts [1].
- *Data management skills* are perhaps even more important to analytical professionals than statistical and mathematical expertise. It is often commented that such professionals spend the majority of their time manipulating data–finding, integrating, cleaning, matching, and so on. And the most commonly sought software skill by employers of data scientists is not a statistical program, but rather SQL–a query language for data management [2]. There is little doubt

that analytical professionals need skills in managing and manipulating data, and for some this activity will constitute a major component of their jobs.

- *Relationship and consulting skills* enable analysts to work effectively with their business counterparts to conceive, specify, pilot, and implement analytical applications. Relationship skills–advising, negotiating, and managing expectations–are vital to the success of all analytical projects. Furthermore, an analyst needs to communicate the results of analytical work: either within the business to share best practices and to emphasize the value of analytical projects or outside the business to shape working relationships with customers and suppliers, or to explain the role of analytics in meeting regulatory requirements (e.g., utility company rate cases). This skill has been described as "telling a story with data [3]."
- *Coaching and staff development skills* are essential to an analytical organization, particularly when a company has a large or fast-growing pool of analysts, or when its analytical talent is spread across business units and geographies. All analytical professionals may not need them, but they are certainly required for supervisors and managers of large teams. When analytical talent is not centralized, coaching can ensure that best practices are shared across the company. Good coaching not only builds quantitative skills but also helps people understand how data-driven insights can drive business value.

One survey of quantitative analysts' activities suggested that there are really several categories of the role [4]. Based on their self-reported time allocation across 11 different analytical activities, the analytical professionals surveyed were clustered into four groups: generalists, data preparation specialists, programmers, and managers. Every participant indicated they did a little of each activity; however, managers mostly managed, programmers mostly programmed, and data prep folks mostly worked on data acquisition and preparation. The generalists do all these activities, of course, but focus more on analysis, interpretation, and presentation than other activities. Across all four categories, the least amount of time was spent on data mining and visualization.

Of course, few individuals come equipped with the full array of skills I've described; this is where teaming comes in. To constitute effective teams, a company needs the right mix of analytical talent in its teams of analysts. For example, it is often a good idea to balance hard-core quantitative experts–who focus on more advanced analytical techniques–with business-oriented "translators"–who have a broader skill set, combining strong analytics with business design and management skills to link professionals to their customers.

3.2.1 More Advanced or Recent Analytical and Data Science Skills

The practice of analytics has changed substantially over several different "eras [5]." However, the skills I've described for basic analytical work don't go away over time. That's in part because companies still have a need for descriptive analytics and the other activities performed in early analytics periods, and also because the skills required for that era still apply in later eras of analytics. However, as analytical practice has evolved, new skills are added. That is, the skills for doing analytics across the different eras of analytical practice, unfortunately, are cumulative. To be more specific, none of the statistical, business acumen, relationship, data management, and coaching capabilities required for traditional quantitative analysis go away when organizations move into the era of "big data." This occurred around the turn of the twenty-first century in Silicon Valley, when organizations needed new data management and analytical approaches for the rise of online business.

But there are new skills required in the big data era. Data scientists–the new term for people doing high-skill analytical and data management work in this environment–typically have advanced degrees in technical and scientific fields [6]. Because they are testing many different approaches to online operations and commerce, they need experimentation skills, as well as the ability to transform unstructured data into structures suitable for analysis. In Silicon Valley, performing these tasks also requires a familiarity with open-source development tools. If the data scientists are going to help develop "data products"–products and services based on data and analytics–they need to know something about product development and engineering. Perhaps because visual displays are a good way to comprehend a very large data set, the time that big data took off was also the time that visual analytics became widely practiced in large organizations, so a familiarity with visual display of data and analytics also became important during this period.

The next era, which I would argue began around 2012 or 2013 in the most sophisticated companies, involved the combination of both big and small data for analytics within large organizations. What skills got added at this point? In addition to mastering the new technologies used in combining big and small data, there's a lot of organizational and process change to be undertaken. If operational analytics means that data and analytics will be embedded into key business processes, there's going to be a great need for change management skills. At UPS, for example, which initiated a large real-time driver routing initiative called ORION during this period, the most expensive and time-consuming factor by far in the project was change management–teaching about and getting drivers to accept the new way of routing. This period was also marked by the early use of statistical machine learning approaches, which were necessary to handle the large and fast-changing data environment of the period.

The current era, which started perhaps 5 years ago in online businesses and 2 years ago in other industries, involves extensive use of artificial intelligence or cognitive technologies. This means that analysts and data scientists need a heavy dose of new technical skills–machine and deep learning, natural language processing, and so forth. There is also a need for work design skills to determine what tasks can be done by smart machines, and which ones can be performed by (hopefully) smart humans.

The cumulative nature of these additional skills over time means that it is even more important to take a team-based approach to analytical and data science projects. It is impossible to imagine, for example, that someone who possesses the rare skill of deep learning analytics would also have all the other skills I've mentioned thus far in this chapter. The only way to have all the necessary skills on a team is to staff projects with people who hold different–and hopefully complementary–skill sets.

3.2.2 The Larger Team

Analytics were initially created to improve human decision-making. But there are many circumstances in organizations in which analytics aren't enough to ensure an effective decision, even when orchestrated by a human analyst. In order for analytics to be of any use, a decision-maker has to assess the analytical outcomes, make a decision on the basis of them, and take action. Since decision-makers may not have the time or ability to perform analyses themselves, such interpersonal attributes as trust and credibility between analysts and decision-makers come into play. If the decision-maker doesn't trust the analyst or simply doesn't pay attention to the results of the analysis, nothing will result from the analytical work, and the statistics might as well never have been computed.

I cited one such example in my first book on analytics [7]. In the course of research for that book, I talked to analysts at a large New York bank who were studying the profitability of the bank's branch network. The analysts went through a detailed and highly analytical study in the New York area–identifying and collecting activity-based costs, allocating overheads, and even projecting current cost and revenue trends for each branch in the near future. The outcome of the analysis was an ordered list of all branches and their current and future profitability, with a clear red line drawn to separate the branches that should be left open from those that should be closed.

The actual outcome, however, was that not a single branch was shut down. The retail banking executive who sponsored the study was mostly just curious about the profitability issue, and he hardly knew the analysts. He probably wasn't aware of all the work that would go into the analysis process. He knew–but the analysts didn't–that there were many political considerations involved in, say, closing the branch in Brooklyn near where the borough president had grown up, no matter where it ranked on the ordered list of branches. Basing actions on analytics often require a close, trusting relationship between analyst and decision-maker, and that was missing at this bank. Because of the missing relationship, the analysts didn't ask the right questions about the analysis, and the executive didn't frame the question for them correctly.

Instead of just analysts and data scientists, there are really three groups whose analytical skills and orientations are at issue within organizations. One is the senior management team–including the CEO–that sets the tone for the organization's analytical culture and makes the most important decisions. Then there are the professional analysts and data scientists, who gather and analyze the data, interpret the results, and report them to decision-makers. The third group is a diverse collection I have referred to as analytical amateurs. They comprise a large category of "everybody else," whose daily use of the outputs of analytical processes is critical to their job performance. These could range from frontline manufacturing workers, who have to make multiple small decisions on quality and speed, to middle managers, who also have to make decisions with respect to their functions and units–which products to continue or discontinue, for example, or what price to charge for them. IT employees who put in place the software and hardware for analytics also need some familiarity with analytical topics, and also qualify as analytical amateurs.

To really succeed with analytics, a company will need to acquaint a wide variety of employees with at least some aspects of analytics. Managers and business analysts are increasingly being called on to conduct data-driven experiments, interpret data, and create innovative data-based products and services [8]. Many companies have concluded that their employees require additional skills to thrive in a more analytical environment. One survey found that more than 63% of respondents said their employees need to develop new skills to translate big data analytics into insights and business value [9]. Bob McDonald, at one point CEO of Procter & Gamble and then head of the U.S. Veterans Administration, said about the topic of analytics (and business intelligence more broadly) within P&G:

> We see business intelligence as a key way to drive innovation, fueled by productivity, in everything we do. To do this, we must move business intelligence from the periphery of operations to the center of how business gets done.

With regard to the people who would do the analysis, McDonald stated:

> I gather there are still some MBAs who believe that all the data work will be done for them by subordinates. That won't fly at P&G. It's every manager's job here to understand the nature of statistical forecasting and Monte Carlo simulation. You have to train them in the technology and techniques, but you also have to train them in the transformation of their behavior [10].

Of course, all senior executives are not as aggressive as McDonald in their goals for well-trained analytical amateurs. But in even moderately sophisticated companies with analytics, there will be some expectations for analytical skills among amateurs of various types. As Jeanne Harris and I wrote in the new edition of our book *Competing on Analytics,*

To succeed at an analytical competitor, information workers and decision-makers need to become adept at three core skills [11]:

Experimental: Managers and business analysts must be able to apply the principles of scientific experimentation to their business. They must know how to construct intelligent hypotheses. They also need to understand the principles of experimental testing and design, including population selection and sampling, in order to evaluate the validity of data analyses. As randomized testing and experimentation become more commonplace in the financial services, retail, and telecommunications industries, a background in scientific experimental design will be particularly valued. Google's recruiters know that experimentation and testing are integral parts of their culture and business processes. So job applicants are asked questions such as "How many tennis balls would fit in a school bus?" or "How many sewer covers are there in Brooklyn?" The point isn't to find the right answer but to test the applicant's skills in experimental design, logic, and quantitative analysis.

Numerate: Analytical leaders tell us that an increasingly critical for their workforce is to become more adept in the interpretation and use of numeric data. VELUX's [Anders] Reinhardt [until recently global head of business intelligence at the Danish window company] explains that "Business users don't need to be statisticians, but they need to understand the proper usage of statistical methods. We want our business users to understand how to interpret data, metrics, and the results of statistical models." Some companies, out of necessity, make sure that their employees are already highly adept at mathematical reasoning when they are hired. Capital One's hiring practices are geared toward hiring highly analytical and numerate employees into every aspect of the business. Prospective employees, including senior executives, go through a rigorous interview process, including tests of their mathematical reasoning, logic, and problem-solving abilities.

Data literate: Managers increasingly need to be adept at finding, manipulating, managing, and interpreting data, including not just numbers but also text and images. Data literacy is rapidly becoming an integral aspect of every business function and activity. Procter & Gamble's former chairman and CEO Bob McDonald is convinced that "data modeling, simulation, and other digital tools are reshaping how we innovate." And that changed the skills needed by his employees. To meet this challenge, P&G created "a baseline digital-skills inventory that's tailored to every level of advancement in the organization." The current CEO, David Taylor, also supports and has continued this policy. At VELUX, data literacy training for business users is a priority. Managers need to understand what data are available, and to use data visualization techniques to process and interpret them. "Perhaps most importantly, we need to help them to imagine how new types of data can lead to new insights," notes Reinhardt [12].

As with analytical professionals, additional function unit- or business unit-specific expertise in analytics may be needed by amateurs. In the case of IT professionals, for example, those who provision and support data warehouses and lakes should have some sense of what analyses are being performed on data, so that they can ensure that stored data are in the right formats for analysis. HR workers need to understand something about analytics so that they can hire people with the right kinds of analytical skills–and how analytics can be employed to identify promising employees, or those likely to leave the company soon. With the rise of artificial intelligence, even the corporate legal staff may need to understand the implications of a firm's approach to automated decision-making in case something goes awry in the process. There are also an increasing number of AI applications in corporate litigation as well.

Interview With Greta Roberts

When asked for her thoughts on the essentials of assembling an analytics team, cofounder and CEO of Talent Analytics Greta Roberts responded:

One thing that has been a curiosity for us has been the question of whether there a quintessential data scientist. If you had a list of all the necessary attributes, could you find it all in one person? I never believed that for an instant, but it was anecdotal. We wanted to study it quantitatively. It was interesting because, since it was on the analytics side, you would think the first thing that analytics people would do is to use a quantitative approach to understand the people who are actually doing the analysis. Harlan Harris, Marck Vaisman, and Sean Murphy wrote a book–*Analyzing the Analyzers*–that makes me ask, "Why not just apply analytics to understand the analyzers?" That's when we really said that, instead of just saying anecdotally, "We think this is what a data scientist is," we would actually see if there is a way to really understand them.

Just as in IT there is not just one kind of IT person, there is not one kind of analytics person. I think because analytics is still relatively forming in its new iteration–you do have people that do the entire analytics workflow. There are people that need to do everything from forming the question to data acquisition and collection to visualization, programming, interpretation, presentation, communication . . . you name it. I think we're starting to see that some of that is breaking into specialization. We've seen anecdotally that sometimes you have data scientists who now just do the data acquisition and collection (and maybe preparation) side, and then turn the work over to people to do the programming and the design of the algorithms. The programmers, in turn, then turn it over to people who actually do the presentation. We've been very interested because of the work that we do here, always really understanding the people that do the analytics work. We've been interested in moving beyond anecdotes around analytics professionals.

There is not a single kind of marketer or one kind of sales rep or one kind of IT person . . . or one kind of anything, really. For any role, there is always a lot of granularity inside the role. We are really only interested in the analyzers and whether there is a way to analyze the analyzers. Is there a way to categorize a little better to make it easier for people to identify the people who are doing the work? I think the passion comes from the interest in being involved. It is interesting that there is a quantitative approach, and it's particularly interesting to use the same approach–analysis–to understand the analyzers. That is fun on all kinds of different levels!

This is an excerpt from one of a series of interviews with analytics professionals and educators commissioned by the *INFORMS Analytics Body of Knowledge* Committee.

3.3 Managing Analytical Talent

In addition to hiring people with the right kinds and levels of skills, there are a number of activities that are involved in ongoing management of analytical talent [13]. One such activity is to conduct an assessment of the analytical skills within your organization and a "gap analysis" of the differences between the current state and the desired future state. The level of details in the assessment may vary by the purposes of the organization, but may include a roles inventory of analytical jobs and their locations within the organization, a skills inventory, and an analytics talent map. The skills inventory might include a listing of the analytical skills required, and a comparison to the desired skill levels and numbers of people possessing them. A talent map is a high-level mapping of current roles and skills, comparisons to desired future objectives, and elements of plans to close the gap–all ideally in some visual format that is easily comprehensible by busy executives.

The factors measured in the assessment will also differ by the strategies and priorities of the organization conducting the assessment. Some typical examples of factors include the following:

- How many people are there in each of the major analytics functions?
- What percentage of analytical professionals are capable of predictive and prescriptive analytics, as opposed to descriptive analytics?
- How many data scientists are able to use machine learning to build models?
- What percentage of data management-oriented employees have any experience with Hadoop and other big data tools?
- How many employees are familiar with each of the software tools in our approved portfolio for analytics?

- How many analytics staff have close and trusting relationships with the business leaders in the units and functions they serve?
- What analytics/technical skills exist within the current staff by type and number of years of experience?
- What percentage of analytics team members have more than 3 years, or less than 1 year, of experience in analytics?
- Which software/tools have the most and least number of skilled resources available for development and support activities?

Answering these types of questions can allow analytics leaders to build the initial foundation of their organization's talent strategy and roadmap. In order to ensure that the information is relevant to the entire enterprise, it is important to involve all analytics leaders within the company and to include questions or decision points that address the unique nature of the organization and industry. In addition to providing important information, for highly decentralized analytics groups such an inventory can also be a first step toward building a greater level of cohesion.

After doing the assessment, a company will normally want to formulate some objectives and plans for what to do about the results, with a time frame for planned changes. One company, for example, determined that only 5% of its analytics staff were comfortable with predictive analytics, and it wanted to shift to 95% with that skill over 5 years. Another organization determined that its staff lacked close relationships with business leaders, so it developed clearer assignments of analysts to business units, and asked business leaders to participate in annual performance assessments for analytics staff.

A one-time talent assessment is of limited value. People, their skills, and objectives for new capabilities change all the time. Organizations should reassess their analytical roles and skills every year or two. Once an assessment process is in place, it can be repeated relatively easily.

3.3.1 Developing Talent

Many analytics organizations primarily think about hiring people with needed skills. But it is often less expensive and more effective to develop skills through education and training programs, either in-house or in partnership with universities. If there is a critical analytical skill that an organization identifies that is particularly important, it is not difficult to arrange a training program for it. There are, for example, training programs available for organizations that want their analysts to achieve CAP certification from INFORMS.

For another example within a specific firm, Cisco Systems has been expanding for several years into advanced services that analyze the data thrown off by devices like routers and switches. In addition, Cisco has been using analytics extensively for internal purposes, such as sales propensity modeling and demand/production forecasting.

However, managers within Cisco felt that they lacked the data science skills to effectively perform all these activities. Desmond Murray, a Senior Director for IT at Cisco, was running Enterprise Data and Data Security for the company in 2012. His team was adopting new big data technologies (Hadoop, SAP HANA, etc.) for the company to use, but demand within the business was limited. He concluded that a set of educational offerings around data science would build awareness and stoke demand for these new technologies.

Murray designed a distance education (an obvious approach, given Cisco's distance conferencing business offerings) program on data science with two different universities. The program would last for 9 months and results in a certificate in data science from the university. Students attend virtual classes a couple of nights a week, and of course had homework as well. Cisco is now on its sixth student cohort with 40 students in each. About 300 data scientists have been trained and certified, and are now based in a variety of different functions and business units at Cisco.

But Murray, by now head of the Enterprise Data Science Office within the IT organization, didn't stop there. He realized that the newly trained data scientists needed some support from their managers if they were going to be satisfied in their new roles. So Cisco also created a 2-day executive program led by business school professors on what analytics and data science are and how they are typically applied to business problems. The program also covers how to manage analysts and data scientists, and how to know whether their work is effective. Cisco's initiatives to develop its employees' analytics and data science skills are relatively unusual, but they don't have to be. Any company that is serious about analytics and data science could undertake similar steps.

3.3.2 Working with the HR Organization

Analytics and data science organizations in companies can do a lot to identify and inculcate needed skills. At some point, however, it will probably be wise to collaborate with a company's human resources (HR) function to institutionalize talent management processes. HR groups can help to establish formal job titles, create linkages between skill and seniority levels and compensation, and provide internal and external resources for training. If analytics and data science skills are considered strategic, HR groups can help to source, nurture, and manage them. Many HR organizations are themselves interested in doing more with analytics in their own functions, so a partnership with analytics groups can be mutually beneficial.

HR organizations can provide guidance about the type of future skills that the organization will need. Additionally, HR leadership can describe the types of nontechnical skills that they are planning to develop or already have available to support the analytics function (e.g., business acumen, relationship, or communication skills).

At Cisco, the creation of data science skill development programs revealed that there was no standard at Cisco–or at many other firms, for that matter–for who is a serious data scientist and who isn't. So they created a "Pyramid of Analytic Knowledge" to classify different levels of expertise and establish a career track. Murray and his successor worked with Cisco's HR organization to incorporate these into official job classifications and compensation bands.

Interview With Russell Walker

When asked about soft skills and the analytics professional, Kellogg School of Management Clinical Associate Professor Russell Walker responded:

Many of us have personality traits and interests that we cannot divorce ourselves from very easily. However, I suspect that with appropriate awareness and training, there are probably effective tools for doing so. I also suspect that this that would be an enormous undertaking and perhaps even an exercise in great frustration for many people.

I would not necessarily expect employees to be someone other than who they are. But in a business setting, employees should be aware of the impact of their work on others and the contribution of others to their work. This is best achieved by creating some sort of collaborative environment, and this is an approach that more analytical professionals should embrace. Perhaps you have some tasks at work that you can–or even should–do alone. However, many tasks cannot be accomplished alone. Your manager is involved, your coworkers are involved, your customers are involved–someone pulls the data, someone performs the analyses, someone creates the PowerPoint deck, etc. In an analytics project, there is generally an enormous amount of work to do across the team, and the team is often rather disparate. So being mindful and respectful of this aspect of analytics is critical.

In a seemingly simple exercise, I ask students I have assigned to work on a project in teams to go to dinner as a team and interview each other. The objective is to gain a better understanding of the specific strengths and interests of their teammates. I think this goes a very long way in simply helping everybody communicate. There are people who are much better at so-called soft skills, just as there are people who are better at drawing than others, and again this is probably driven by personality. Can we teach everybody to draw? With sufficient effort, we probably could make everybody good–not necessarily an artist, but reasonably accomplished at drawing. But could we make everybody a grand master craftsman or an artist? I do not think so!

So we should all recognize limitations and strengths in others and in ourselves, and by doing so see that a team can be better if someone takes the lead role in the presentation, someone takes the lead role in model building, etc. Someone should not go to work and be expected to dwell on and toil in their weaknesses. Corporations are actually pretty good at addressing this; at the end of the year, your manager will perform an annual review and say, "Here are all the things you didn't do well, and this is why we're not going to give you a full bonus." But if we look at sports as an example, no coach or general manager goes to the quarterback at the end of a season and says, "You didn't kick any field goals this year, so we're going to take money away from you." You ask the quarterback to do his job, and you let the kicker do his job.

If you are acutely aware of your strengths and your limitations, you can avoid being assigned to a role in which you are set up to fail by having a frank discussion with your employer. And employers can avoid assigning employees–their human capital and valuable assets–roles for which they might not have the requisite traits or interests.

This is an excerpt from one of a series of interviews with analytics professionals and educators commissioned by the *INFORMS Analytics Body of Knowledge* Committee.

3.4 Organizing Analytics [14]

One of the key questions to address in managing analytical teams is "How should we best organize our analysts and data scientists?" It is a common question arising from a common situation: Analysts and analytics/big data projects are often scattered across the organization. That is how companies get started with analytics–here and there as pockets of interest arise. However, when an organization starts to get serious about analytics and data science, it often adopts an enterprise perspective in order to develop analysts effectively and deploy them where they create the greatest business value. In order to achieve these objectives, pockets of analytics and data science usually need to be coordinated, consolidated, or centralized.

The trend over the past decade has clearly been toward centralization of analysts, and that makes sense for several good reasons. If a company wants to differentiate itself in the marketplace through its analytical capabilities, it doesn't make sense to manage analytics locally. Skilled and experienced analysts and data scientists are a scarce and high-demand resource. A central function can deploy them on the most important projects, including cross-functional and enterprise-wide projects that may be otherwise difficult to staff. Centralization also facilitates analyst development because people have more opportunity to connect with and learn from one another. In addition, a central group with a critical mass of people helps with recruiting analysts by demonstrating the organization's commitment to analytics and providing new hires with a community. Finally, research led by my frequent coauthor Jeanne Harris [15] suggests that analysts in centralized and well-coordinated structures are more engaged and less likely to leave their employer than their decentralized counterparts.

However, recent trends suggest that analytics and data science teams are not immune from the normal pressures that move centralized functions in a more decentralized direction. Previously centralized analytics groups have been decentralized and dispersed in several different companies over the past year or two. The leaders of these groups cite several reasons for the decentralization, including the visibility of centralized budgets, complaints of lack of responsiveness by business unit and function leaders, and perceptions of excessive bureaucracy in large analytics groups. It seems likely, then, that despite the efficiency and effectiveness benefits of centralization, there will be the usual oscillation between centralization and decentralization in analytics and data science groups.

Another common situation among organizations I encounter is a significant analytics presence in one or two business functions, plus small pockets of analytics across the rest of the organization. The lead functions vary by industry–risk management and trading in financial services, engineering and supply chain in manufacturing, and marketing in consumer businesses. The challenge here is simultaneously to connect the pockets of analytics and spread the wealth of expertise resident in the advanced units. In these cases, full centralization could be unnecessarily disruptive, so the organization needs other mechanisms to coordinate analyst talent supply.

In the book *Analytics at Work*, Jeanne Harris, Bob Morison, and I discuss five common organizational models [16]. They are a useful place to start, but organizing your analysts isn't as simple as just picking one. There are different organizational circumstances, with many variables and mitigating factors in play, and many variations on these five options. This section attempts to decompose the organizational models for analysts and data scientists, and provides tools for developing and tuning your own model.

3.4.1 Goals of a Particular Analytics Organization

When debating alternative organizational structures for analytical and data science groups, it is important to keep the overriding goals for the organization in mind. Typically, the following are some of the goals of analytical groups and their leadership within companies:

- Supporting business decision-makers with analytical capabilities
- Helping to develop new customer-oriented products and features involving data and analytics
- Providing leadership and a "critical mass" home for analytical and data science-oriented people, and the ability to easily share ideas and collaborate on projects across analysts

- Fostering visibility for analytics and big data throughout the organization, and ease in finding help with analytical problems and decisions
- Creating standardized methodological approaches, tools, and processes
- Researching and adopting new analytical and data science practices
- Reducing the cost to deliver analytical outcomes
- Building and monitoring analytical capabilities and expertise

Different priorities for these goals may lead to different organizational models. For example, the goal of supporting business decision-makers with analytics may be best served by locating analysts directly in business units and functions that those decision-makers lead. That decentralized approach may also be the most effective one for development of products and services based on analytics and data. However, such decentralization may work against the goal of giving analysts and data scientists the ability to easily share ideas and collaborate.

Note that throughout this section (and the chapter in general) I have generally mentioned analysts and data scientists in the same breath. This usage is intentional; I believe that it was always difficult to clearly differentiate between "traditional" quantitative analysts and data scientists, and it is becoming increasingly difficult over time. At one point, data scientists tended to be more experimentally focused than traditional analysts, and also were likely to write code to transform unstructured data into structured formats for analysis. But now the tasks that these two groups perform certainly overlap, and the cachet of the data scientist title means that it is being applied to more jobs. My assumption is whatever organizational structure makes sense for one group also makes sense for both; that is, analysts and data scientists should be part of the same larger group. Of course, there are always exceptions to any organizational structure rule.

As I suggested above, no set of organizational structures and processes is perfect or permanent, so organizations must decide what particular goals are most important at any point in their analytical life cycles. For example, if an organization has had a centralized group of analysts and data scientists for a while and it has become unresponsive to business unit needs, it may be time to establish stronger ties between analysts and specified business units and leaders. A company with highly localized analytics may need to switch, at least for a while, to a more centralized structure. If possible, however, organizations should avoid rapid swings from highly centralized structures to highly decentralized structures, and back again. There are usually less disruptive ways to achieve the desired goals.

3.4.2 Basic Models for Organizing Analytics

Figure 3.1 shows the common organizational models described in *Analytics at Work.*

In a *centralized* model, all analyst groups are part of one corporate organization. Even if located in or primarily assigned to business units or functions, all

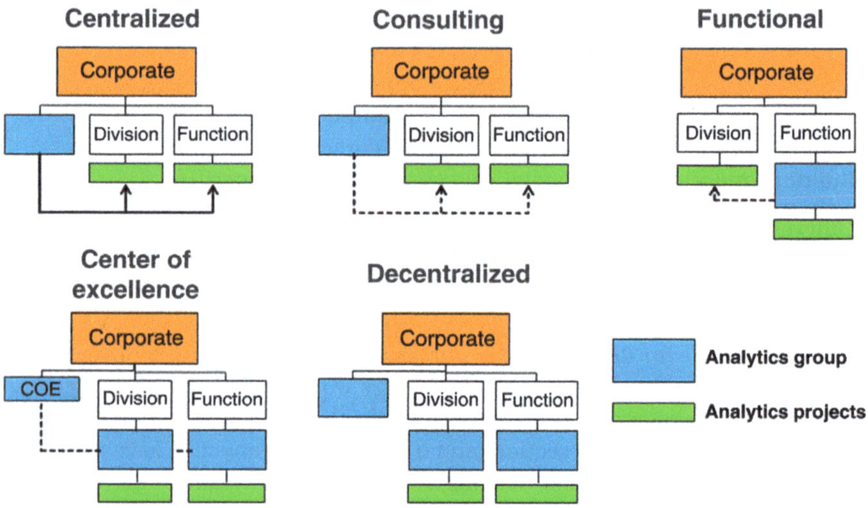

Figure 3.1 Common organizational models described in *Analytics at Work*. (Adapted from Davenport, Harris, and Morison, 2010.)

analysts report to the corporate unit. This obviously makes it easier to deploy analysts on projects with strategic priority, as well as to develop skills and build community. However, especially if the analysts and data scientists are all housed in the corporate location, it can create distance between them and the business (although this can be mitigated by other factors, as I describe below). Implementing a centralized model for analytics is easiest where there is successful precedent for operating other functions or managing scarce resources as centralized shared services.

In a *consulting* model, all analysts/data scientists are part of one central organization, but instead of being deployed from corporate to business unit projects, the business units "hire" analysts for their analytical projects. This model is more market driven, and especially important here is the analyst/consultants' ability to educate and advise their customers on how to utilize analytical services– in other words, to make the market demand smart. This model can be troublesome if enterprise focus and targeting mechanisms are weak, because analysts may end up working on whatever business units choose to pay for (or whatever wheel is squeakiest) rather than what delivers the most business value.

In a *functional* or "best home" model, there is one major analyst/data scientist unit that reports to the business unit or function that is the primary consumer of analyst services. This analyst unit typically also provides services in a consulting fashion (or even better, strategic prioritization) to the rest of the corporation. As already mentioned, many financial services and manufacturing firms have, in effect, a functional model today, with one or two well-established analyst groups in functions like marketing or risk management. The best home may migrate as analytical applications are completed and the analytical orientation of the corporation changes, typically from operations to marketing.

A *center of excellence* model is a somewhat less centralized approach that still incorporates some enterprise-level coordination. In this structure, analysts are based primarily in business functions and units, but their activities are coordinated by a small central group. The CoEs are typically responsible for issues such as training, adoption of analytical tools, and facilitating communication among analysts. The CoE builds a community of analysts/data scientists and can organize or influence their development and their sharing across units. It is most appropriate for large, diverse businesses with a variety of analytical needs and issues, but that still would benefit from central coordination. This is perhaps the most popular of the five models. In the era of business intelligence, this model was sometimes called a "business intelligence competency center."

There are many variations on this model, depending on the powers of the CoE. Do analysts report to it dotted line? Does it control the staff development agenda and resources? Does it double as a Program Management Office (PMO), with powers to coordinate priorities and resources across business units? Or are the business units solidly in charge of their analysts?

In a *decentralized* model, analyst groups are associated with business units and functions, and there is likely an analytics group or groups for corporate functions, but there is no corporate reporting or consolidating structure. This model makes it difficult to set enterprise priorities and difficult to develop and deploy staff effectively through borrowing and rotation of staff. It is most appropriate in a diversified multibusiness corporation where the businesses have little in common. But even then it makes sense to build a cross-business community of analysts so that they can share experience. As a result, this is the model I (and my *Analytics at Work* coauthors) am least likely to endorse.

Beneath the surface, each of these models is essentially either centralized or decentralized. The consulting and functional models are variations on centralization–the consulting model has different funding and deployment methods, and the functional model is centralized, just not at corporate. The CoE model is an overlay on a decentralized structure. So are other hybrid models, most commonly a combination of decentralized analyst groups in business units plus a central group at corporate that focuses on cross-functional, cross-organizational, and enterprise-wide initiatives.

These five models have pros and cons and trade-offs in terms of deployment and development and other objectives. Figure 3.2 indicates the strengths of each in terms of four specific goals.

3.4.3 Coordination Approaches

One basic structure may be the best general fit, but no model will be best in terms of meeting all goals. Whatever the basic model, there will be a need to coordinate

	Centralized	Consulting	Functional	CoE	Decentralized
Deploy analysts on the most important business initiatives	●		◑	◑	
Deploy analysts close to the business for purposes of service, responsiveness, and local learning		◑	◑	●	●
Maximize the overall learning and development of analysts	●	◑		◑	
Maximize the engagement of analysts	●		◑	◑	

● = very strong ◑ = somewhat strong

Figure 3.2 The strength of the five models. (Adapted from Davenport, Harris, and Morison, 2010.)

across analyst groups or across different parts of the business that are consuming analyst services. In a sense, all models are hybrids. Even if all analysts and data scientists work in one centralized corporate unit, the customers for their services are spread across the enterprise. You need coordination mechanisms to manage and meet demand for analytics.

There are a variety of common coordination mechanisms, some of which we've already mentioned. The mechanisms can supplement the formal reporting structure for purposes of enabling groups to plan and work together, and developing an enterprise view of priorities and resources. Think of them as ways of supplementing and fine-tuning a basic centralized or decentralized model, or of compensating for its inherent weaknesses. And note that all present challenges.

Program Management Office

This is a formal corporate unit for setting enterprise priorities, coordinating analytics and big data initiatives, influencing resource deployment on key initiatives, and facilitating the borrowing of staff across analytics groups. As mentioned above, it may be a function within a center of excellence. PMOs are especially useful where potential business value from analytics is high and resources are scarce and distributed. Under a PMO, the deployment process must be sophisticated to meet the dual needs of project staffing and analyst development.

Federation

Analyst groups and their associated business units work together on priorities, coordination of initiatives, resource deployment, and analyst development under a set of "guidelines of federation." The most basic form of federation is a clearly chartered enterprise governance or steering committee. These committees add an immediate enterprise view, but they sometimes lack clout and even commitment. Some firms have considered federation as a sixth type of organization model.

Community

Decentralized analysts can be encouraged to share ideas and analytical approaches in a community. Such a community would typically involve occasional meetings, seminars, written communications, or electronic discussions or portals. It may be facilitated by a community organizer, and typically benefits from some budget. In most cases, this is a relatively weak coordination mechanism.

Matrix

Analyst groups report both to their associated business units and to a corporate analytics unit, with one line solid and the other dotted. Establishing dotted-line reporting to a central organization injects an imperative to get coordinated, but dotted-lines can lose their force over time if they're not regularly exercised.

Rotation

Some of the analysts in a centralized model are physically located in and dedicated to business units on a rotational basis. Or there is an enterprise-wide program facilitating the lending and migration of analysts across decentralized units. The strength and success of rotation programs are easy to gauge–analysts really do have mobility across the enterprise.

Assigned Customers

Some centralized analytics groups, such as the one at Procter & Gamble, have assigned or "embedded" analysts to work exclusively with particular business units and the leaders of those units. The assignments fall short of a matrixed tie in the organizational structure, but they help to ensure that the analytical needs of the units and their leaders are met. Recently, however, some of the embedded analysts at P&G have been put into a matrix structure; business units and functions were more comfortable having their analysts report to them.

For purposes of deploying analysts on the most important business initiatives, the PMO is the strongest mechanism. For purposes of developing analysts, all of the mechanisms can help the cause, but rotation programs may have the most profound effect. The coordination mechanisms can be used in combination–for example, a PMO focused on deployment and a community focused on development, or a federation focused on coordination and a matrix focused on ensuring alignment with business needs.

What Model Fits Your Business?

Any basic organizational design for analyst may look good on paper, but it's got to work in the context of how the business already operates. To evaluate, design, implement, and refine organizational structures, you've got to look behind the organization chart and consider some basic variables that have to be working together for any organizational model to succeed. These factors can either mitigate or strengthen the effects of any particular organizational structure. Figure 3.3 lists six key variables [17].

Home location is the geographical location where analyst groups officially reside for administrative purposes. Home base and formal reporting lines have been the dominant variables in organizational design, especially in companies where more headcount has indicated more power. However, in today's more fluid and collaborative organizations, home location means less and less (especially if coordination mechanisms are effective). Home location is a matter of convenience, with the goals of limiting travel to work locations, accommodating employees' preferences, and getting enough people in one place regularly to sustain a community. In many firms today, analysts are based offshore, either as employees or contractors.

Work location is where the work of business analytics is performed, typically a mix of in the field (wherever the business customers of analytical models and services may be) and in regional or corporate analytics centers (where colleagues and support services are readily available). It is generally best to locate analytics work, insofar as possible, where the corresponding business work is. This greatly

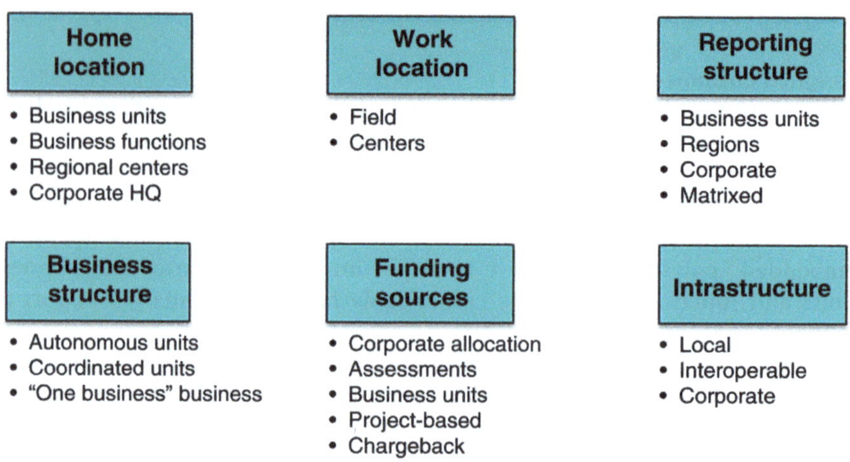

Figure 3.3 Six key variables for tuning organizational designs.

facilitates communication with business leaders and those who perform the work process under analysis. Make sure that home location and reporting structure don't erect barriers to analysts' working close to the business.

Reporting structure is the formal lines of connection, direction, and administration. Analysts and their groups typically report to local business units, to corporate, or to an intermediate unit (e.g., business sector or region) if the corporation is so structured. Some reporting structures are matrixed, with analysts reporting solid-line to business units and dotted-line to the corporate analytics organization, or vice versa. Reporting structure may be predetermined if analytics is part of another organization, such as marketing or IT. Reporting lines should not be so rigid as to impede the flexible staffing and development of analysts. Given the advantages of enterprise coordination of analytics, a least a dotted line to a central group or CoE makes sense in most organizations.

Business structure is the shape of the enterprise. Are its business units highly autonomous? Or are they closely coordinated? To what extent do business units already share functions, services, and important-but-scarce resources? Is power concentrated at the regional level? Centralizing analysts and data scientists may seem the logical thing to do, but then prove very impractical if that flies in the face of a locally autonomous or regionalized business structure.

Centralized analytics groups are a natural match for an integrated "one business" business. If business units are intertwined and must work with and rely on one another regularly, you need a centralized or consulting model, or else a strong federation. If business units are autonomous with little interconnection, analysts may stay decentralized, but a center of excellence helps in sharing experience and building the analyst community. And if the enterprise relies extensively on business partners to perform major processes, you may need a centralized structure, especially if there's need or opportunity to coordinate analytics with partners.

Funding sources are seldom considered in the context of organizational design, even though paralysis is guaranteed if organizational structure and funding sources are at odds. Friction is minimal if funding follows the lines of formal reporting, but matters are seldom that simple because business services like analytics often have multiple funding sources. These may include funds from corporate, business unit assessments, direct funds from business units, chargeback to business units for analyst time, and project-based funding from the sponsoring business unit or units. The organizational questions are as follows: To what extent does the basic model under consideration align with funding sources? How does funding need to be revised or influenced by coordination mechanisms to support the analytics organization and its work?

Project-based funding is the most market and demand driven, but it requires a certain level of maturity among business customers in setting analytics ambitions and priorities, and among analyst groups in advising customers and marketing their services. Project-based funding (or other funding for services performed) should in most cases be supplemented by seed funding (to foster innovation) and infrastructure funding (to build capability), usually from corporate.

Infrastructure includes the configuration and ownership of other essential resources, especially technology and data. This variable is similar to funding sources–alignment is essential to the success, but the variable is seldom considered in organizational design. Analysts cannot work across business processes and units if local systems and databases, inconsistent tools, and fragmented infrastructure prevent it. And business units cannot incorporate new technologies and techniques for analytical applications of corporate standards prevent it. To capitalize on analytics, the infrastructure must be local-but-interoperable or corporate-but-flexible.

As a practical matter, those six variables are never perfectly aligned, and organizations will have to experiment with and adjust the coordination mechanisms over time. As a common example, if data and technical infrastructure are fragmented, a company might phase an organizational consolidation alongside (or slightly in advance of) the rationalization and consolidation of those resources.

3.4.4 Organizational Structures for Specific Analytics Strategies and Scenarios

There are at least seven scenarios [18] for how enterprises approach and employ analytics (Table 3.1). These different emphases suggest different basic organizational models.

3.4.5 Analytical Leadership and the Chief Analytics Officer

Another key organizational question is the leadership role for analytics within organizations. There are already a substantial number of "Chief Analytics Officers (CAO)", and I expect that more will emerge. The role may not always have that title (it may, for example, be combined with Chief Data Officer–particularly in financial services), but there is a need–at least for each of the three centrally coordinated models described above–for someone to lead the analytics organization. The CAO could be either a permanent role, or a transitional role for an organization wanting to improve its analytical capabilities. There are a few Chief Data Scientists in organizations, but often these roles are combined with Chief Analytics Officer titles.

The roles of a Chief Analytics Officer could include any or all of the following:

- Mobilizing the needed data, people, and systems to make analytics succeed within an organization.
- Working closely with executives to inject analytics into company strategies and important decisions.

Table 3.1 Other factors driving effectiveness of analytics organizational structures.

Scenario	Definition	Basic model
Traditional analytics and BI	Make analytics tools and resources available to meet a broad variety of business needs	Centralized
Analytics for the masses	"Democratize" analytics and spread their use broadly across the organization	Centralized, with considerable effort to create self-service approaches
Big data	Tap the analytical potential of unstructured and nonquantitative data	Functional if one unit is in the lead leveraging these data; otherwise, consulting or centralized
Decision-centered	Enable the rapid and accurate execution of business decisions– both frequent/structured and infrequent/new	Model relatively unimportant if analysts can work closely with decision-makers, with a means of sharing methods and experience
Embedded analytics	Make real-time, automated analytical decisions part of core business processes and systems	Centralized or consulting, and close relationship with IT
Function- or process-specific analytics	Use specialized analytical technologies and applications to excel at a differentiating business process	Functional if there's an organization focused on the process; otherwise, consulting or centralized
Industry-specific analytics	Use specialized analytical technologies and applications to excel at processes common to an industry	Centralized or consulting, or functional if focus is on very specific applications

- Supervising the activities and careers of analytical people.
- Consulting with business functions and units on how to take advantage of analytics in their business processes.
- Surveying and contracting with external providers of analytical capabilities.

One key issue for the CAO role is whether analytical people across the organization should report to it. While an indirect reporting relationship (as one dimension of a matrixed organization) may be feasible, a CAO without any direct or indirect reports seems unlikely to be effective.

In one insurance firm, for example, the CEO was passionate about the role of analytics, and named a CAO as a direct report. But the CAO had only a couple of staff; all other analytics people in the organization did not report to him. The CEO did not want to "rock the organizational boat" by having such traditional analytical functions in insurance as actuaries and underwriters report to the CAO. As a result, the CAO felt he had no ability to carry out his objectives; he resigned from the role, and the CEO did not replace him.

3.5 To Where Should Analytical Functions Report?

There are a variety of different places in the organization to which centralized analytical/data science groups and their CAO leaders can report. While there is no ideal reporting relationship, each one has its strengths and weaknesses. In the following section each alternative is discussed.

Information Technology
Some organizations, such as a leading consumer products firm, have built analytical capabilities within the IT organization, or transferred them there. There are several reasons why this reporting relationship makes sense:

- Analytics are heavily dependent upon both data and software, and expertise on both of these is mostly likely to reside in an IT function.
- The IT function is used to serving a wide variety of organizational functions and business units.
- Analytics are closely aligned with some other typical IT functions, for example, business intelligence and data warehousing.

Of course, there are some disadvantages as well. IT organizations are sometimes slow to deliver analytical capabilities, and may have a poor reputation as a result. They may also overemphasize the technical components of analytics, and not focus sufficiently on business, organization, behavior, skill, and culture-related issues. Finally, IT organizations typically want to produce standardized and common solutions, and this may inhibit one-off analytical projects. In principle, however, there is no reason why IT organizations cannot overcome these problems.

Strategy
A few analytical groups, including those at a large retailer, report to a corporate strategy organization. This relationship allows analysts to become privy to the key strategic initiatives and objectives of the organization. Another virtue is that strategy groups are often staffed by analytically focused MBAs who may understand and appreciate analytical work, even if they cannot perform it themselves. The possible downsides to this reporting relationship are that strategy groups may not be able to marshal the technical and data resources to make analytical projects succeed, and strategy groups are usually relatively small.

Shared Services
In organizations with a shared administrative services organization, an analytics group can simply be part of that capability. The primary benefit of such a

reporting structure is that analysts can serve anyone in the company–and often there are charging and resource allocation mechanisms in place for doing so. The downside is that analytics may be viewed as a low-value, nonstrategic resource like some other shared service functions. With the appropriate mechanisms in place, this problem can surely be avoided.

Finance

Being a numbers-focused function, finance organizations have the potential to be a home for business analytics groups. The obvious virtue of this arrangement would be the ability to focus analytics on the issues that matter most to business performance, including enterprise performance management itself. For some unknown reason, however, most CFOs have not embraced analytics, and the finance function remains a logical, if uncommon, home for analytical groups. At some firms, however, including Deloitte (for internal analytics) and Ford, the finance function is beginning to play a much stronger role in championing analytical projects and perspectives.

Marketing or Other Specific Function

As noted above, if an organization's primary analytical activities are concentrated on marketing or some other specific function, then it makes sense to incorporate the analytical group within it. The resulting structure would allow a close focus on the analytics applications and issues in the functional area. Caesars Entertainment, for example, has put analytics in a reporting relationship to marketing. Obviously, it would also make it more difficult for analytical initiatives outside those functional areas to be pursued.

Product Development

The most likely industries for having analytics (and data science) reporting to product development are those–like online businesses–where there are a substantial number of "data products," or products and services based on analytics and data. There are, for example, analytics groups at LinkedIn, Facebook, and Google who report into product development organizations.

3.5.1 Building an Analytical Ecosystem

Most of the foregoing discussion about analytical capabilities has been focused on organizing and developing internal analytical capabilities. But there is a broad set of analytical offerings that are available from a wide variety of external providers as well. The providers include consultants, IT (primarily software) vendors, offshore analytical outsourcers, data providers, and other categories of assistance. Some provide general analytical help across industries, but in almost every industry there are also specialized analytics and data providers. Many firms can benefit from working with such "analytical ecosystems" to improve their capabilities.

The key in constructing an effective analytical ecosystem is not to let it grow at random, but to identify the analytical capabilities the organization needs overall. Then a decision should be made as to whether internal or external capabilities are most appropriate to fill a specific need. In general, external capabilities make sense when the need is highly specialized, not likely to be needed frequently, and not critical to the organization's ongoing analytical capabilities.

A major pharmaceutical firm's Commercial Analytics group, for example, has a well-developed ecosystem. There is a large group–more than 30–of internal analysts, but their capabilities are supplemented by outside help when necessary. The group has worked with specialized consultants on analysis of physician targeting, for example. The company's primary prescription data provider also works with it on analytics issues. Software vendors have consulted on analytical methods and techniques. Finally, the group supplements its work with help from an offshore analytics vendor in India.

3.5.2 Developing the Analytical Organization over Time

A final point is that analytical organization structures should develop and evolve over time. An internal structure and ecosystem that makes sense at the beginning of developing analytical capabilities will become obsolete later on. For example, it may be very reasonable to have a highly decentralized organizational model early on, but most firms create mechanisms for coordination and collaboration around analytics as they mature in their analytical orientations. It may also make sense to "borrow" a number of external resources in a firm's early stages of analytical maturity before making the commitment to build internal capabilities. In addition, companies may want to add data science capabilities to existing analytics groups to take advantage of the potential of big data.

The best way to adapt organizational capabilities to current needs is with a strategy or plan. Admittedly, in the early stages, there may not be anyone with the formal authority to even create a plan. However, if it appears that analytics are going to be key to an organization's future, it may make sense for a small group of analysts or data scientists to get together and create a bottom-up one.

At a large U.S. bank, for example, the head of the distribution organization (including physical branches, call centers, ATMs, and online channels) realized that she had a large number of analysts in her organization, but they weren't providing the value of which they were capable. She met with the managers of the diverse analytics and reporting groups in her business unit, and asked one of them to take the lead in assessing the problem. His work determined that the vast majority of the groups worked on reports rather than more predictive analytics, and that there were virtually no resources devoted to cross-channel analytics. With this start, the group began to develop a plan to remedy the situation and shift the balance toward predictive analytics and a cross-channel perspective. There was also a major focus on reducing the amount and frequency of reports.

Later, this same sponsor moved a different business unit toward heavier use of machine learning technologies.

Plans should probably be revised every year or so, or with major changes in the demand or supply around analytics. There are usually clear signs–if anyone is looking–that the current model has become dysfunctional. It is a key step in an organization's analytical development that someone takes responsibility–either informally or formally–for assessing the organization of analytical resources, and for creating a better model.

No set of skills, plans, or organizational structures is perfect–even for a given time and situation–and every structure or skillset, if taken beyond its limits, will become a limitation. The leaders of contemporary organizations will need to become conversant with their analytical capabilities and how they are organized. Most importantly, they will need to realize when their current organizational approach and team no longer functions effectively, and needs to be restructured and/or reskilled.

References

1 Harris J, Craig E, Egan H (2010) *Counting on Analytical Talent* (Accenture Institute for High Performance), March.
2 Muenchen RA (2017) *The Popularity of Data Science Software*, R4Stats website. Available at http://r4stats.com/articles/popularity/. June 19 version.
3 Davenport TH (2014) *Ten Kinds of Stories to Tell with Data*. Harvard Business Review blog post, May 5. Available at https://hbr.org/2014/05/10-kinds-of-stories-to-tell-with-data.
4 Roberts P, Roberts G (2013) *Research Brief: Four Functional Clusters of Analytics Professionals*, Talent Analytics Corp, July. Available at http://www.talentanalytics.com/wp-content/uploads/2012/05/ResearchBriefFunctionalClusters.pdf.
5 Davenport TH (2013) *Analytics 3.0*. Harvard Business Review, December. Available at https://hbr.org/2013/12/analytics-30.
6 Davenport TH, Patil DJ (2012) Data scientist: the sexiest job of the 21st century. *Harvard Bus. Rev.* 90(10): 70–76.
7 Davenport TH, Harris JG (2007) *Competing on Analytics* (Harvard Business Review Press).
8 Davenport TH, Kudyba S (2016) Designing and developing analytics-based data products. *MIT Sloan Management Review*, Fall 2016. Available at http://sloanreview.mit.edu/article/designing-and-developing-analytics-based-data-products/.
9 Hartman T (2012) *Is Big Data Producing Big Returns?* Avanade Insights blog post, June 5. Available at http://blog.avanade.com/avanade-insights/data-analytics/is-big-data-producing-big-returns/.

10 McDonald quotations from Davenport TH, Iansiti M, Serels A (2013) *Managing with Analytics at Procter & Gamble*, Harvard Business School case study, April.

11 The three core skills needed for analytical managers is adapted research originally published in Harris J (2012) Data is useless without the skills to analyze it. *Harvard Business Review*, September 13. Available at https://hbr.org/2012/09/data-is-useless-without-the-skills.

12 Davenport TH, Harris JG (2017) *Competing on Analytics* (Harvard Business Review Press, revised edition).

13 This section draws on content from a research brief by Harrington E (2014) *Building an Analytics Team for Your Organization*, International Institute for Analytics, September. Available at http://iianalytics.com/research/building-an-analytics-team-for-your-organization-part-i.

14 This section is a revised and updated version of a chapter by Morison R, Davenport TH (2012) Organizing analysts, in *Enterprise Analytics*, Davenport TH, ed. (Prentice Hall).

15 Accenture Institute for High Performance (2010) *Counting on Analytical Talent*.

16 Davenport TH, Harris JG, and Morison R (2010) *Analytics at Work: Smarter Decisions, Better Results* (Harvard Business Press), pp. 104–109.

17 Framework is based on *Building an Analytical Organization*, Business Analytics Concours and nGenera Corporation, 2008.

18 Framework is based on Morison R, Davenport TH (2008) *Mastering the Technologies of Business Analytics*, Business Analytics Concours and nGenera Corporation.

4

The Data

Brian T. Downs

Accenture Digital, Data Science Center of Excellence, Dallas, TX, USA

4.1 Introduction

Regardless of one's area of specialization or interest, it is true that most analytics students and professionals devote most of the effort and energy that goes into training to learning analytics methods and algorithms. A review of a typical curriculum in business analytics will reveal a sharp focus on the tools and techniques required, often in a specific context such as marketing or operations, to be a successful analytics practitioner. Therefore, it is often a surprise for people starting out in the field to discover that on most analytics projects, most of one's time is not spent on using the algorithms recently mastered with such great effort and determination. Rather, it is the lot of an analytics professional to spend most of their time messing with data. This chapter provides a practitioner's view of the different types of data, and some of the challenges in identifying, collecting, and preparing data for analysis.

4.2 Data Collection

4.2.1 Data Types

Before exploring data collection, a review of the various types of data will be useful. Figure 4.1 shows a useful hierarchy for describing these.

Qualitative data result from classifying something or labeling its attributes. There are three main types of qualitative data. *Nominal* data results when we identify things with named categories that do not have any natural or intrinsic value associated with them. For example, the wooden poles a utility company uses to transmit power to its customers can be classified by the species of tree from which they are made. Pine, fir, and cedar are meaningful categories in that each

INFORMS Analytics Body of Knowledge, First Edition. Edited by James J. Cochran.

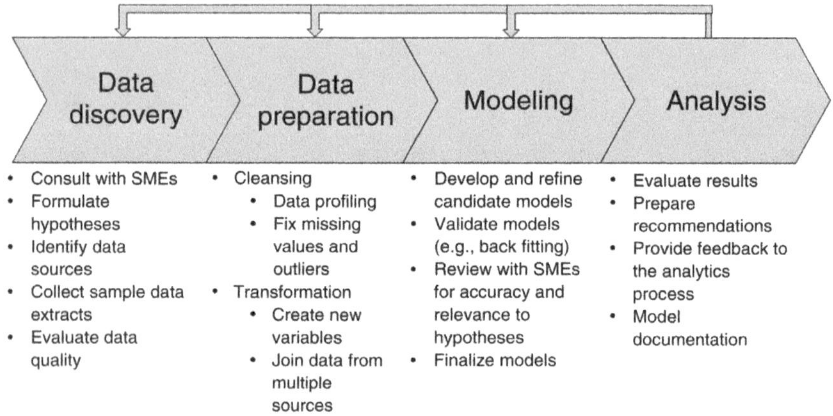

Figure 4.1 The analytics process.

has intrinsic properties that affect their performance in this application, but there is no obvious way to rank them based on the nominal classification alone. One could use this classification to perform an analysis to see if there are statistically significant differences in the lifespan of wood poles made from each species.

An important special case of nominal data is *binary* data. This type of data places something into one of two mutually exclusive and collectively exhaustive categories, often implying opposite states. A quality inspection of an item on a production line can result in a pass/fail. A production process can be in control/out of control. A magazine subscriber can renew/not renew. This type of data has become increasingly important as methods for predicting how likely an event of interest is to occur have seen widespread use in a variety of contexts. A manufacturing company may wish to predict how likely a machine is to fail given current operating conditions using data that can be collected from the production process. A cable television provider may wish to predict how likely a customer is to drop their cable service given demographic information and their history of problems and complaints. There are many number of classification methods that can predict events with binary outcomes effectively. The challenge in many cases is that historical data that contain multiple instances of each outcome may not be available, or will require some time to collect.

Ordinal data are created when one classifies things into categories where there is an implicit relationship between categories. The use of small, medium, and large to describe the size of things has an implicit meaning in many contexts. We expect a medium drink to contain more than a small drink, and a large drink to contain more than a medium drink. In the context of completing a survey, one might be asked to rank something from worst to best on a scale of 1–10, with the expectation that 5 is better than 2, 10 is better than 6, and so on. The problem with both of these examples is that the rank ordering does not tell us the magnitude of the difference between each category. There is no way to know from the classification that a medium drink is 33% bigger than a small drink, or how much better in absolute terms a rank of 10 is than a rank of 5.

Quantitative data are created when things are counted or measured. *Discrete* data result from counting things, and therefore is typically expressed as an integer value. The number of nights one has stayed with their preferred hotel chain is an example of discrete data. The number of warranty claims received on a model of smart phone is another. Neither of these things are recorded as fractional values as they refer to discrete events that have occurred an integer number of times.

Continuous data are generally anything that can be measured, and as such may have fractional values depending on how fine of a measurement is made. The flow rate of crude oil through a pipeline, the exhaust temperature of a diesel engine, and the daily output of a chemical process are all things that can be measured and the result will generally be a real number. One thing to be cautious about when using continuous data is that the quality and reliability if the data can be affected by the method of collection. Devices such as electronic sensors can be unreliable or influenced by the surrounding environment. Data recorded by human interaction are naturally prone to errors.

Time is also a potential consideration. Data collected from several subjects at approximately the same point in time are referred to as *cross-sectional* data. Examples of cross-sectional data are candidate preferences of voters immediately prior to an election, the high temperature on a given date in the 100 most populated U.S. cities, or the sizes of donations given to a charity during a fund raising drive. The most common purpose for collecting cross-sectional data is to develop an understanding of characteristics of a population at a particular point in time. Data collected from a single subject at several approximately equally spaced points in time are referred to as *time series* data. Examples of time series data are weekly sales for an item, the daily number of visitors to a museum, or the monthly rainfall measured at a weather station. The most common purpose for collecting time series data is to predict future values of the time series, such as when a sales history is used to predict future sales for planning purposes. In many cases, a time series will have measurable components that can be estimated using appropriate analytical methods. A trend indicates a long-term shift in the overall level of the time series, while seasonality is a cyclic pattern that repeats within a specific time interval such as a day or a year. If a single subject is observed over several points in space, the data are referred to as *spatial data*. Spatial data are similar to time series data and are often analyzed with methods designed for time series data. Finally, data are increasingly collected from several subjects at several approximately equally spaced points. These data, which have characteristics of both cross-sectional and time series data, are referred to as *panel data* (or *longitudinal data* or *cross-sectional time series data*). Time series and cross-sectional data are each a special case of panel data in which either the number of periods of time or the number of subjects is one.

Another type of data that has proliferated in recent years is unstructured *text data*. New technologies have been developed to collect and store this type of data, which can be collected from Web sites, social media, and discussion groups in the form of comments, reviews, and opinions. This type of data is stored as documents and may be used for text mining and sentiment analysis. As will be discussed later, the lack of structure in this type of data makes it difficult to store in a traditional database. Therefore, it has been a driving force in the evolution of nonrelational database technologies in recent years.

Interview With Harrison Schramm

Sometimes clients have data that could be useful to an analytics project. Harrison Schramm, who recently retired from a 20-year career as a helicopter pilot and an operations research analyst in the U.S. Navy, shares his thoughts on obtaining data for an analytics project from a client:

There are two ways to approach this, and the choice depends on the stakeholder. One way to go about it is to get stakeholders excited about what you are doing and make them want to help you by giving you their data. This is the preferred route.

The other route is to make stakeholders utterly terrified of what you are going to do if they don't give you data. This is a horrible route to take, but sometimes you have to go down this path. If you are working with a large organization, you cannot expect every segment of that organization to be excited about what you're going to do. So if one department is recalcitrant, you just end up having to say, "If you don't give us the data, then we're going to assume this, this, and this . . ." and you pathologically craft those assumptions so all of a sudden that giving you their data looks a lot better to them than those assumptions you're threatening to make. It's a varsity move–it's not for freshman.

This is an excerpt from one of a series of interviews with analytics professionals and educators commissioned by the *INFORMS Analytics Body of Knowledge* Committee.

4.2.2 Data Discovery

There are two types of analytics projects that are often encountered in practice. *Management Consulting*-type projects involve the use of analytics to solve a problem or answer a particular set of questions. These types of projects deal with one-time decisions and the "leave behind" from the effort is a report that contains analysis and recommendations. The questions addressed can be relatively simple, such as "Should I add storage capacity at a facility?", or they can be complex such as "How can I reduce my variable conversion cost while pursuing a high variety, highly customized make-to-order production strategy?" The analytics practitioner may use a single technique, or a combination of predictive analytics, simulation, and optimization. The data may come from a variety of internal and external sources, but are generally discarded after the project is complete. As such, there may not be a need to collect and merge the data into a permanent and sustainable environment. Data often will be collected in spreadsheet format, from a variety of sources, and will require considerable manual effort to prepare. Detective work may be required to locate some data elements as there may not be a system of record that contains what is needed, or even worse the data that are in the system of record may not be accurate. These types of projects can often be completed by people with analytics as their primary skill set as such people usually have some basic data management skills as well. Larger projects with high volumes of data may require data integration specialists to assist with data preparation.

The other common type of analytics project is *Application Development.* In these projects, analytics tools and algorithms are imbedded in an information technology system to support a set of business processes. Examples of such processes include forecasting and demand planning, sales and operations planning, and production process monitoring and control. In these applications, the analytics component is executed on a periodic basis. This can be anywhere from fractions of a second to monthly, depending on the type of process. Supporting a recurring process requires that the data needed by the analytics models be current and complete. This generally requires the development of a data warehouse, or at least an operational data store (ODS), which may involve combining data from a variety of sources in a single environment, and developing extract, transform, and load (ETL) procedures that capture data from source systems and move it into the analytics environment. ETL processes will also clean and transform data, creating tables and views that can be loaded directly into analytics applications. For these types of projects, the "leave behind" will be a functioning application, as well as the data infrastructure necessary to support it. In addition, there will be documented procedures in place to maintain the integrity of both the analytics models and the data. Application Development projects usually require skill sets beyond analytics, including data integration, system architecture, and data visualization.

At a high level, most analytics projects follow a similar process flow, although the amount of effort and complexity can vary widely depending on the scale of the initiative. Figure 4.2 shows a high-level view of typical activities that an analytics practitioner will undertake to complete a project. Note that a sizable proportion of the activities are data related. Data discovery is a critical first step as it is necessary to define the objectives of the work, as well as to determine the likelihood of a successful outcome.

Data discovery must begin with discussions with subject matter experts (SME) to understand the research question to be addressed with analytics. In a business

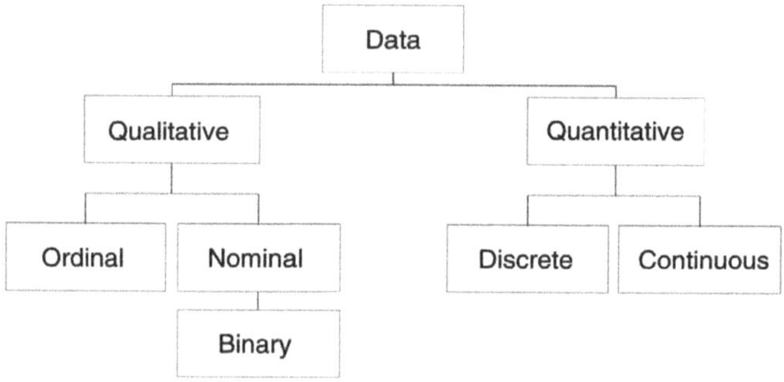

Figure 4.2 Types of data.

context, these are typically people who work in a client business and who are familiar with the attendant operational and business processes. It is the job of the analytics practitioner to understand the business issues at stake, and to frame those issues as testable hypotheses or to propose an approach for addressing a business need. An example of the former is "The results of our supplier audits can be used to predict which of them are most likely to be noncompliant in the next audit cycle." An example of the latter could be "We can use simulation modeling to understand the effect of different lot size policies on our manufacturing conversion costs."

Once the problem and analytics approach have been identified, it is necessary to determine whether the proposed approach is feasible. Critical elements for answering this question are the availability and quality of the necessary data. This requires carefully listing all of the required data elements for the analytics work and identifying possible sources for each. A data source will have an owner, whether it resides in a corporate information system or in a spreadsheet on a personal computer. Enlisting the cooperation of a data source owner is a key to success in analytics projects, as access to data and help understanding its format and structure are essential.

Figure 4.3 provides a way to categorize potential data sources. Along one dimension, one can think of data that are collected manually versus data that are collected by an automated process. Any process that involves a human being recording data on paper or through a form, electronic or paper, is manual data entry. Automated data collection does not require human intervention. Along the other dimension, there are data that are collected specifically in support of the analytics project at hand versus data that have been collected to support another business process but which can be used for the analytics project at hand. This last can be problematic since the specifications of the data being collected were designed to support a different objective, and there is a good chance that it will not be an exact fit for the needs of the current effort. It will likely require additional effort to augment and transform such data into a form fitting the current objective.

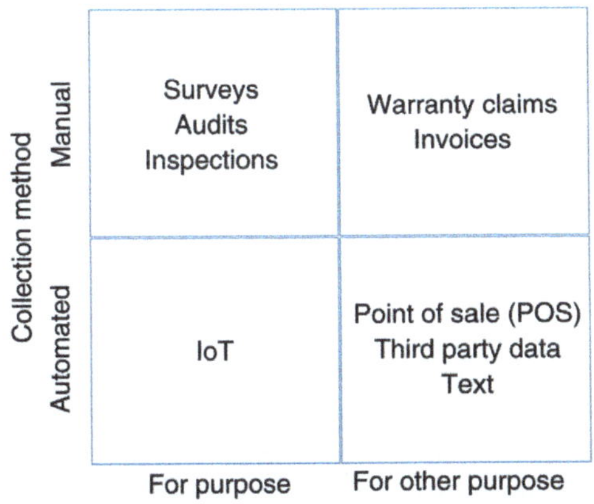

Figure 4.3 Data sources.

Surveys, audits, and inspections are all examples of methods requiring manual data collection that are designed to investigate specific questions using analytics. Surveys use statistical methods to identify a representative sample of a target population to measure their response to a set of research questions. Audits measure compliance to a set of standards, usually along several dimensions. Inspections entail a point-by-point examination of specific operating criteria, usually with a binary (pass/fail) outcome. For each of these, people are directly involved in the gathering and recording of the data, and thus there are opportunities for errors to occur in the process. These can take the form of simple keystroke errors, known as "fat fingering," or the failure to correctly record a response or observation. Other concerns relate to the effect of human judgment on the data collection process. Survey respondents may not answer truthfully because of a reluctance to express an unpopular viewpoint. Different auditors may evaluate the same situation as having different levels of compliance. Different inspectors may employ different thresholds as the standard for a pass/fail recommendation, or even fail to complete the entire inspection process. It is essential that an analytics practitioner be mindful of these potential sources of trouble when using such data to analyze the populations on which these tools are used.

There are transaction-oriented systems that rely upon manual data entry as the means by which data are digitized. Manual processes are often used to create and process invoices, creating records of customer, product, pricing, and shipment information. Such records are initially created for accounting purposes as a financial record of the transaction, but the same data can be used for other analysis such as sales forecasting and production planning. An industrial equipment manufacturer had a system that relied on data entered manually by technicians at their dealer sites to create warranty claims. The system had an electronic form that needed to be completed with data such as the product serial number, time of failure, the parts replaced, and a failure code to classify the nature of the claim. While initially used as a way to receive and process warranty claims so that dealers can be reimbursed for warranty service, over time this system creates a history of warranty claims that can be used by other analyses such as predictive maintenance, root cause analysis, and fraud detection.

An anecdote from the last example highlights the importance of a thoughtful design when creating tools for manual data entry. The field in the form that requested a code to classify the specific failure mode was free text, rather than a pull-down list that provided specific choices. Often the technicians would be in a hurry, or would not have the right code handy, and would enter an invalid code "99" just to complete the form and get the claim submitted. They would write short details in a free text field to describe the work performed. While in most cases this was enough information to get the claim reimbursement, it created a history of warranty claims that did not have the correct failure mode associated with many of the records. This made the data almost unusable for deeper analysis without someone trying to manually review the free text fields to recode the problem records, a task that proved impractical from both time and accuracy perspectives.

Some basic guidelines for design of forms for manual data collection can alleviate some of the data quality risks inherent in this method of data collection:

- Automate the workflow as much as possible, eliminating intermediate steps that use paper or spreadsheets. The use of tablets or other mobile devices for data collection in the field will improve both the accuracy and completeness of data.
- Limit the use of text fields.
- Use pull-down lists, radio boxes, and check boxes wherever possible instead of text fields to limit the potential for errors.
- Make clear which fields are required for the form to be submitted.
- Be sure that required data formats (e.g., dates, currency) are clearly indicated on the form.
- Validate the data and correct errors before allowing the form to be submitted.

This list is not exhaustive and there are many resources available on the Internet to assist in designing forms for data collection. It is important that an analytics practitioner be mindful of these issues when designing tools or applications for these types of applications.

A common lament from many companies is that they are drowning in data, but do not know how to extract value from all the data they possess. This can be attributed in part to factors such as the low cost of data storage, and the proliferation of inexpensive sensors that can be used to collect data from equipment at intervals as small as a fraction of a second. In many industries, the collection of process data from sensors and other systems such as SCADA (Supervisory Control and Data Acquisition) is a prevalent form of automated data collection that is used to monitor and control processes, as well for other analytics-driven applications such as predictive maintenance. This type of data is the life blood of the Internet of Things (IoT), as the automation of the data collection process enables the automation of monitoring and control processes. This type of data consists of high-frequency time series, typically measurements of process parameters such as pressure, temperature, or velocity. It is usually captured and stored in a *data historian*, such as AspenTech's InfoPlus.21 or OSISoft's PI, that is designed for the efficient storage and retrieval of time series data. Examples of this type of automated data collection can be found in a variety of industries. According to *Aviation Week* (http://aviationweek.com/connected-aerospace/internet-aircraft-things-industry-set-be-transformed), there are now jet engines that have over 5000 sensors, and produce over10 GB of data a second. Industrial equipment manufacturers that serve the mining industry have developed sensors and software that collect operational data from shovels and haul trucks. Utilities collect process data from power generation equipment such as turbines. And processes throughout the oil and gas industry are closely monitored using automated data collection, from upstream drilling and extraction through refining and downstream chemical production. Value can be extracted from this enormous volume of data, but not without considerable effort in the data preparation step of the analytics process.

Other types of automated data collection occur in systems that are used for transaction management purposes such as point of sale (POS) systems. Such systems are used to process transactions in retail operations and perform critical tasks such as invoice preparation, payment and membership discount processing, inventory management, and promotion processing. Since tools such as bar code and credit card readers are a part of the system, the need for manual data entry is nearly eliminated and errors are minimized. The resulting sales records, which contain information about both customers and products, are captured in a data base that can be used for a variety of analytics applications. These include customer segmentation to allow targeted promotions, supply chain segmentation to enable segment-specific strategies such as make to order (MTO) and make to stock (MTS), and forecasting and demand sensing to allow in-season adjustments to production quantities and inventory placement.

Another important data source to be considered is called third party data. These are data provided by an external source that will collect, cleanse, and transform data into usable form, typically on a subscription basis. This includes industry-focused companies such as S&P Global Platts, which provides energy and commodities information, including pricing. Experian is known as a credit reporting corporation with a global footprint. The U.S. Bureau of Labor Statistics compiles a variety of price and production indices, which range from the aggregate level to the industry or commodity specific such as construction, natural gas, and electric utilities. These indices are time series, and are based on good and services specific to the sector that they measure. They are often available in a seasonally adjusted form, with seasonal variation removed, making it easier for analytics models to identify correlations between different time series. This is especially useful when an index is a leading indicator, giving it predictive power for other related time series.

4.3 Data Preparation

Referring to Figure 4.1, one can see that the next step in the analytics process is data preparation. This step can be divided into two parts: data cleansing and data transformation. The objective of the data preparation step is to collect the data that have been gathered from various sources into a single location, and transform it into a form that can be consumed by analytical tools and software. Figure 4.4 shows the flow of data through the process, and the important activities conducted at each step.

Data profiling involves a univariate analysis of each of the variables in a data source, as well as a record-by-record evaluation of the completeness of the data. This is to allow the analyst to evaluate the suitability of the data for the project at hand. For quantitative data, this analysis will involve plotting the distribution of the variable, and identifying measures of central tendency such as the mean and median, as well as measure of dispersion including maximum, minimum, range, variance, and skewness. In addition to providing a sense of the overall shape of the data, this analysis provides insight about the possible probability distributions that may apply to the data, including whether the assumption of normality is warranted. In addition, profiling can help with the identification of missing values, extreme values, or problems with scaling.

Data profiling of qualitative data will involve the creation of frequency histograms to confirm that the data values are valid and complete. This aids in the identification of common data errors, such as missing or inconsistent values, or problems such as high cardinality of values in a categorical variable.

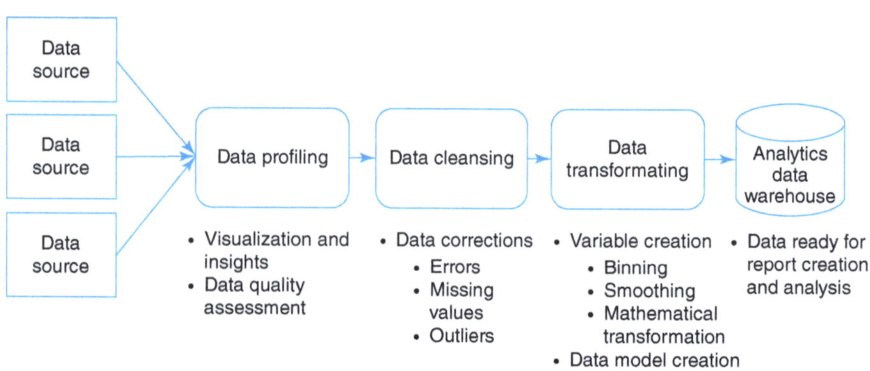

Figure 4.4 Data preparation.

Most commercially available analytics software has standard routines that are available for data profiling. These can greatly reduce the amount of effort involved in the process. However, this task can be completed using the tools that are available in a typical spreadsheet program, although the time and effort required to do so will be much greater than using a tool designed specifically for that purpose. No matter what the tool employed, a key output of data profiling is a list of data issues that need to be remediated to proceed with the analysis. What follows is a discussion of some common data problems that need to be addressed at this stage of the process.

Missing values are endemic to many data sources, and they can occur in a variety of ways often as the result of human involvement in the data collection process. Operational data collected in the field are particularly prone to missing values. Technicians may neglect to enter key information in a form such as identification codes for assets that they are inspecting simply because they cannot see through obstructions such as vegetation. Survey data may have missing values. It is typical that high-income respondents are reluctant to answer questions about their income level and may not respond to them. Whatever the source of the missing values, the critical question to answer is whether the missing values affect the representativeness of the data relative to the population from which it comes. If the sample size is large and the number of missing values is few, the missing values can be discarded without altering the results of the analysis. However, if the number of missing values is large, or if the incidence of the missing values is due to some systemic cause as described in the second example above, it will be necessary to attempt the estimation of the missing values.

In cases where data are gathered from operational systems, it may be necessary to pull data together from multiple sources to create records for analysis. For example, in the case described above, suppose there are technicians performing inspections in the field, and some of the data elements in the inspection form are missing. If there is a master list of assets, it may be possible to fill gaps such as missing identification codes by comparing timestamps from inspection records, and ascertaining the location of the crew at the time the inspection was performed. Comparing this type of data with the geographic coordinate data contained in the master list of assets may make it possible to identify the assets for which the identification codes are missing. In practice, this type of forensic approach to missing value correction is quite common, although it is labor-intensive and usually requires the assistance of someone who has a profound understanding of the data.

When there is a sensitivity to responding to certain questions, perhaps about topics such as income or politics, survey data may contain missing values that indicate response bias, called not missing at random (NMAR). One way to identify this is to create a new binary variable coded as response/no response, and to compare mean values of response variables between the two groups. If there are significant differences, this indicates a nonresponse bias and one should be careful about making inferences using such data.

In other situations, there will not be significant differences between the response and no response group for variables of interest. In this case, the data are missing at random (MAR). One can proceed with the analysis without fear of nonresponse bias, although there will be smaller sample sizes for the questions where there are no responses. If we repeat the comparison of means for all response variables and find no significant differences between the response and no response groups, then we have the best outcome and the data are said to be missing completely at random (MCAR).

There are two approaches often employed in situations where there are missing data due to no response on survey instruments. Pairwise deletion occurs when the responses to each question are summarized individually and the missing value is just excluded from the analysis. Similarly, one can perform correlation analysis on such data, but with a smaller sample size due to the missing values. List-wise deletion is used when using tools such as multiple regression or classification models. Since these methods seek to determine the relative influence of each of a group of predictor variables, any missing value requires that the entire record be excluded from the analysis.

Other techniques for handing missing values fall into the category of imputation. This is the substitution of some value for the missing values using mathematical methods of estimation. The simplest is to use the mean value of all observations for a missing value. This has the desirable property of not changing the mean of the variable, although it will dilute the correlation between that variable and any other. There are many other methods available for imputation of missing values. The reader is advised that many of them are quite advanced and will require some experience and skill to properly implement.

Another common problem with data comes in the form of *nonstandard values*. This happens when the same categorical data value is represented in the data with more than one set of characters. For example, the category "not applicable" may be represented as NA, N/A, n/a, N_A, and so on. This often results where there is manual data entry and nothing in the data entry process enforces the standardization of the response. This is common with abbreviation as well, including those for state and country names. This can also happen when sharing data between countries. A recent example involved data being shared between different groups within a multinational company. Certain special characters such as the @ and % that appear to be the same character have different values in Chinese and English character sets, creating instances of variables that looked the same but were different. While nonstandard values appearing in data sets are very common, it is a straightforward issue to fix. A frequency histogram of all observed values makes it easy to identify such cases, and a best practice is to automate the replacement of nonstandard values as part of an ongoing ETL process.

A related issue with categorical data can occur when a variable has a high cardinality. This occurs with data such as zip codes where the number of unique values is very high. These could also be variables such as e-mail addresses, user names, and social security numbers. The high number of unique values makes these variables impossible to use in tools such as linear models as such variables will not have enough observations per level of the variable to create a model. In practice, this type of data will either be discarded or transformed into a new variable using a technique called binning. For example, a long list of phone numbers may be mapped to a new variable made of just the area code. Zip codes could be mapped to a new variable called region, made up of a small number of geographical areas in the country. Binning is one example of data transformations that will be discussed in a later section of the chapter.

When dealing with quantitative data, the problem of outliers will often occur. An outlier is said to occur when a value is observed for a quantitative variable that is more than three standard deviations away from its mean. Outliers happen for many reasons, and understanding the reason for their occurrence is essential to knowing the proper remedy. Often an outlier is due to a mistake or malfunction. For example, heavy mining trucks have many sensors that monitor critical systems. These track important operating parameters such as the temperature and pressure of oil, fuel, and cooling systems, as well as tire pressure and payload, at intervals of a fraction of a second. However, the normal operating condition of a mining truck is to be carrying hundreds of tons of payload over rough terrain, or to have 40 ton loads dropped into the bed of the truck while loading, often in extremes of altitude and temperature. Under such harsh conditions, sensors can malfunction, as can the software used to collect the data. Data collected from operating assets will often have extreme values that are known to be erroneous. In such environments, observations with extreme values are discarded.

If the outliers in a data set are recurring and predictable, discarding those observations can mean a loss of valuable information that can cause bias in statistical models. However, many statistical modeling techniques are sensitive to extreme values. In such cases, an appropriate variable translation may be necessary to keep the predictive power of the variable reducing the influence of the outliers. An example of such a transformation is the creation of a new variable that is the base 10 logarithm of the observed variable, often after adding a constant to eliminate zeros and negative values.

Another problematic characteristic of some data is that it can be very *noisy*. This occurs when there is a high level of random variability in the data. The data collected from mining trucks described above is an example of data that are noisy. The shocks and vibration endured by the equipment result in data that have a high variance, obscuring the directional changes in the key operating parameters that may be occurring. Another example of data that is noisy is the intra-day price for a stock. Over the course of the day, there may be substantial variation in the price of individual transactions. This obscures directional changes that are of interest. High variability in time series data can be handled by transformations such as smoothing. One approach is to create a new variable that is based on a rolling moving average of a fixed number of the observed values. Since the high and low extreme values are effectively negated by the averaging, the result will be a new time series that has lower variability that will more transparently reveal any trends in the overall level of the time series. Another approach is to reduce the frequency of the time series by taking the minimum, maximum, and average of the time series over fixed intervals such as an hour. This creates three new time series that can be used to monitor not only the average value but also the range and variability.

Another issue with data that may require transformation is *skewness*. This happens when the distribution of continuous data has a long tail on one side, often because the values of the variable are bounded by zero on one side, leading to a long upper tail in the distribution. Such distributions are not consistent with the assumption of normality that is required for many parametric methods, and a transformation is used to mitigate this difficulty. Such transformations involve using a function that will impact the long upper tail the most. Examples of these transformations include logarithmic ($\ln(x)$, $\log10(x)$), inverse ($1/x$), and square root ($\text{sqr}(x)$).

Data sets with multiple observations taken on the same population may experience high correlation between variables. While this correlation can be informative and a useful output of the data profiling process, one is advised to be cautious about using correlated variables in predictive models. These collinear variables contain redundant information, and can make it difficult to estimate the parameters of the models. Often this is due to a hierarchical relationship among variables, such as supplier name and source country. The analyst is encouraged to limit the inclusion of all but the most descriptive variables when modeling.

In the previous discussion, the methods for data transformation that have been discussed included *binning, smoothing,* and *fitting.* Binning divides the values of a continuous variable into intervals. Binnig discretizes the data, turning quantitative data into categorical data. Most statistical software will have a capability to create new categorical variable from bins using any number of methods such as assigning an equal number of observations to each bin, or creating bins of equal width and assigning a record to a bin if its value falls within the defined interval. Binnig can also be done based upon prior knowledge of the data.

There are mixed views about the use of binning. It does involve the loss of information, and the use of too few bins can hide information such as a multiple modality in the continuous data. However, it does have advantages. Binning reduces the influence of outliers on the model by converting them to a level of a categorical variable. It can also help with the interpretation of the coefficients of the final predictive models as it has the effect of scaling variables that are of different magnitudes. It can also increase the number of degrees of freedom of a model.

Smoothing is a technique most often used to reduce the volatility of a time series. As already mentioned, simple multi-period moving averages can be effective for reducing volatility. In this approach, a new time series is created from the old one by taking successive averages of a fixed number of periods. For each new point in the new time series, the oldest observation in the previous average is dropped and the next one in the series is added to the calculation. Another popular method for smoothing a time series is called Loess Regression. This is a nonparametric method that performs least-squares regression on a local neighborhood of the time series. The new time series is predicted within a specified range, or span, and may include other predictor variables. The result is a new time series with a smoothness that increases with the width of the span, although this does not minimize the sum of squared errors of the Loess Regression. This functionality is available in commercial and open-source tools.

Several approaches for transforming variables by fitting functions are mentioned above. Another technique of interest for transforming data is *normalization*. Not to be confused with normalization in a database context, this refers to scaling data to eliminate differences of magnitude between continuous variables that can create numerical issues solving for model coefficients, as well as difficulties in comparing and interpreting the estimated coefficients. Typical methods include the following:

- Min–max: The value is scaled by subtracting the minimum from the value, and dividing by the range (max–min) of the observed values.
- *Z*-score: The value is scaled by subtracting the mean from the value, and dividing by the standard deviation of the observed values.
- Decimal scaling: The value is divided by some power of 10, to adjust the range of the observed values.

Another transformation that can be used to discretize time series data is to count the number of events that occur within a specific time interval. For example, suppose data are collected from a diesel engine. A sensor collects data for the temperature of the engine coolant at regular intervals. The engine has a protection system that causes the engine to be derated (the power reduced) when the coolant temperature exceeds 225°F. The raw time series will be noisy and difficult to use. One approach is to consider an event of interest to occur whenever the temperature exceeds this threshold. A new variable can be created that will count the number of these events that occur within a specified interval. The transformed variable can now be used to examine the relationship between these events, which may be transitory in nature, and the occurrence of other events such as unplanned maintenance. This is a valuable transformation as the collection of events and alarms is quite common in asset monitoring systems.

A final consideration on the topic of data transformation is *data reduction*. With the advent of inexpensive data storage and inexpensive devices that can collect data at high frequencies, it is not uncommon for data warehouses to become quite large. Analyses that run using data sets with terabytes of data can become impractical due to the processing time required. Data reduction seeks to reduce the size of the data warehouse while preserving the information contained in the data. There are many techniques used to perform data reduction. This discussion will focus on two examples.

The first is called *principal components analysis* (PCA). PCA finds new variables, called components, that represent the data in a lower dimensional space. PCA reduces the dimensions by an orthogonal transformation of the data that is achieved through the following process:

- Start with a data matrix of m observations of n variables.
- Subtract the mean of each variable from each observation.
- Calculate the $n \times n$ covariance matrix.
- Calculate the eigenvalues and eigenvectors of the covariance matrix.
- The principal component is the eigenvector with the largest eigenvalue.
- Select some subset of p eigenvectors with the p largest eigenvalues.
- Derive the new data by creating a matrix of p eigenvectors and transposing it. Multiply this by the mean adjusted to complete the transformation.

If the correlation between the original n variables is high, the difference between n and p will be significant and there will be substantial reduction in the size of the data. These new data can be used for model development in a fashion like the original data. However, much of the redundancy and unimportant information is removed by the projection into the lower dimensional space.

Another commonly used technique for data reduction is data sampling. There are a variety of sampling methods that can be used to reduce the number of instances submitted to an algorithm while retaining the original characteristics of the data. Simple random sampling without replacement (SRSWOR) is used to select n records from a set of m records, where $n < m$ and every record has an equal probability of being selected. Simple random sampling with replacement (SRSWR) is similar, except that each record that is selected is replaced and may be selected again on the next draw. If the population from which the sample is not homogeneous, then a stratified sample may be taken. Suppose that a sample of individuals consists of three groups or strata: youth, adults, and seniors. A simple random sample (SRS) may be taken from each stratum to accurately reflect the data of the entire population. One challenge with using SRS methods for data reduction is that while they do reduce the size of the data, which will improve computational performance and memory usage, they also increase the sample variance. This will make it more difficult to detect small differences between groups, and will generally reduce the effectiveness of statistical algorithms. More complex algorithms are available for data reduction, also sometimes called *data squashing*. These methods select n records from a set of m records, where n is much smaller than m, and add an additional column that contains a weight that is representative of the frequency of occurrence of that record in the original population. Numerous references to data squashing methods and their application and effectiveness can be found in the statistical literature.

4.4 Data Modeling

4.4.1 Relational Databases

After the data have been cleaned and transformed, the ETL process will deposit them into a data warehouse. The most common type of data warehouse is built using a *relational database*. The software underlying the structure of relational databases is called a relational database management system (RDBMS). Originally proposed by an IBM researcher named E.F. Codd in 1970, a relational database stores data in tables. Each table consists of rows called records that usually represent one entry of the content of the table. For example, each record can contain information about a customer, an asset, or a purchase order. Each record consists of columns or fields that contain data related to that instance. So, a customer record might contain account number, first name, last name, phone number, and e-mail address.

Figure 4.5 illustrates the critical concept in a relational database. Each table will have a *primary key* that serves as a unique identifier for that record in the table. In this example, Asset Type, Asset ID, Work Order Number, and Task Code all serve as a primary key in a table. When they appear in other tables, they are called *foreign keys*. The relationships between the tables are highlighted by the connections. For example, in the Assets table, Asset ID is the primary key as each asset will have a unique Asset ID. In the Work Orders table, Asset ID is a foreign key and the relationship between the two keys is said to be *one to many*. The Asset ID may appear many times in the Work Orders table as the asset may have been serviced many times. However, the information uniquely related to the asset appears just once in the Asset table. This allows the relational database to store the data in a more compact form, and master data such as we see here regarding Assets, Asset Types, and Tasks needs to be retrieved only if we have a need to associate it with transactional data such as we see in the Work Orders table.

Essentially all relational databases use structured query language (SQL) to write queries and maintain the database. A query allows the user to extract data from several different tables to create a new record format specific to a required

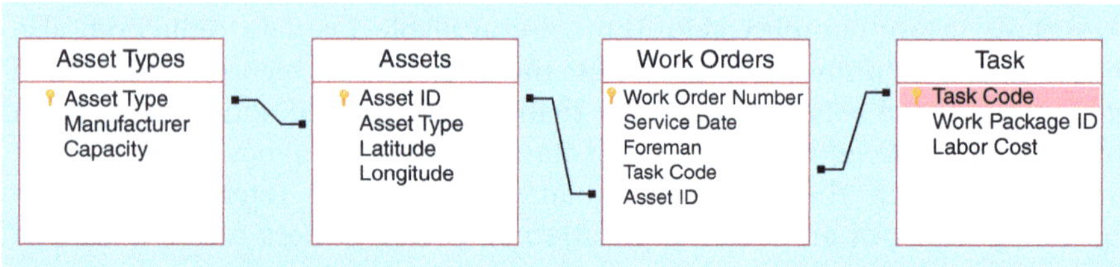

Figure 4.5 Relational database.

purpose, such as being processed by an algorithm. Suppose that one wanted to examine the work order history and compare the maintenance costs by manufacturer and by work crew foreman. The following example of SQL code is called a *query*. It will create a new *view* of the data that will be useful for the analysis:

```
SELECT
    [Work Orders].[Work Order Number],
    [Work Orders].Foreman,
    [Work Orders].[Asset ID],
    [Asset Types].[Asset Type],
    [Asset Types].Manufacturer,
    [Task].[Labor Cost]
FROM
    [Task],
    [Assets],
    [Asset Types],
    [Work Orders]
WHERE
    [Assets].[Asset ID] = [Work Orders].[Asset ID]
    AND
    [Task].[Task Code] = [Work Orders].[Task Code]
    AND
    [Asset Types].[Asset Type] = Assets.[Asset Type]
ORDER BY
    [Work Orders].[Service Date];
```

The query will select individual fields from records within the Work Orders, Asset Types, and Task tables, and join them together using the relationships defined between the primary and foreign keys. The SELECT portion of the query lists the fields that are to go into the new view. The FROM portion of the query lists the target tables or views from which records are to be selected. The WHERE portion of the query lists conditions that must be satisfied for selected records to be displayed in the view. The ORDER BY portion of the query allows the user to sort the new records in the view based upon one or more specified fields. Note that the fields specified in either the WHERE or ORDER BY sections need not appear in the view itself. The resulting data record will contain the following fields:

- Work Order Number
- Foreman
- Asset ID
- Asset Type
- Manufacturer
- Labor Cost

In addition, the records in the view will be sorted by the Service Date contained in the Work Orders table. A variety of operators are available in SQL that can be used to create complex views from many tables or views, giving it tremendous power for query development and data maintenance.

Another powerful feature of a relational database is the ability to enforce specific data types for each field, as well as adding constraints on the values allowed for each field. For example, it is possible to restrict a field to only integer values within an allowed range, or to require that a date/time value be in a specified format. This facilitates the enforcing of necessary business rules and prevents introduction of incorrect or erroneous data into the database. This structured approach to maintaining *data integrity* is one of the primary advantages of a relational database.

For efficiency and security, *stored procedures* are sometimes used to perform complex or frequently repeated tasks. A stored procedure is a block of code, either SQL or some other language such as Java or C++, that can be used to implement business logic. Because they are stored in the database and run on the database server, they typically run with less overhead and better security than when applications send dynamic queries to the database from outside the database server.

4.4.2 Nonrelational Databases

A nonrelational database, also sometimes called a NoSQL database, is any database that does not rely on the tabular structures and primary and foreign key relationships supported by a traditional RDBMS. These databases have become popular in recent years as part of the so-called *big data* explosion that has been driven in part by the sheer volumes of data that an Internet-connected world can create. But it has also been driven by the fact that these data are also more unstructured as social media-driven content has proliferated.

Interview With Robert Clark

When asked his work with big data, *RTI International Senior Research Biologist Robert Clark provided the following example:*

On the LungMAP project, we are considering the normal development of the human lung. On that project, we are looking at imaging and we are looking at genomics, transcriptomics, proteomics, metabolomics, lipidomics -and all have very large data sets. The human cell has 3.3 billion base pairs and 20,000 genes. Then you probably have 100,000 different proteins in each of your cells, and then the RNA is probably, oh, 300,000 RNAs per cell, 12,000 different RNAs, things like that. Trying to align all that information and then analyze it separately and then together is a huge feat. We use all kinds of data analysis tools as well as imaging and machine learning to draw on and annotate images so that people understand what's on these images rather than having people go in and manually draw them. And that is just one mapping project.

On the LungMAP project, we are currently storing everything in the cloud using tools developed for use in the cloud. But, for example, we had a great deal of 3D image data that came from one of our research centers–It was 150 Terabytes of data, and it took a whole day to download it on a special drive within the Amazon cloud that we're using. Then it took another entire day to retrieve it from the cloud so that we could examine it. And that's just a first few steps in a complex analytics project.

This is an excerpt from one of a series of interviews with analytics professionals and educators commissioned by the *INFORMS Analytics Body of Knowledge* Committee.

Nonrelational databases encompass a variety of different technologies, but tend to share some characteristics. Since they do not rely upon a relational model, no predefined schema must be constructed before data can be loaded into them. In addition to having a flexible schema, they can handle unstructured data that do not fit into the tabular structures of the relational model. Most of them *scale horizontally*, meaning they can be increased in size by adding additional clusters of inexpensive, commodity servers. And with a few exceptions, they follow an open-source model and do not require expensive license and maintenance fees to get started.

Columnar databases such as Redshift and Cassandra organize data by columns, not rows as in the traditional model. Queries are still processed using traditional SQL, but for many applications the efficiency is greater as the input/output process is more efficient for many common types of queries used in analytics applications. This is because such queries often touch on many rows but only a few columns in the data; and in the row-oriented structure of a traditional relational database, this means scanning across each row to retrieve the required columns. In a columnar database, only the relevant columns need to be scanned for increasing performance.

A *key-value store* is a database without a schema that stores all the data in a single blob. Each value of data can have a different form, and will have a unique key identified with it. The fundamental structure of the key-value store is called an associative array that contains what are called key-value pairs. This approach does not allow the processing of complex queries using SQL. The only way to retrieve data is using the unique key associated with it. Since this means a direct request to the data object in memory or on disk, it will be very fast. However, since the operations typically performed with SQL such as joins are not available in the database, they will need to be done in the code calling the database. A key reason to use a key-value store is scalability as the simple architecture makes this easy to do. Aerospike and Cassandra are examples of products that offer key-value stores.

A *document-oriented* database is designed to store semistructured data, typically using Java Script Object Notation (JSON) or XML. MongoDB is an example of this type of database, which can be thought of as a subset of a key-value store. This is because document-oriented databases use a key to document look-up like the key-value relationship. The difference is that while the characteristics of the data in a key-value store are not visible, a document-oriented database will typically have an API for developing queries based upon the structure of the documents.

A *graph* database is used to map complex relationships between objects such as people, things, and locations. In a graph database, objects are stored as vertices and directed edges. For example, the vertices may represent people and activities, and the directed edges may represent relationships such as "friend" and "likes." Graph databases are useful for analyzing data where there are complex relationships between entities. Examples of this can be found in customer relationship management applications, such as identifying product bundles to suggest to a customer. Another example is market segmentation, where one might seek to identify interests of customers that are strongly related to preferences for certain products. Titan is an example of a graph database.

4.5 Data Management

Any company or organization that has physical assets or human resources is likely to have processes in place to ensure that those are maintained and protected, recognizing the value they create. An organization's data are another asset that should be actively managed and maintained just like any other asset. In recent years, most organizations have acquired data at rates far exceeding those at which they acquire new assets or employees. To ensure that their data are reliable, usable, and available to create value for the consumers of data within the organization, it needs to have a thoughtful and systematic approach to data management. The literature about data management and data governance is extensive, and has been the subject of entire volumes. It is useful to highlight some concerns that a successful data management strategy will address.

As described earlier in this chapter, a great deal of effort can be expended to clean data when gathered from its original source. Therefore, it is important that methods for *data capture* are implemented to prevent the entering of incomplete or incorrect data into the system. This should also incorporate the necessary business rules to ensure that the data collected matches the requirements for the business processes it is intended to support.

Effective data management programs will create roles for *data stewards* who are responsible for monitoring the quality of specific elements of data on an ongoing basis. Data stewards may use dashboards and reports combined with their unique understanding what constitutes data quality for their data, and will not allow changes to be made to the data structure without approval from the appropriate governance body within the organization. Data stewards make sure that usable data stay usable.

Metadata is often called "data about data." It serves as the documentation for the data and can contain information such as when and by whom the data were created, as well as the structural elements of the data such as tables, fields, and relationships. It can also contain information about who has access to the data and at what security level. Metadata can also describe the various elements of the data in relation to the business processes that generated it so that the user can have a practical understanding to complement the technical description. Collection and maintenance of metadata is an important data management function, especially if the data are to be archived, or retained and published and made available to other users within the organization. This is especially the case as organizations seek to build large data platforms to support self-service analytics on demand across the entire enterprise.

Master data consists of the key objects necessary to describe and run a business. These are typically lists of people, places, or things core to the business such as customers, regions, and products. They may be scattered across multiple systems in the organization. *Master data management* (MDM) is a process by which a single view of these core objects can be presented to users across the entire enterprise in a consistent and current format. Each of these objects has a unique life cycle, sometimes called the CRUD cycle (create, read, update, and delete), that must be managed according to defined procedures for each step in the objects life cycle. Effective MDM will require frequent updates to the master data store, and will often rely upon specialized applications such as customer data integration (CDI) or product information management (PIM) tools.

5

Solution Methodologies

Mary E. Helander

Data Science Department, IBM T. J. Watson Research Center, Yorktown Heights, NY, USA

Modern solution methodology offers a set of macro- and micro-practices that help a practitioner systematically maximize the odds of a successful analytics project outcome.

5.1 Introduction

Methodology is all about approach. Every discipline, whether it be applied or theoretical in nature, has methodologies. While there is no one standard analytics solution methodology, common denominators of solution methodologies are the shared purposes of being systematic, creating believable results, and being repeatable. That is to say, a solution methodology helps practitioners and researchers alike to progress efficiently toward credible results that can be reproduced.

Whether we mean a methodology at a macro- or microlevel, analytics practitioners at all stages of experience generally rely on some form of methodology to help ensure successful project outcomes. The goal of this chapter is to provide an organized view of solution methodologies for the analytics practitioner. We begin by observing that, in today's practice, there does not appear to be a shared understanding of what is meant by the word *solution*.

5.1.1 What Exactly Do We Mean by "Solution," "Problem," and "Methodology?"

In its purest form, a *solution* is an answer to a *problem*. A *problem* is a situation in need of a repair, improvement, or replacement. A *problem statement* is a concise description of that situation. *Problem definition* is the activity of coming up with the *problem statement*. *Problem-solving*, in its most practical sense, involves the

INFORMS Analytics Body of Knowledge, First Edition. Edited by James J. Cochran.

collective actions that start with identifying and describing the problematic situation, followed by systematically identifying potential solution paths, selecting a best course of action (i.e., the *solution*), and then developing and implementing the solution. *Problem-solving* is, by far, one of the most valuable skills an analytics practitioner can hone, and is even an important life skill!

Most of us first encountered problem-solving as students exposed to mathematics at primary, secondary, and collegiate education levels, where a problem–for example, *given two points in a plane,* (x_1, y_1) *and* (x_2, y_2), *find the midpoint*–is more often than not stated explicitly. The solution $\left(\frac{x_1 + x_2}{2}, \frac{y_1 + y_2}{2}\right)$ can be found with some geometry and algebra wrangling. See Eves [1]. If asked to solve this problem for homework or on an exam, we probably did not get full credit unless we showed our work. This shown work we can think of as the *solution methodology* for the problem. This sample math problem can be used to illustrate the fact that there are different ways to solve a problem: For example, to use those same methodology steps to find a midpoint solution, if presented with the two points in polar coordinates, one may proceed using an entirely different approach by applying methods from trigonometry.

Similarly, in analytics practice, the path to a solution is generally not unique. For example, Ref. [2] describes a study of the variation in approach (and results) by 29 independent analytics teams working on the same data and problem statement. The path to a solution may involve a straightforward set of steps, or it may need some clever new twist; the method chosen may depend on the form of the available data, the assumptions, and context. A big difference between the problems that we encounter in school and the problems that we encounter in real life is usually that in real life, we are rarely presented with a clean problem statement with, for example, the *given information*. Still, writing down the steps we use to get from the problem statement to the solution is generally a good idea. In most cases, we can write down steps that are general enough so that we're able to find solutions to new and challenging problems.

What do we mean by a "solution"? To the purist, a solution is this: The correct answer to a problem. It is what you write down on your exam in response to a problem statement. If you get the answer right, and if you have adequately satisfied the requirement of showing your work, you earn full credit for the solution. In some cases, you may get the wrong answer, but if some of your shown work is okay, you may still earn partial credit. Similarly, in practice, analytics that produce unexpected or flawed results may earn their creators recognition for solid work that has gone into the project, and practitioners may get the opportunity to revise these analytics, just as authors of peer-reviewed papers may get the opportunity to make major revisions to their work during the review process. Without a transparent methodology, however, it is more difficult for evaluators of a project to appreciate the practitioners' findings and effort when they are presented with results that are unexpected or questionable.

Methodological steps are analogous to what we mean more generally by a solution methodology or approach. When we're starting out, the steps give us an approximate roadmap to follow in our analytics project. When we're done, if we've followed a roadmap and have perhaps even documented the steps, then it is easier to trace these steps, to repeat them, and to explain to stakeholders and potential users or sponsors how our solution was derived. It might be that the steps themselves are so innovative that we patent some aspect of the approach, or perhaps we find that publishing about some aspect of the project, the technology, or the outcome is useful for sharing the experience with others and promoting best practices. In any of these cases, having followed some methodology helps tremendously in describing and building credibility into whatever it was that we did to reach the solution.

5.1.2 It's All About the Problem

Experienced analytics professionals already know this too well: In practice, new projects rarely, if ever, start out with a well-defined problem statement. The precision of a problem statement in the real world will never be as clearly articulated as it was in our math classes in grade school, high school, and college. Indeed, there may be contrasting and even conflicting versions of the underlying problem statement for a complex system in a real-world analytics project, particularly when teams of people with varying experiences, backgrounds, opinions, and observations come together to collaborate. Using our sample math problem to illustrate, this would be equivalent to some people's thinking that the problem is to find a point solution (x, y), while others might think that the solution should be defined by the intersection of two or more lines, or perhaps that it should be defined by a circle with a very small radius that covers the point of intersection, and so on. The point is that the solution can be relevant to the interpretation of the problem, and thus, when the problem is not defined for us precisely–and even sometimes when the problem is–people may interpret it in different ways, which may lead to entirely different solution approaches.

An important message here is that time and effort up-front on a problem statement is time well spent, as it will help clarify a direction and create consistent understanding of the practitioners' end goals.

5.1.3 Solutions versus Products

In today's commercial world of software and services, the word *solution* may be used to describe a whole collection of technologies that address an entire class of problems. The problems being solved by these commercial technologies may not be specifically defined in the ways we have been used to seeing problems defined in school. For example, a commercial supply chain software provider may have a suite of *solutions* that claim to address all the needs of a retail business.

In other words, in today's world of commercial software and services, the word *solution* has become synonymous with the word *product*. In fact, in some circles, it is **not cool** to say that the solution solves a problem because this suggests that there *is* a problem. *Problems*, at least in our our modern Western capitalist culture, are no big deal. Therefore, we don't really have them. However, we do have plenty of solutions, especially when it comes to commercial products. So, we begin this chapter by pointing out the elephant in many project conference rooms: Problems are not sexy, but solutions are! While this line of thinking is indeed the more positive and inspiring outlook, and while it makes selling solutions easier, unfortunately, it often leads to implementing the wrong solutions, or to failing altogether at solution implementation. Why? There are many reasons, but one of the most obvious and common reasons is that ill-defined, poorly understood, or denied problems are difficult–if not impossible–to actually solve.

Interview With Eric Stephens

When asked to consider how the analytics professional determines whether to pursue an analytic solution to a problem, the Vanderbilt University Medical Center's Manager of Population Health Analytics Eric Stephens offered the following thoughts:

This is a process. The overall problem should be defined before analytics is brought into the equation. That is, you should first have a thorough understanding of what the business user is trying to solve, what the context is, and what other approaches might have been tried previously. In many cases, people who are not analytics practitioners think that an analytic solution is required just because a problem or an issue has data available or associated with it. This may sound strange coming from someone who works in analytics, but I do not think it's true that every single issue that has data associated with it is necessarily an analytics problem.

There has always been a tendency for business users and executives to assume that if there are data associated with or related to a problem, then it is automatically an analytics problem; this is especially true in this age of analytics, in which the capability is getting so much attention and people are employing it more and more frequently. However, determining if an analytic solution is truly called for requires the analytics practitioner to gain an understanding of the problem context and the associated business rationale.

Suppose I am presented with a problem and have already done the work necessary to define it properly. I then think, "OK, how might I apply analytics to solve this?" Perhaps I then realize that this is not the appropriate way to think about the problem, and thus may consider a different approach. Maybe that will entail only reporting some information, so ultimately I am not using an analytic solution per se. In other words, if I feel like I am really reaching or really working hard to find an analytics approach to a problem, this should raise an intuitive red flag that the problem might not necessarily be one that requires an analytic solution.

At the end of the day, it really comes down to your ability to dissociate the problem from whatever data may be available, so that you are thinking about the problem simply on its own and not automatically jumping into the related data. If you do this first, you have accomplished a couple of things. First, if it is an analytics problem, it helps you think about the problem a bit more creatively, because you do not want to be restricted by whatever data may or may not be readily available. At the same time, it forces you to consider the problem from a broader perspective: You may realize not only that this may not be an analytics problem but also that there may actually be a better nonanalytic approach to solving it. You may not have considered the nonanalytic approach had you locked into your mind that an analytic solution would be necessary before you fully understood the problem. It comes down to having a strong overall business understanding, and then being able to determine the likely impact of the problem and the potential solutions before considering possible analytics approaches. Oftentimes, thinking in this way can help you realize either that you are not dealing with an analytics problem or there may be a better approach that is not analytic in nature.

This is an excerpt from one of a series of interviews with analytics professionals and educators commissioned by the *INFORMS Analytics Body of Knowledge* Committee.

5.1.4 How This Chapter Is Organized

The previous section, hopefully, has left the reader with a strong impression that recognizing the underlying problem is a first step toward solving it. It is in this spirit that this chapter introduces the notions of *macro-* and *micro*solution methodologies for analytics projects and organizes their content around them. Macro-methodologies, as we shall see in a section devoted to their description, provide the more general project path and structure. Four alternative macro-methodologies will be described in that section with this important *caveat*: Any one of them is good for practitioners to use; the most important thing is for practitioners to follow *some* macro-methodology, even if it is a hybrid.

Micro-methodology, on the other hand, is the collection of approaches used to apply specific techniques to solve very specific aspects of a problem. For every specific technique, there are numerous textbooks (and sometimes countless papers) describing its theory and detailed application. There is no way we will be able to cover all possible problem-solving techniques, which is not the purpose of this chapter. Instead, this chapter covers an array of historically common techniques that are relevant to INFORMS and to analytics practitioners in order to illustrate micro-solution methodology, that is, to expose, compare–and in some cases, contrast–the approaches used.

Figure 5.1 provides an illustration of the chapter topic breakdown. Note that all solution methodology descriptions in this chapter, both at macro- and

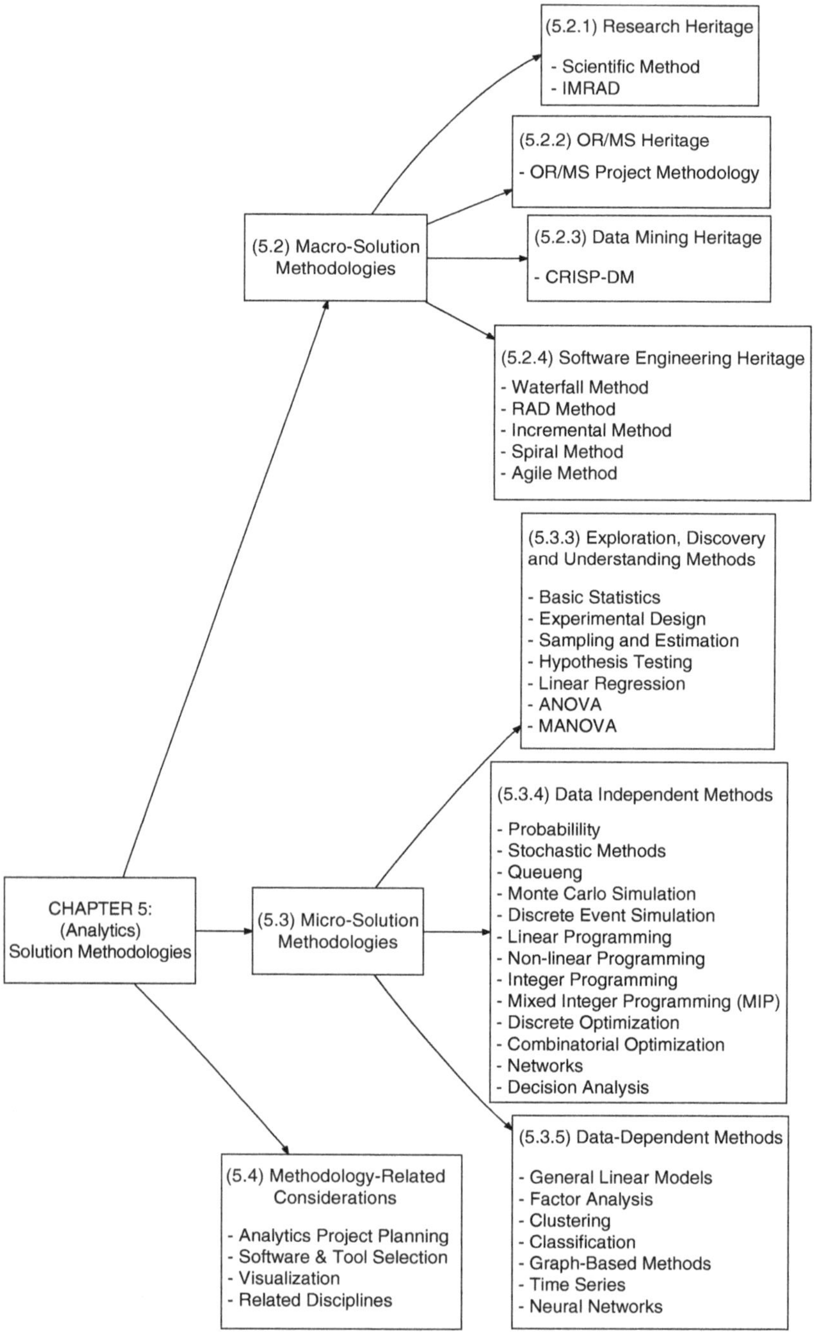

Figure 5.1 A breakdown of analytics solution methodologies (and related topics) covered in this chapter.

microlevels, are significantly biased in favor of operations research and management sciences. This is so because this chapter appears in an analytics book published in affiliation with INFORMS, the international professional organization aimed at promoting operations research and management science. The stated purpose of INFORMS is "to improve operational processes, decision-making, and management by individuals and organizations through operations research, the management sciences, and related scientific methods." (see the INFORMS Constitution [3].)

5.1.5 The "Descriptive–Predictive–Prescriptive" Analytics Paradigm

With the rise in the use of quantitative methods, particularly OR and MS, to solve problems in the business world, the business analytics community has adopted a paradigm that classifies analytics in terms of descriptive, predictive, and prescriptive categories. These correspond respectively to analytics that help practitioners to understand the past (i.e., describe things), to make recommendations about the present (i.e., prescribe things), and to understand the future (i.e., predict things). The author of this chapter believes that the paradigm originated at SAS [4], one of the most well-known analytics software and solutions companies today.

Granted, many disciplines today are using analytics, and the descriptive–predictive–prescriptive analytics paradigm has no doubt helped evangelize analytics to disciplines. However, it should be noted that we explicitly have chosen to organize this chapter directly around macro- and micro-methodologies, and within the micro-category, exploratory, data-independent, and data-dependent technique categories. While intending to complement the"descriptive–predictive–prescriptive" analytics paradigm, this orgnaization emphasizes that solution techniques do not necessily fall neatly into one of the paradigm bins. Instead, techniques in common categories tend to have threads based on underlying problem structure, model characteristics, and relationships to data, as opposed to what that the analytics project outcome may drive (i.e., to describe, to predict, or to prescribe). From the perspective of analytics solutions methodologies, this can also help avoid an unintentional marginalization of techniques that fall into the descriptive analytics category.

5.1.6 The Goals of This Chapter

After reading this chapter, a practitioner will

1) be able to distinguish macro- versus micro-solution methodologies,
2) be ready to design a high-level analytics project plan according to some macro-level solution methodology,
3) be better at assessing and selecting appropriate microlevel solution methodologies appropriate for a new analytics project, based on a general understanding of the project objectives, type and approximate amounts of data

available to the project, and various types of resources (e.g., people and skills, computing, time, and funding),
4) be armed with a few pearls of wisdom and lessons learned in order to help maximize the success of her or his next analytics project,
5) understand the significance of methodology to the practice of analytics within operations research and other disciplines.

5.2 Macro-Solution Methodologies for the Analytics Practitioner

As described in the Introduction, a macro-solution methodology is comprised of general steps for an analytics project, while a micro-methodology is specific to a particular type of technical solution. In this section, we describe macro-methodology options available to the analytics practictioner.

Since a macro-methodology provides a high-level project path and structure, that is, steps and a potential sequence for practitioners to follow, practitioners can use it as an aid to project planning and activity estimation. Within the steps of a macro-methodology, specific micro-methodologies may be identified and planned, aiding practitioners in the identification of specific technical skills and even named resources that they will need in order to solve the problem.

Four general macro-methodology categories are covered in this section:

A. The scientific research methodology
B. The operations research project methodology
C. The cross-industry standard process for data mining (CRISP-DM) methodology
D. The software engineering methodology

We reiterate here that there is some overlap in these methodologies and that the most important message for the practitioner is to follow a macro-solution methodology. In fact, even a hybrid will do.

5.2.1 The Scientific Research Methodology

The scientific research methodology, also known as the *scientific method* [5], has very early roots in science and inquiry. While formally credited to Francis Bacon, its inspiration likely dates back to the time of the ancient Greeks and the famous scholar and philosopher Aristotle [6].

This methodology has served humankind well over the years, in one form or another, and has been particularly embraced by the scientific disciplines where theories often are born from interesting initial observations. In the early days, and even until more recently (i.e., within the last 20 years–merely a blip in historical time!), a plethora of digital data was not available for researchers to study; data were a scarce resource and were expensive to obtain. Most data were planned, that is, collected from human observation, and then treated as a limited, valuable resource. Because of its value both to researchers' eventual conclusions and to the generalizations that they are able to make based upon their findings, the scientific methodology related to data collection has evolved into a specialty in and of itself within applied statistics: experimental design. In fact, many modern-day graduate education programs in the United States require that students take a course related to research methodology either as a prerequisite for graduate admission or as part of their graduate coursework so that graduate students learn well-established systematic steps for research, sometimes specifically for setting up experiments and handling data, to support their MS or PHD thesis. Often, this type of requirement is not uncommon in social sciences, education, engineering, mathematics, computer science, and so on–that is, these requirements are not limited strictly to the sciences.

The general steps of the scientific method, with annotations to show their alignment with a typical analytics project, are the following:

A.1. *Form the Research Question(s).* This step is the one that usually kicks off a project involving the scientific method. However, as already noted, these types of projects may be inspired by some interesting initial observation. In applying this step to an analytics project in practice, the research questions may also relate to an underlying problem statement, which typically forms the preface for the project.

A.2. *State One or More Hypotheses.* In its most specific form, this step may involve stating the actual statistic that will be estimated and tested: for example, $H_0 : \mu_1 = \mu_2$, that is, that two treatment means are the same. (Note that a *treatment mean* is the average of observations from an experiment with a set of common inputs, that is, fixed independent variable values are the *treatment*.) Interpreted more broadly, the hypotheses to test imply the specific techniques that will be applied. For example, the hypothesis that two means are identical implies that some specific techniques of experimental design, data collection, statistical estimation, and hypothesis testing will be applied. However, one might also consider more general project hypotheses: for example, we suspect that cost, quality of service, and peer pressure are the most significant reasons that cell phone customers change their service providers frequently. These types of hypotheses imply specific techniques in churn modeling.

A.3. *Examine and Refine the Research Question and Hypotheses.* In this step, the investigating team tries to tune up the output of the first two steps of the scientific method. Historically, this is done to make sure that the planning going forward is done in the most efficient and credible way, so that ultimately, the costly manual data collection leads to usable data and scientifically sound conclusions–otherwise, the entire research project becomes suspect and a waste of time (not to mention, the discrediting of any conclusions or general theory that the team is trying to prove). This step is not much different in today's data-rich world: Practitioners should still want to make sure they are asking the right questions, that is, setting up the hypotheses to test so that the results they hope to get will not be challenged, while trying to ensure that this is all done as cost-effectively and in as timely a manner as possible. In today's world, because of the abundance of digital data, this sometimes means exploration on small or representative data sets. This can lead to the identification of additional data needed (including derivatives of the available data), as well as adjustments to the questions and hypotheses based on improved understanding of the underlying problem and the addition of preliminary insights. Notice the carry forward of "problem understanding" that happens naturally in this step. In fact, it is good to consider the acceptable conclusion of this step as one where the underlying problem being addressed can be well enough articulated that stakeholders, sponsors, and project personnel all agree. Some preliminary model building, to support the "examination" aspect of this step, may occur here.

A.4. *Investigate, Collect Data, and Test the Hypotheses.* In traditional science and application of the scientific method, this meant the actual steps of performing experiments, collecting and recording observations, and actually performing the tests (which were usually statistically based). Applied to analytics projects, this macro-methodology step means preparing the final data, modeling, and observing the results of the model.

A.5. *Perform Analysis and Conclude the General Result.* In this step, we perform the final analysis. In traditional science, does the analysis support the hypotheses? Can we draw general conclusions such as the statement of a theory? In analytics projects, this is the actual application of the techniques to the data and the drawing of general conclusions.

As is evident here, the scientific method is a naturally iterative process designed to be adaptive and to support systematic progress that gets more and more specific as new knowledge is learned. When followed and documented, it allows others to replicate a study in an attempt to validate (or refute) its results. Note that reproducibility is a critical issue in scientific discovery and is emerging as an important concern with respect to data-dependent methods in analytics (see Refs [7,8]).

Peer review in research publication often assumes that some derivative of the scientific method has been followed. In fact, some research journals mandate that submitted papers follow a specific outline that coincides closely with the scientific method steps. For example, see Ref. [9], which recommends the following outline: Introduction, Methods, Results, and Discussion (IMRAD). While the scientific method and IMRAD for reporting may not eliminate the problem of false discovery (see, for example, Refs [10,11]), they can increase the chances of a study being replicated, which in turn seems to reduce the probability of false findings as argued by Ioannidis [12].

Because of this relationship to scientific publishing, and to research in general, the scientific method is recommended for analytics professionals who plan eventually to present the findings of their work at a professional conference or who might like the option of eventually publishing in a peer-reviewed journal. This methodology is also recommended for analytics projects that are embedded within research, particularly those where masters and doctoral theses are required, or in any research project where a significant amount of exploration (on data) is expected and a new theory is anticipated. In summary, the scientific method is a solid choice for research-and-discovery-leaning analytics projects as well as any engagement that is data exploratory in nature.

5.2.2 The Operations Research Project Methodology

Throughout this chapter, analytics solution methodology is taken to mean the approach used to solve a problem that involves the use of data. It is worth bringing this point up in this section again because, as mentioned in the Introduction, our perspective assumes an INFORMS audience. Thus, we are biased toward these methodology descriptions for analytics projects that will be applying some operations research/management science techniques. While it was natural to start this macro-section with the oldest, most established, mother of all exploratory methodologies (the scientific method of the last section), it is natural to turn our attention next to the macro-method established in the OR/MS practitioner community.

In general, one may find some variant of this project structure in introductory chapters of just about any OR/MS textbook, such as Ref. [13], which is in its fourth edition, or Ref. [14], which was in its seventh edition in 2002. (There have been later editions, which Dr. Hillier published alone and with other authors after the passing of Dr. Lieberman.)

Most generally, the OR project methodology steps include some form of the following progression:

B.1. *Define the Problem and Collect Data.* As most seasoned analytics and OR practitioners know, problem statements are generally not crisply articulated in the way we have been used to seeing them in school math classes. In fact, as noted earlier, sponsors and stakeholders may have disparate and sometimes conflicting views on what the problem really *is*. Sometimes, some exploratory study of existing data, observing the real-world system (if it exists), and interviewing actors and users of the system helps researchers to gain the system and data understanding needed for them to clarify what the problem is that should be solved by the project. The work involved in this step should not be underestimated, as it can be crucial to later steps in the validation and in the acceptance/adoption/implementation of the project's results. It is a good idea to document assumptions, system and data understanding, exploratory analyses, and even conversations with actors, sponsors, and other stakeholders. Finding consensus about a written problem statement, or a collection of statements, can be critical to the success of the project and the study, so it is worth it to spend time on this, review it, and attempt to build broad consensus for a documented problem statement.

Collecting data is a key part of early OR project methodology, and is intriciately coupled with the problem definitition step, as noted in Ref. [14]. In modern analytics projects, data collection generally means identifying and unifying digital data sources, such as transactional (event) data (e.g., from an SAP system), entity attribute data, process description data, and so on. Moving data from the system of record and transforming it into direct insights or reforming it for model input parameters are important steps that may be overlooked or under-estimated in terms of effort needed.

As noted earlier, we live in a world where "solutions" are sexy and "problems" are not–further adding to the challenge and importance of this step. In comparison with the scientific method of the previous section, this step intersects most closely with the activities and purposes of **A.1**, **A.2**, and **A.3**.

B.2. *Build a Model.* There are many options for this step, depending on the type of problem being solved and on the objective behind solving it. For example, if we are seeking improved understanding, the model may be descriptive in nature, and the techniques may be those of statistical inference. If we are trying to support a complex decision, such as where to build a new firehouse and how to staff it, then we may build descriptive models to analyze current urban demand patterns; we may build predictive models that take those outputs to project future demand; and then we may build an optimization model to locate the facility so that future demand is best served. Much of this step is based on available data, as well as on available tools and skills, which sometimes means we choose to build the models that we are most familiar with or that we have the skills to support. This step most intersects with the activities and purposes of **A.3**, although it is not an exact mapping.

B.3. *Find and Develop a Solution.* In OR, this traditionally has meant the work of solving the equations or doing the math that finds the solution, designing the algorithm, and coming up with a computer code to implement the algorithm. There are many variants of this step today because the models may be derived fully from data or logic, and the micro-methods for finding the solutions can be specific to the technique. However, the common denominator here has to do with the algorithm, or in some cases, the heuristic: It is the recipe for taking the data, assumptions, and so on. and converting it to a useable result, however that is done. Computer code just helps us to do that most efficiently. This step intersects most closely with the activities and purposes of **A.4**, although it is only partial in mapping. As we shall see in a later section, this step interlocks with micro-solution methodologies that can constitute the details of this macro-step.

B.4. *Test (Verify) and Validate.* This step is actually a whole bunch of activities. Testing and verifying are often used interchangeably in software development, and since we often program (i.e., "implement") our model solution (algorithm, heuristic, process, model, etc.), the interchange works here in the OR project methodology. The act of testing, or verifying, is making sure that whatever it is you made and are calling the model or solution is actually doing what you think it is doing. This is different from validation, which is making sure a model is representative of whatever you are trying to mimic, for example, a real-world system or process and a decision-making scenario. Validation asks the following question: Does the model behave as if it were the real system? There are entire areas of research devoted to these topics, not just in the analytics and OR fields, but in statistics and software engineering as well. They all are better because of the cross learning that has happened. For example, statistical methods can be used to generate and verify test cases. In validation, statistical methods are often used in rigorous simulation studies–which are basically statistical experiments done with a computer program, and as such lend themselves very nicely to things such as pairwise comparison with historical observations from the true system. Dr. Robert Sargent is one of the pioneers in computer simulation, output analysis and verification, and validation methodologies–the canonical methods he described in his 2007 paper [15] provide valuable lessons not only for simulation modelers, but also for those doing testing, verification, and validation in other types of analytics and OR projects.

B.5. *Disseminate, Use, or Deploy.* Once the solution is ready to be used, it is rolled out (disseminated, deployed), and the work is still not done! Usually, at this stage, there needs to be training, advocacy, sometimes adjustment, and virtually always maintenance (fixing things that are wrong, or adding new features as the users and stakeholders hopefully become enthralled with the work and have new ideas for it). At this stage, it is usually useful to have baked in some monitoring–that is, if you can think ahead to put in metrics that automatically observe value that is being derived from using the solution, that's awesome foresight. In too many analytics and OR projects, deployment and dissemination merely means a final presentation and report. In some cases, those recommendations are good enough! In others, they might signal that the true solution is not really intended to be "used." Sometimes, this leads to an iterative process of refinement and redeployment, allowing practitioners to restart this entire step process. In other cases, you write the report, and perhaps an experience paper gets submitted to a peer-reviewed journal or is presented at an INFORMS conference. Whatever the outcome, practitioners need to keep in mind that all projects are worthy learning experiences–even the ones that are not deployed in the manner in which we were hoping.

It is not surprising that the OR project method, being exploratory in nature, is somewhat of a derivative of the scientific method. As Hillier and Lieberman point out in the introductory material of Ref. [16], operations research has a fairly broad definition, but in fact gets its name from research on operations. The *study objects* of the research are "operations," or sometimes "systems." These operations and systems are often digital in their planning and execution, and so tons of data now exist to model, recreate them, and model/experiment with them. In other words, these observable digital histories mean they are rich in data (analytics) that can be used to model very quickly. Unfortunately, the ability to jump right into modeling, analysis, and conclusions often means skipping over early methodological steps, particularly in the area of problem definition.

5.2.3 The Cross-Industry Standard Process for Data Mining (CRISP-DM) Methodology

"The cross-industry standard process for data mining methodology," [17,18] known as CRISP or CRISP-DM, is credited to Colin Shear, who is considered to be a pioneeer in data mining and business analytics [19]. This methodology heavily influences the current practical use of SPSS (Statistical Package for the Social Sciences), a software package with its roots in the late 1960s that was acquired by IBM in 2009 and that is currently sold as IBM's main analytics "solution" [18].

As an aside, note that SAS and SPSS are commercial packages that were born in about the same era and that were designed to do roughly the same sort of thing–the computation of statistics. SAS evolved as the choice vehicle of the science and technical world, while SPSS got its start among social scientists. Both have evolved into the data-mining and analytics commercial packages that they are today, heavily influencing the field. As mentioned earlier, the "descriptive–predictive–prescriptive" paradigm appears to have its roots in SAS. As noted above, CRISP is heavily peddled as the methodology of choice for SPSS. However, we note that this methodology is a viable one for data-mining methods that use any package, including R and SAS.

The steps of the CRISP-DM macro-methodology, from Ref. [17], are the following:

C.1. *Business Understanding.* This step is, essentially, the *domain understanding plus problem definition* step. In the business analytics context, CRISP calls out specific activities in this step, such as stating background, defining the business objectives, defining data-mining goals, and defining the success criteria. Within this step, traditional project planning (cost/benefits, risk assessment, and project plan) are included. This step also involves assessment of tools and techniques. Note that this step aligns with **B.1** of the OR project methodology.

C.2. *Data Understanding.* This is a step used to judge what data is available, by specifically identifying and describing it (for example, with a data dictionary) and assessing its quality or utility for the project goals. In most cases, actual data is collected and explored/tested.

C.3. *Data Preparation.* This is the step where analysts decide which data to use and why. This step also includes "data cleansing" (roughly, the act of finding and fixing or removing strange or inaccurate data, and in some cases, adding, enhancing, or modifying data to fix incomplete forms), reformatting data, and creating derivative data (i.e., extracting implied or derived attributes from existing data, merging data, etc.). An example of reformatting data would be converting GIS latitude and longitude (i.e., latitude/longitude) data from degree/minute/second format, for example, $41^\circ\ 13'\ 1''$ N, $73^\circ\ 48'\ 27''$ W, to decimal degrees, that is, 41.217, -73.808.An example of enhancing in data cleansing is finding and adding a postal code field to a street, city, state address or geocoding the address (i.e., finding the corresponding latitude/longitude). Data merging is a common activity in this step, and it generally is used to create extended views of data by adding attributes, via match up by some key. Note that a common "mistake" among inexperienced data scientists is to try to merge extremely large *unsorted* data sets. Packages such as SPSS, SAS, and R, and even scripting languages such as Python, allow for these common types of data movement, but without presorting lists, execution to accomplish merge operations can end up taking days instead of a few minutes when the list sizes are in the millions, which is not an unrealistic volume of data to be working with these days.

C.4. *Modeling.* This is the step where models are built and applied. In data mining and knowledge discovery, the models are generally built from the data (e.g., a regression model with a single independent variable is basically a model of a linear relationship where the data is used to derive the slope and y-intercept). Other modeling-related steps include articulating the assumptions, assessing the model, and fitting parameters. Note that this step, together with the previous two steps, aligns with *B.2* and *B.3* of the OR project methdology.

C.5. *Evaluation.* This step is equivalent to the OR project verification and validation step. See *B.4.* Note that Dwork et al. [20] give a well-recognized example of a validation method for data dependent methods.

C.6. *Deployment.* This step is equivalent to the OR project deployment step. See *B.5.*

The CRISP-DM macro-methodology is thought of as an iterative process. In fact, the scientific method and the OR project method can also be embedded in an iterative process. More details of the CRISP-DM macro-methodology can be found in Chapter 7.

5.2.4 Software Engineering-Related Solution Methodologies

Software engineering is relevant to analytics macro-solution methodology because of the frequent expectation of an outcome implemented in a software tool or system. The steps of the most standard software engineering methodology, the waterfall method, are the following:

D.1. *Requirements.* This step is a combination of understanding the business or technical environment in which a system will be used and identifying the behavior (function) and various other attributes (performance, security, usability, etc.) that are needed for a solution. Advisable prerequisites for identifying high quality requirement specifications are problem, business, and data understanding. Thus, this step aligns with *B.1*, *C.1*, and *C.2*.

D.2. *Design.* The design step in software engineering translates the requirements (usually documented in a "specification") into a technical plan that covers, at a higher level, the software components and how they fit together, and at a lower level, how the components are structured. This generally includes plans for databases, queries, data movement, algorithms, modules or objects to be coded, and so on.

D.3. *Implementation.* Implementation refers to the translation of the design into code that can be executed on a computer.

D.4. *Verification.* Similar to previous macro-methodologies, verification means testing. In software, this can be unit testing, system testing, performance testing, reliability testing, and so on. The step is similar to other macro-method verification steps in that it is intended to make sure that the code works as intended.

D.5. *Maintenance.* This is the phase, in software development, that assumes the programs have been deployed and when sometimes either bug fixes will need to be done or else new functions may be added.

A number of other software engineering methodologies exist. See, for example, Ref. [21] for descriptions of rapid application development (comprised of data modeling, process modeling, application generation, testing, and turnover), the incremental model (analysis, design, code, test, etc.; analysis, design, code, test, etc.; analysis, design, code, test, etc.), and the spiral model (customer communication, planning, risk analysis, engineering, construction and release, evaluation). When looking more deeply at these steps, one can see that they can also be mapped to the other macro-methodologies–note that Agile, a popular newer form of software development, is very much like the Incremental model in that it focuses on fast progress with iterative steps.

5.2.5 Summary of Macro-Methodologies

Figure 5.2 shows how the four macro-solution methodologies are comparatively related. It is not difficult to imagine any of these macro-methodologies

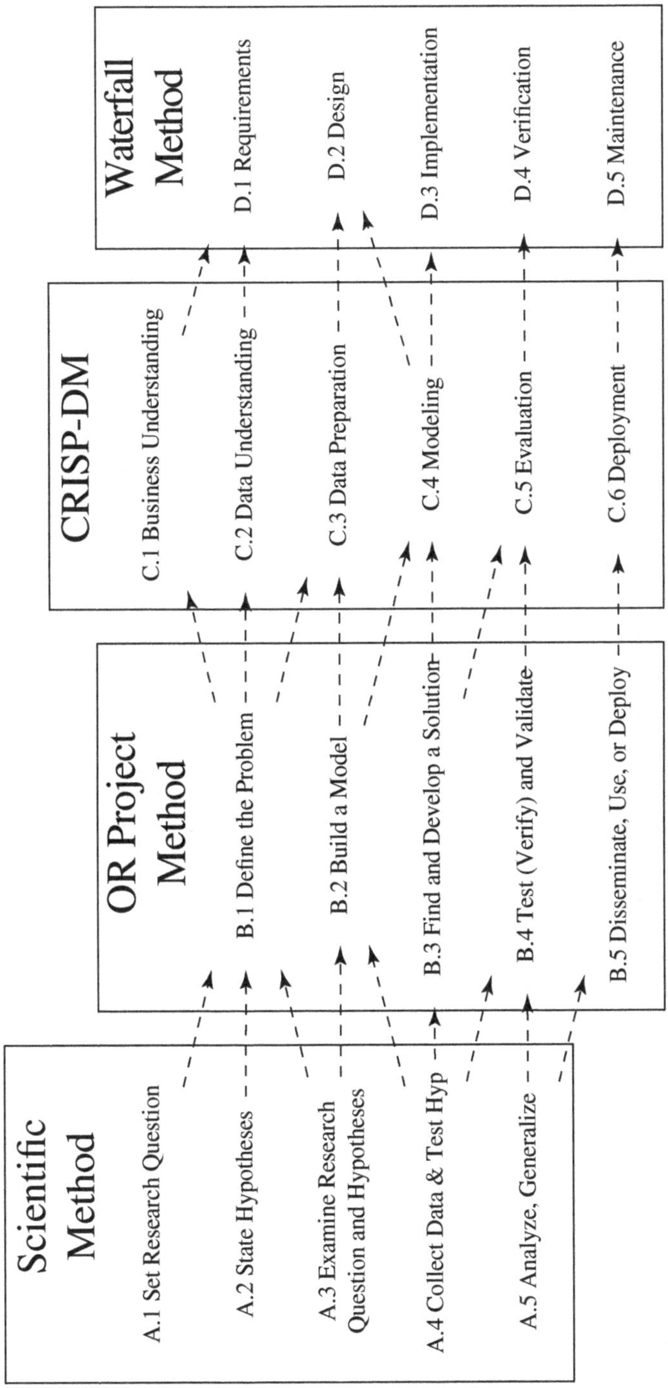

Figure 5.2 Relationship among the macro-methodologies.

embedded in an iterative process. One can also see, through their relationships, how it can be argued that each one, in some way, is derivative of the scientific method.

Every analytics project is unique and can benefit from following a macro-methodology. In fact, a macro-methodology can literally save a troubled project, can help to ensure credibility and repeatability, can provide a structure to an eventual experience paper or documentation, and so on. In fact, veteran practitioners may use a combination of steps from different macro-methodologies without being fully conscious of doing so. (All fine and good, but, in fact, you veterans could contribute to our field significantly if you documented your projects in the form of papers submitted for INFORMS publication consideration and if, in those papers, you described the methodology that you used.)

The take-home message about macro-methodologies is that it is not necessarily important exactly which one of them you use–its just important that you use one (or a hybrid) of them. It is recommended that, for all analytics projects, the steps of *problem definition* and *verification and validation* be inserted and strictly followed, whether the specific macro-methodology used calls them out directly or not.

5.3 Micro-Solution Methodologies for the Analytics Practitioner

In this section, we turn our attention to micro-methodology options available to the analytics practitioner.

5.3.1 Micro-Solution Methodology Preliminaries

In general, for any micro-methodology, two factors are most significant in how one proceeds to "solutioning":

i) The specific modeling approach
ii) The manner in which the data (analytics) are leveraged with respect to model building as well as analysis prior to modeling and using the model

Modeling approaches vary widely, even within the discipline of operations research. For example, data, numerical, mathematical, and logical models are distinguished by their form; stochastic and deterministic models are distinguished by whether they consider random variables or not; linear and nonlinear models are differentiated by assumptions related to the relationship between variables and the mathematical equations that use them, and so on. We note that micro-solution methodology depends on the chosen modeling approach, which in turn depends on domain understanding and problem definition–that is, some of those macro-methodology steps covered in the previous section. Skipping over those foundational steps becomes easier to justify when the methods that are most closely affiliated with them (e.g., descriptive statistics and statistical inference) are side-lined in a rush to use "advanced (prescriptive) analytics."

Thus, we begin this micro-solution methodology section by re-stating the importance of following a macro-solution methodology, and by emphasizing that the selection of appropriate micro-solution methodologies–which could even constitute a collection of techniques–is best accomplished when practitioners integrate their selection considerations into a systematic framework that enforces some degree of precision in *problem definition* and *domain understanding*, that is, macro-method steps in the spirit of *A.1*, *A.2*, *A.3*, *B.1*, *B.2*, *C.1*, *C.2*, *C.3*, and *D.1* (see Figure 5.2).

All of this is not to diminish the importance of the form and purpose of the project analytics, that is, the data, in selection of micro-solution methodologies to be used. In fact,

- how data are created, collected, or acquired,
- how data are mined, transformed, and analyzed,
- how data are used to build and parameterize models, and
- whether general "solutions" to models are dependent or independent of the data

are all consequential in micro-solution methodology. However, it is the *model* that is our representation of the real world for purposes of analysis or decision-making, and as such it gives the *context* for the underlying problem and the understanding of the domain in which "solving the problem" is relevant. This is why consideration of (i) the specific modeling approach should always take precedence over (ii) the manner of leveraging the data. Thus, this section is organized around modeling approaches first, while taking their relationship to analytics into account as a close second.

5.3.2 Micro-Solution Methodology Description Framework

This section presents the micro-solution methodologies in these three general groups:

Group I. Micro-solution methodologies for exploration and discovery

Group II. Micro-solution methodologies using models where techniques to find solutions are independent of data

Group III. Micro-solution methodologies using models where techniques to find solutions are dependent on data

Note that these groups are not directly aligned with the "descriptive–predictive–prescriptive" paradigm but are intended to complement the paradigm. In fact, depending on the nature of the underlying problem being "solved," and as this section shall illustrate, a micro-methodology very often draws from two or three of the three (i.e., "descriptive," "predictive," and "prescriptive") characterizations at a time–sometimes implicitly, and at other times explicitly.

Since it is impractical to cover every conceivable technique, this section covers an array of historically common techniques relevant to the INFORMS and analytics practice with the goals of illustrating how and when to select techniques. (Note that we will use the word *technique* or *method* to describe a specific micro-solution methodology.) While pointers to references are provided for the reader to find details of specific techniques, we use certain model and solution technique details to expose why choosing an approach is appropriate, how the technique relates to micro (and in some cases, macro)-methodology, and to compare and contrast choices in an effort to help the reader differentiate between concepts. And while there are many, many flavors of models and modeling perspectives (e.g., an iconic model is usually a physical representation of the real world, such as a map or a model airplane), we'll generally stay within the types of models most familiar to the operations research discipline. Further reading on the theory of modeling can be found in the foundational work of Zeigler [22], in introductory material of Law and Kelton [23], and of course in our discipline standards such as Hillier and Lieberman [14,16] and Winston [13]. Others, such as Kutner et al. [24], Shearer [25], Hastie et al. [26], Provost and Fawcett [27], and Wilder and Ozgur [28], expose and contrast the practice and theory of modeling led from the perspective of data first. General model building is also the topic of the next chapter of this book.

We turn next to the presentation of each of the above micro-solution methodology groups. Each micro-methodology group is presented using the following framework:

1) What are the general characteristics of *problems* we try to "solve" by micro-solution methodologies of this group? What are some examples?
2) Which *models* are used by the micro-solution methodologies of this group? What are the typical underlying assumptions of the models, and what are their advantages and disadvantages?
3) How are *data* considered by this group? That is, how are data created, collected, acquired or mined, transformed, analyzed, used to build and parameterize models, and so on?
4) What are some of the known *techniques* related to finding solutions to the underlying problem based on use of each model type?
5) What is the *relationship to macro-methodology steps*?
6) What are the *main takeaways* regarding the micro-methodology group?

5.3.3 Group I: Micro-Solution Methodologies for Exploration and Discovery

This group of micro-solution methodologies includes everything we do to explore operations, processes, and systems to increase our understanding of them, to discover new information, and/or to test a theory. Sometimes, the real-world system, which is the main object of our study, exists and is operational so that we can observe it, either directly or through a data history (i.e., indirectly). Sometimes, the operation we are interested in does not exist yet, but there are related data that help us understand the environment in which a new system might operate. The important thread for this group involves *discovery.*

Group I: Problems of Interest

Problems that are addressed by methods in this exploratory group are in this group because they can be generally characterized by, for example, the following questions: How does this work? What is the predominant factor? Are these two things equal? What is the average value? What is the underlying distribution? What proportion of these tests are successful? In fact, it is in this group that the (macro) scientific method has most relevance, because it helps us to formulate research queries and structure the processes of collecting data, estimating, and inferring. Exploration and discovery is often where analytics projects start, both in research and the real world of analytics practice. It is also not uncommon to repeat or return to exploration and discovery steps as a project progresses and new insights are found, even from other forms of micro-solution methodologies. As an example, consider a linear programming model (that will be covered in Group II) that needs cost coefficients for instantiating the parameters of an objective function. In some cases, simple unit costs may exist. In many real-world scenarios, however, costs change over time and have complex dependencies. Thus, estimating the cost coefficients may be considered an exploration and discovery subproblem within a project. In this example, the problems addressed may be finding the valid range for a fixed cost coefficient's value or finding 95% confidence intervals for the cost coefficients. Questioning the assumption that the cost function is indeed linear with respect to its variable for a specified range is another example of a problem here.

Group I: Relevant Models

When considering exploration and discovery, the relevant models are statistical models. Here, we mean statistical models in their most general sense: the underlying distributions, the interplay between the random variables, and so on. In fact, part of the exploration may be to determine the relevant underlying statistical model–for example, determining if an underlying population is normally distributed in some key performance metric, or if a normal-inducing transformation of observations will justify a normality assumption. The importance of recognizing the underlying models formally when doing exploration and discovery is related to the assumptions formed for using subsequent techniques.

Group I: Data Considerations

Data when the micro-methodology group is one of exploration and discovery may be obtained in a number of ways. In the most classic deployment of the scientific method, data are created specifically to answer the exploration questions, by running experiences, observing, and recording the data. In today's world of digital operations and systems, historical data are often available to enable the exploration and discovery process. Data "collection" in these digital cases may take more of the form of identifying digital data sources, exploring the data elements and characterizing their meaning as well as their quality, and so on, and even "mining" large data sets to zero in on the most pertinent forms of the data. In these cases of *already-existing* data, it is equally important to consider the research questions, the underlying problem being solved, and the relevant models. For example, one may have a fairly large volume of data to work with (i.e.,"Big Data"), but despite the generous amount of data, the data cover a time period or geography that is not directly relevant to the problem being studied. For example, if a database contains millions of sales transactions for frozen snacks purchased in Scandinavian countries during the months of January and February, the data may not be relevant to finding the distribution of daily demand for the same population during summer months, or for a population of a different geography at any time, or for the distribution of daily demand for frozen meals (i.e., nonsnacks) for a population of any geography in any time period. In some situations, we may have so much data (i.e., "Big Data") that we decide to take a representative random sample.

In general, for this group of methods, the problem one wishes to solve and the assumptions related to the statistical models considered are the most important data considerations. In certain cases, practitioners may like to think that their exploration process is so preliminary that a true problem statement (that is sometimes stated as a research question plus hypotheses) and any call out of modeling assumptions are considered unnecessary. However preliminary, exploration can usually benefit by introducing some methodological steps, even if the problem statement and modeling assumptions are themselves preliminary.

Group I: Solution Techniques

Keeping in mind that "solving" a problem related to an exploration and discovery process involves trying to answer an investigational question, it should be no surprise that techniques related to descriptive statistical models are at the core of the micro-solution methodologies for this group. Applied statistical analysis and inference have a traditional place in the general research scientific methods related to exploration, and they also carry the discovery needed for the data handling and wrangling required by other "advanced" models and solution techniques. In fact, one of the great ironies of our field is that the statistical models and techniques that constitute "descriptive models and techniques" are the oldest and most well formed in theory and practice of all solution methodologies related to analytics and operations research. Hence, passing them over for "advanced" (e.g., prescriptive or predictive) techniques should elicit at least some derision.

This collection of techniques might be, arguably, the most important subset of the micro-solution methodology techniques. Why? Because even prescriptive and predictive techniques reckon on them.

Techniques here range from deriving descriptive statistics (mean, variance, percentiles, confidence intervals, histograms, distributions, etc.) from data to advanced model fitting, forecasting, and linear regression. Supporting techniques include experimental design, hypothesis testing, analysis of variance, and more–many of which are disciplines and complete fields of expertise in and of themselves.

The methods of descriptive statistics are fairly straightforward, and most analytics professionals likely have their favorite textbooks to use for reference. For example, coming from an engineering background, one may have used Ref. [29]. Reference [30] is the standard for mathematics-anchored folks. Reference [31] is the usual choice for the serious experimenters. For the most part, all of these methods help us to use and peruse data to gain insights about a process or system under study. Usually, that system is observable, either directly or indirectly (e.g., in the form of a digital transaction history, which is often the case today). While not as old as the scientific method, the field of statistics is old enough to have developed a great amount of rigor–but it also has lived through a transformational period over the past 30+ years, as we've moved from methods that rely on observations that needed to be carefully planned (i.e., experimental design) and took great effort to collect (i.e., sampling theory and observations) to a world in which data are ubiquitous. In fact, many Big Data exploratory methods are based on using statistical sampling techniques–even though we may have available to us, in glorious digital format, an exhaustive data set, that is, the entire population!

Histograms, boxplots, scatter plots, and heatmaps (showing the correlation coefficient statistics between pairs of variables) are examples of visualizations that, paired with descriptive statistics and inference, help practitioners to understand data and to check assumptions. See Figures 5.3–5.6, respectively. Histograms and boxplots are powerful means of identifying outliers and anomalies that may lead to avoiding data in certain ranges, identifying missing values, or even spotting evidence of data-transmission errors.

Descriptive statistics are equally powerful for exploring nonquantitative data. Finding the number of unique values of a text field, and finding how frequently these unique values occur in the data, is standard for understanding data. Again,

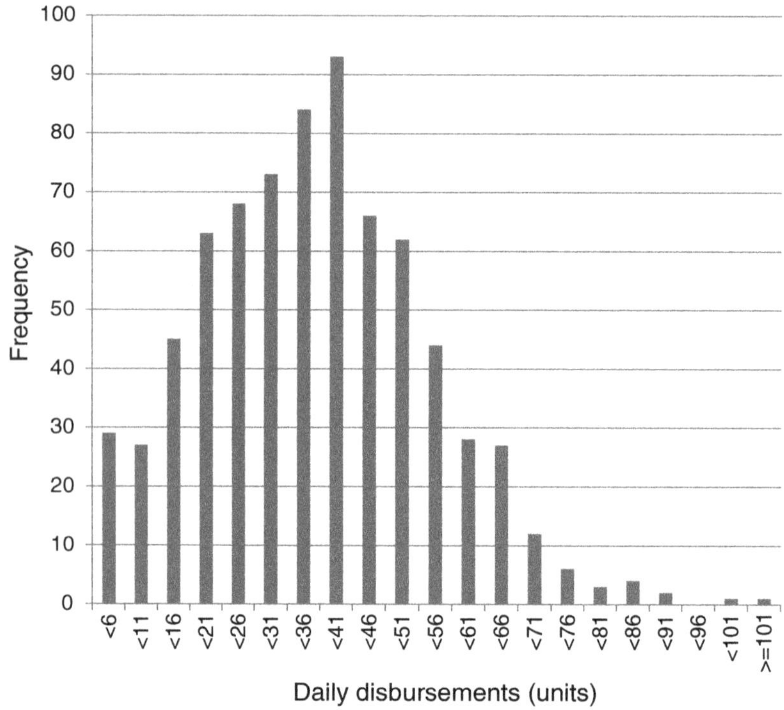

Figure 5.3 An example of a histogram showing the frequency (distribution) for unit disbursements of a single food item at a New York City digital food pantry from January 2, 2013 to April 24, 2017.

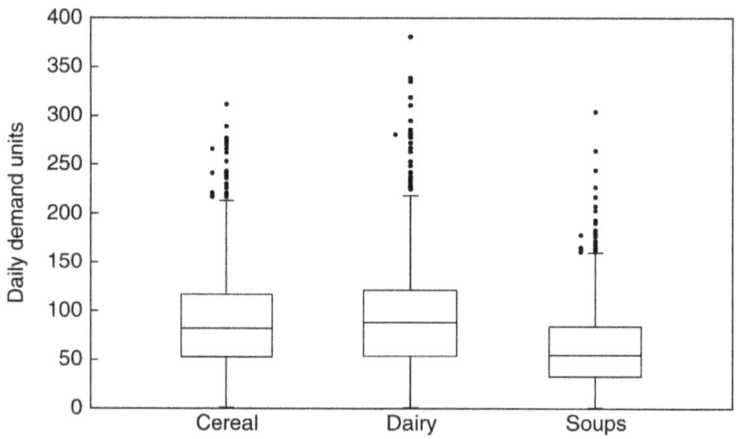

Figure 5.4 An example of a boxplot showing the distribtuion for unit weekday demand of for three food categories at a New York City digital food pantry from January 2, 2013 to April 24, 2017.

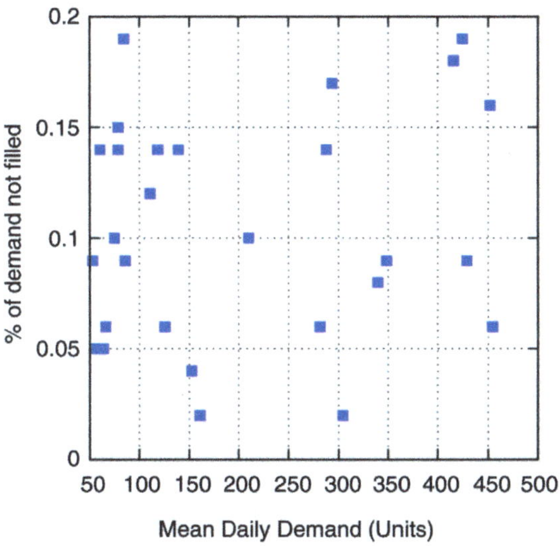

Figure 5.5 Example of a scatter plot visually showing the relationship between the daily (mean) demand and nonfill percentage for a set of stock keeping units. There appears to be no signficant correlation for this product set.

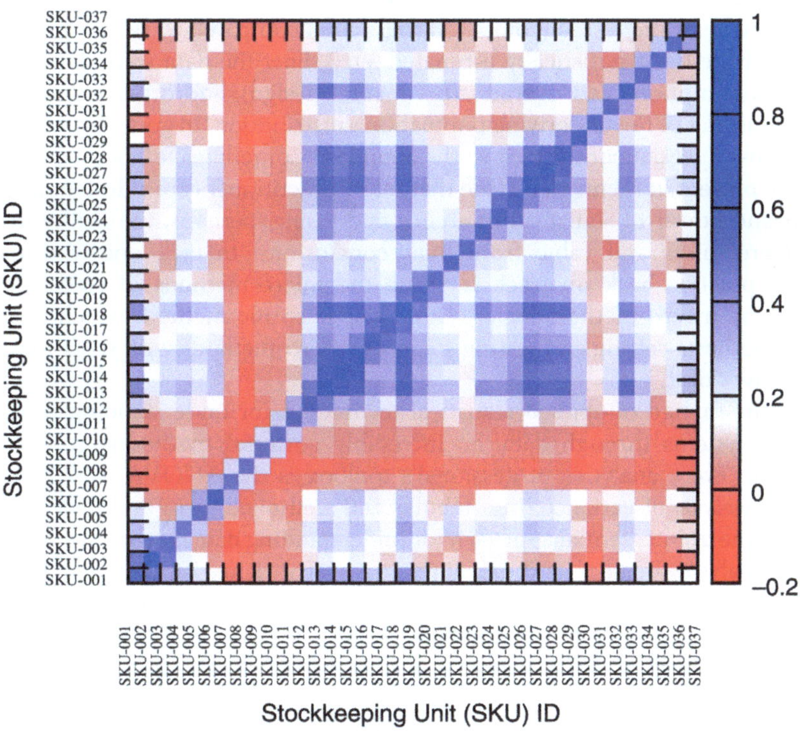

Figure 5.6 Example of a heat plot visually showing the pairwise correlation coefficient for a set of stock keeping units (SKUs). There are several negatively correlated pairs of SKUs, indicated by the dark red, and several positively correlated SKUs, indicated by the blue for those not in the diagonal.

together with scatter plots and heatmaps for data visualization, correlation analysis is usually done during data exploration to help practitioners understand the relationships between different types of data.

Overall, the micro-methodologies formed by the wealth and rigor of statistical analysis provide the analytics professional with tools that are specifically aimed at drawing conclusions in a systematic and fact-based way and at getting the most out of the data available, while also taking into consideration some of the inherent uncertainty of conclusions. For example, computing a confidence interval for an estimated mean not only gives us information about the magnitude of the mean but it also provides a direct methodology for deciding if the true mean is actually equal to some value. We can test to see if the mean is really zero by noticing if the confidence interval includes the value of zero. By virtue of taking variance and sample size into its calculation, the confidence interval, along with the underlying assumption of distribution, gives us a hint about how well we can rely on this type of test.

Hypotheses tests in general are one of the most powerful and rigorous ways to make very solid conclusions based on fact. The methods of hypotheses testing depend on what type of statistic is being used (mean, variance, proportion, etc.), what the nature of the test is (compared to a value, compared to two or more values that have been statistically estimated, etc.), how the data were derived (sampling assumptions and overall experimental design), and other assumptions, such as that of the underlying population's distribution. In going from the sparse, hard-to-get data of the past to the abundant, sometimes full population data of the present, it seems to be true that many practitioners are sidestepping the rigor and power of statistical inference and losing, perhaps, the ability to gain full credibility and value from their conclusions. In fact, one way to bring this practice back on track is to tie the micro-methods of statistics back into the macro-methodologies, either the scientific method, which has natural hypothesis-setting and testing steps, or macro-methods with steps that are derivatives of it.

Within the myriad of applied statistical techniques for understanding processes and systems through data, an incredibly powerful methodology that should be in every analytics professional's toolbox is the ANOVA. ANOVA stands for analysis of variance. In a tabular and well-oiled form and method, ANOVA is the quintessential approach for understanding data by virtue of how they help analysts organize and explain sources of variance (and error). The method gets its name from the fact that the table is an accounting of variance by attributable source, and one way to think about it is really as a bookkeeping practice for explaining what causes variance. ANOVA tables are natural mechanics for performing statistical tests, such as comparison of variance to see which source in a system is more significant. A basic extension of ANOVA is the multi-variate analysis of variance (MANOVA), which extends this methodology by considering the presence of multiple dependent variables at once.

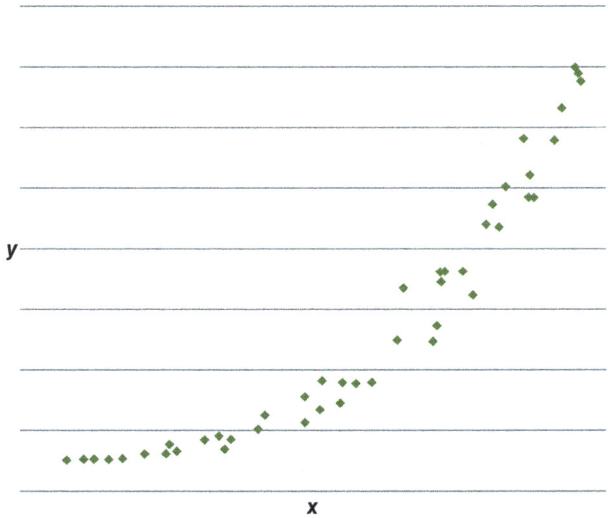

Figure 5.7 An independent and a response variable before transformation-induced linearity.

Any statistics textbook of worth should have at least one chapter devoted to ANOVA computations and applications, including tests. Reference [32] is a favorite text for analysts who frequently use regression analysis, which is closely tied to the methodology of ANOVA–they basically go hand-in-hand. Regression is the stepping stone for analytics and in particular modeling that is derived from data–it is the essential method when one wishes to find a relationship, generally a linear equation, between one or more independent variables and a response variable. The mechanics of this method involve estimating the values of a y-intercept and slope (for a single independent variable). This is called the method of least squares, and it is basically the solution to an embedded optimization problem. Solution methodology for the least squares problem, for example, Ref. [33], is also an illustration showing that the the techniques of micro-methodologies often depend on one another–in this case, a statistical modeling technique dependent on an underlying optimization method. Figure 5.7 exhibits a range of observations before applying a transformation to linearize the data and fit a linear regression (see Figure 5.8), illustrating another common form of common and complimenting techniques (i.e., mathematical data transformation prior to applying of a micro-methodology).

In summary, micro-methodologies for exploration and discovery rely on the following core techniques:

- Basic statistics
- Experimental design

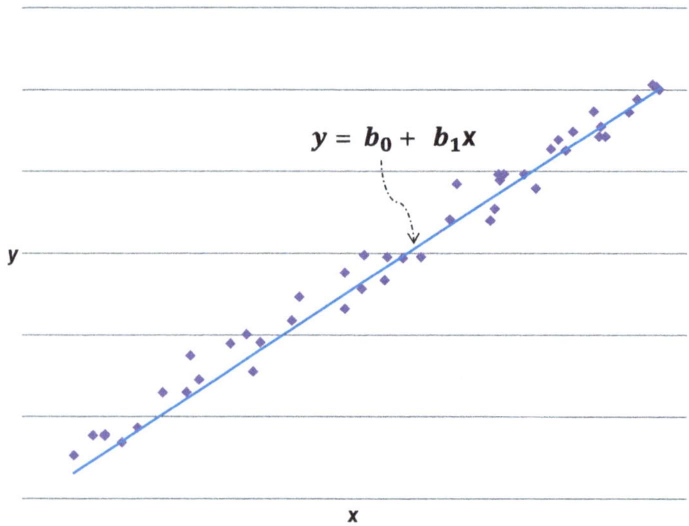

Figure 5.8 An independent and a response variable after transformation-induced linearity, with linear regression line.

- Sampling and estimation
- Hypothesis testing
- Linear regression
- ANOVA and MANOVA

Group I: Relationship to Macro-Methodologies

This area of analytics and OR is most closely and traditionally related to the scientific method and to the discovery and research processes in general, and it is not surprising that there are hundreds, maybe thousands, of textbooks devoted to this statistical topic, since virtually every field of study and research in science, social sciences, education, engineering, and technology relies on these methods as the underlying basis for testing research questions and drawing conclusions from data.

Group I: Takeaways

An important function of applied statistics in the analytics world today is in preparing data for other methods, for example, creating the parameters for the math programming techniques described in the previous section. In this case, and in the case of methods covered in the subsequent sections, statistical inference is the important methodology for providing the systematic process and rigor behind data-preparation steps, for just about any other method in analytics and OR that relies on any data. Thus, in virtually every analytics project involving data, statistical analysis and particularly inference methods will always have a role.

5.3.4 Group II: Micro-Solution Methodologies Using Models Where Techniques to Find Solutions Are Independent of Data

Next, we consider micro-methodologies where the models for which the techniques used to "solve" problems are independent of data. Note that this does not mean that the models and techniques do not use data. On the contrary! Here, the assumption of "independence of data" means that we can find a general solution path whether or not we know the data. In other words, we can find a solution and then plug the data in later so that we can then say something about that particular instance of the problem and its solution.

Group II: Problems of Interest

This group is distinguished by the fact that data, that is, our analytics, create an instance of the problem through parameters such as coefficients, right-hand-side values, interarrival time distributions, and so on. Problems of interest in this group are those in which we seek a modeling context that allows for either experimentation (as an alternative to experimenting on the real-world system) or decision support (i.e., optimization). The problem statements that characterize this group are of one of two forms: *experimental* (i.e., *what-if* analysis) or *prescriptive* (e.g., what should I do to optimize?).

As discussed in the Introduction of this chapter, problem statements are often elusive, particularly in the early phases of a real-world project. In that spirit, it is not uncommon to have a problem statement formulated somewhat generally for this group: How can I make improvements to the system (or operation) of interest? Or, how can I build the best new system given some set of operating assumptions?

Group II: Relevant Models

Some of the modeling options relevant to this group include the following:

- Probability models
- Queueing models
- Simulation and stochastic models
- Mathematical and optimization models
- Network models

Indeed, these modeling options include many viable modeling paths. The most significant factor in determining the modeling path relates back to questions that are fundamental to the problem statement, which may also characterize the analytics project objective: Do I want to model an existing or new system? Am I trying to build a new system or improve an existing one? How complex are the dynamics of the system? Are there clear decisions to be made that can be captured with decision variables and mathematical equations (or equalities) that constrain the variables and may also be used to drive an objective function that minimizes or maximizes something?

Group II: Data Considerations

In this group, data serve the purpose of creating parameters for the models. For simulation, probability, and queueing models, this may mean data that help to fit distributions for describing interarrival or service times or any other random variables in a system. For optimization models, we generally seek data for parameterizing right-hand-side values, technical coefficients within constraint equations, objective function cost coefficients, and so on.

Traditionally, operations researchers developed models with scant or hoped-for data. In some cases, practitioners may have compensated for unavailable data by making inferences from logic and/or using sensitivity analysis to test the robustness of solutions with respect to specfic parameter input values. Indeed, that models with solution techniques became the original core of operations research modeling is not entirely surprising, given the preanalytics era challenge of data availability.

In today's world of analytics, a new challenge is that the data needed to parameterize models in this class may be too much (versus the old problem of too little). In this case, the micro-methods of *Group I* come in handy and should be used, for example, for everything from the estimation of point estimates to finding confidence interval estimates that specify interesting ranges for sensitivity analyses to distribution fitting and hypotheses testing.

Group II: Solution Techniques

- *Basic Probability.* Practitioners should use these techniques following the choice of models that yield descriptions about the inherent uncertainty of events in a system. These are techniques used to estimate discrete choice probabilities or to fit probability distribution parameters. The quintessential example is estimating the probabilities of simple events such as those in a decision tree (see Figure 5.9). Comprehensive treatment of probability models and their solution techniques can be found in Ref. [34].
- *Stochastic Processes.* In general, one moves to stochastic processes (from basic probability models and techniques) when there is a dynamic aspect of systems being studied. Processes are often described by states and transitions, either discrete or continuous in nature. Comprehensive treatment of solution techniques can be found in Ref. [35].
- *Queueing Theory.* A queueing system is basically any system where waiting in line may occur when there is contention for one or more limited resources. These systems occur almost everywhere! For example, they occur when people are waiting in line for a cashier at a grocery store, a bank teller, or an ATM; they occur when manufacturing subassemblies (i.e., partially finished products) wait for the attention of machines and operators; and they occur virtually in call centers and communications systems (e.g., see Ref. [36]). An example of a queueing system configuration in a manufacturing system is

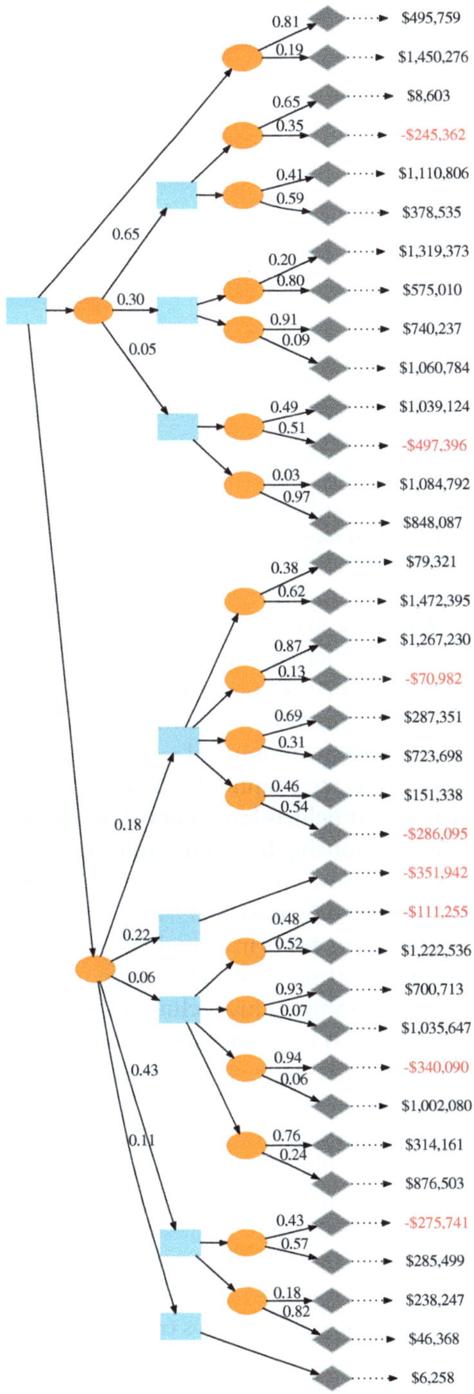

Figure 5.9 Illustration of a decision tree. Square (blue) nodes represent decision points with choice arcs emanating from them. Ovals (orange) represent external events, with uncertainties captured in adjacent probabilistic arcs. Diamonds (gray) illustrate the space of all possible outcomes, each with an associated value.

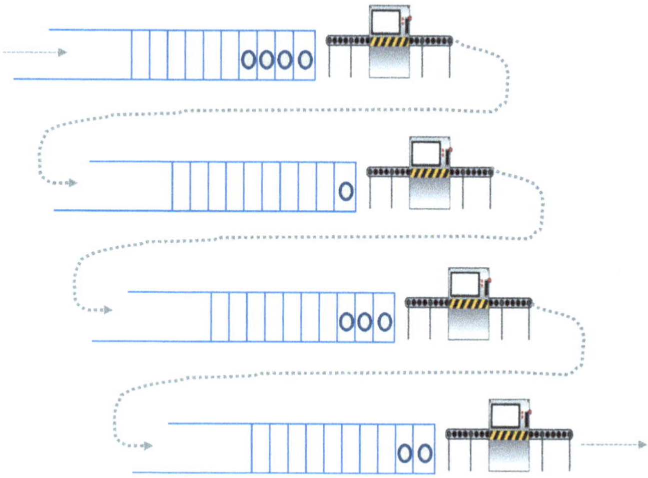

Figure 5.10 Example of a complex queueing system involving four servers in sequence. Work items arrive for service from the top server, then move sequentially downward, queue to queue. When completed by the fourth server, they leave the system.

given in Figure 5.10. Techniques in this area are derivatives of probability, stochastic processes, systems theory, differential equations, and calculus. In simpler systems, a closed form solution (i.e., a well-formed equation) may exist, and in more complicated systems, an approximation or bounding method is used because the equations to "solve" (i.e., find the number of servers to ensure the expected waiting time is no more than x, etc.) cannot be derived. One of the most important results in this area, for our field, is Little's law ($L = \lambda W$, see Refs [37,38]), which basically tells us the relationship between waiting time and queue length in a system with arrivals related to rate λ. For a comprehensive treatment of this area, see the foundational work in Refs [39,40].

- *Monte Carlo Simulation.* This technique has its roots in numerical methods–the canonical application is computing an estimate for the definite integral, that is, the area under a function within a range. This technique works by converting random numbers (between 0 and 1) into points that land proportionally under or over the function. The area approximation is found by counting the number of points generated under the curve and comparing that number to the number of generated pointed containing the function over the range. Today, this method forms the basis for the acceptance–rejection method of random variate generation (see Ref. [23]) and for estimating performance metrics of a system when time advance is not sophisticated.
- *Discrete Event Simulation.* This technique extends the techniques of Monte Carlo by considering the advance of time in a more sophisticated fashion, that is, the *time flow mechanism* and an *event calendar* that keeps track of discrete events to be processed. Discrete events provide the logic for updating system

state variables, which dynamically represent the system and are used to capture performance variables of interest such as (for a queueing system): *resource utilization, waiting time, number in line*, and others.

When random variate generation is used to create, for example, interarrival and service times, these models are considered stochastic. In general, discrete event simulation models rely heavily on statistical and probability models and techniques for preparing inputs. Stochastic simulation models in general, once implemented in computer code (either high level or a language or package designed explicitly for simulation) basically form experimental systems in that they attempt to mimic the real-world system (or some scoped portion) for the purpose of performing what-if analyses. For example, when simulating an inventory-control system, *how are stock-outs impacted if the daily demand doubles but the inventory replenishment and ordering policies stay the same*? In simulating the traffic flowing through an intersection between two major roads, *what is the impact on average time waiting for a red light to turn green, if the timing of the light changing is changed from 45 to 60 seconds?* In simulating cashier lanes in a popular grocery store, *will five cashier lanes be sufficient to ensure that all check-out lanes have fewer than three customers at least 95% of the time?*

Simulation modeling is one of the most malleable techniques in our analytics toolbox. It is also one of the easiest to abuse (e.g., when results from unverified or unvalidated simulation models are proclaimed as "right"). From an analytics solution methodology perspective, it is important to note that simulation output data should be statistically analyzed, that is, appropriate statistical techniques should be deployed. In fact, the techniques (and macro- plus micro-solution methodologies) can and should be applied to the output of simulations. A comprehensive treatment of system simulation is provided in Ref. [23]. In general, this subfield of OR has led the way in methodological innovations, as exemplified by the aforementioned work in model verification and validation by Sargent [15].

- *Mathematical Programming and Optimization.* Mathematical programming and discrete optimization models and techniques are at the core of the operations research discipline. These techniques form what has become known as the *prescriptive* category. At this point, it is worth bringing up that prescriptive approaches provide the classic notion of *context* for the decisions they are designed to support–that is, they define how to *prescribe in general*–while *data*, in the form of model input or output, gives the *instantiation*–that is, they help use the model to prescribe for a *specific* problem instance. For a more in-depth discussion of the INFORMS definition of analytics, that is, aligned with the notion of making *better decisions*, see Ref. [41].

This collection of techniques includes linear programming, nonlinear programming, integer programming, mixed-integer programming, and discrete and combinatorial optimization. A set of specialty algorithms and methods related to network flows and network optimization is often included with these models and techniques.

These methods all begin similarly: There is a decision to be made, where the decision can be described through values of a number of variable settings (called *decision variables*). Feasibility (i.e., that at least one solution represented as values of the decision variable settings can be found) is generally determined by a set of mathematical equations or inequalities (thus, the name *mathematical programming*). The selection of a best solution to the decision variables, if one exists, is guided by one or more equations, usually prefaced by the word *maximize* or *minimize*.

Which solution method to choose among these techniques is generally determined by the forms of variables, constraints, and objective function. Thus, some "modeling" (stating what the variables are, describing the decisions, describing the system and decision problem in terms of the variables, that is, the objective and constraint functions) must usually take place in order for practitioners to determine the appropriate micro-solution methodology. For example, if all constraint and objective functions are linear with respect to the the decision variables, then linear programming micro-methodologies are appropriate. Linear programming is usually the starting point for most undergraduate textbooks and courses in introductory operations research; see, for example, Ref. [14]. The standard micro-solution methodology for linear programming is the simplex method, which dates back to the early origins of operations research (see Ref. [42]).

The simplex method, invented by George Dantzig (considered to be one of the pioneers of operations research [43]), is a methodology that systematically advances and inspects solutions at corner points of a feasible region, effectively moving along the exterior frame of the region. In April 1985, operations research history was made again when Karmarkar presented the interior point method to a standing-room-only crowd at the ORSA/TIMS conference in Boston, Massachusetts [44,45]. The new method proposed moving through the interior of the feasible region instead of striding along from extreme point to extreme point [46]. It held implications not only for solving linear programming models, but also for solving nonlinear programming models, which are distinguished by the fact that one or more of the constraints or the objective function(s) is nonlinear with respect to decision variables.

As the number of decision variables and constraints become large, large-scale optimization techniques become important to all forms of math programs–these micro-methodologies involve solution strategies such as relaxation (i.e., removing one or more constraints to attempt to make the problem "easier" to solve), decomposition (i.e., breaking the problem up into smaller, easier-to-solve versions), and so on. Finding more efficient techniques for larger problem sizes (i.e., problems that have more variables and constraints, perhaps in the thousands or millions) has become the topic of many research theses and dissertations by graduate students in operations research and management science.

Among the most challenging problems in this space are the models where variables are required to be integers (i.e., integer programming or mixed-integer programming) or discrete (leading to various combinatorial optimization methods). While many specialty techniques exist for integer and mixed-integer (combinatorial/discrete) models, the branch-and-bound technique remains the *de facto* general standard for attempting to solve the most difficult, that is, NP (nondeterministic polynomial time) decision problems (see Refs [47,48]), Branch and bound is an example of *implicit enumeration*, and, while not as old as the simplex method, is one of the oldest (and perhaps most general) solution techniques in operations research.

To summarize, mathematical programming techniques span the following:

- *Linear Programming (LP).* These models are characterized by constraints and an objective function, which are linear with respect to decision variables. The canonical reference is Ref. [49]. Introductory operations research textbooks by Hillier and Lieberman [16] and by Winston [13] provide anchoring chapters on linear programming. While most textbook coverage of linear programming focuses on the simplex method, Ref. [33] provides an entry-level version of the interior point method that students and practitioners may find helpful before turning to more complex descriptions, such as those found in Refs [44,46].
- *Nonlinear Programming (NLP).* These models are characterized by constraints or objective functions that are nonlinear with respect to decision variables. Comprehensive treatment can be found in Refs [50,51]. References [13,16] provide introductory material covering the most widely used methods and optimality conditions (i.e., Karush–Kuhn–Tucker, or KKT).

Examining the structure of a nonlinear programming model reveals that there are times when an NLP may be transformed to an LP formulation, which is preferrable because of the general availability of off-the-shelf LP packages. However, it should be noted that one of the most common mistakes by practitioners is to try to use an LP solution package outright for an NLP formulation.

Figure 5.11 shows a classic visualization of a feasible region for math programming, in this case with a linearized feasible region (with two decision variables) and either a linear or nonlinear objective function. In this case, it was possible to achieve a valid linear feasible region for the example by converting a nonlinear inequality (system reliability as a linear function of decision variables that are its component failure intensities, λ_1 and λ_2) using a natural logarithm transformation.

In contrast to linear programming, the methods deployed by nonlinear programming generally follow an if-then-else-if-then-else-if- and so on

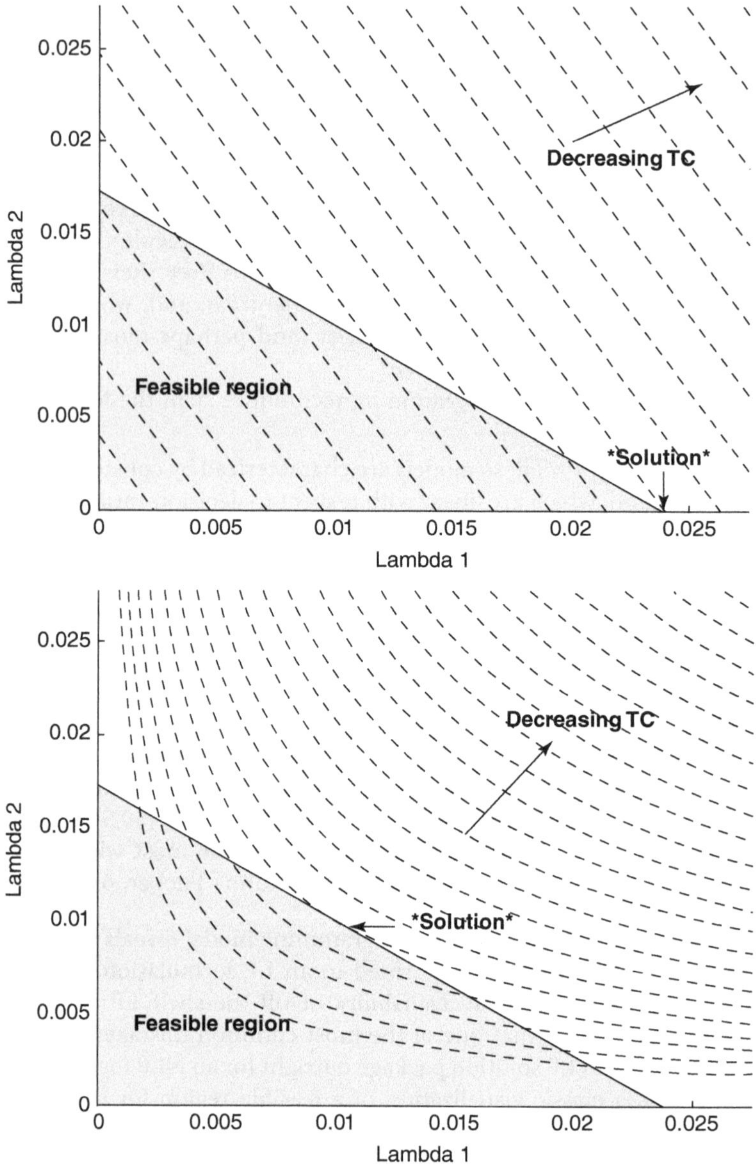

Figure 5.11 Example from Ref. [52] showing a linearized (system reliability) feasible region as a function of two decision variables, LAMBDA1 (λ_1) and LAMBDA2 (λ_2). The contours of the cost-to-attain linear cost function (top), or nonlinear function (bottem), show the optimized solution either at a corner point (top) or at a constraint midpoint (bottem), respectively.

deduction, where one chooses a micro-solution methodology based on the convexity or concavity (or pseudo- or quasi-) forms of the feasible region and objective function. The best way to determine which micro-methodology to use for a nonlinear program is actually to write down the model variables, constraints, and objective function, then mathematically characterize the forms, and then consult one of the classic textbooks, such as Refs [50,51] as a guide to choosing the most appropriate solution technique.

- *Integer and Mixed Programming.* These models are characterized by some or all of the decision variables required to be integer in value. Introductory operations research textbooks by Hillier and Lieberman [16] and Winston [13] both provide excellent chapters on this topic. More in-depth treatment of techniques for handling these types of decision models can be found in Ref. [53].
- *Discrete, Combinatorial, and Network Optimization.* These models are characterized by some or all of the decision variables required to be discrete in nature. Techniques for handling these types of combinatorial decisions can be found in classics by Bertsimas and Tsitsiklis [54] and Papadimitriou and Steiglitz [47]. In some cases, decision problems of the discrete or combinatorial forms (i.e., where the feasible region is generally countable, consisting of discrete solution options as opposed to being in continuous space), we may choose a method that is tailored for the specific problem instead of working with the mathematical programming form directly. Discrete and combinatorial problems usually involve some kind of searching through a space, and often, that space is best represented by a complex data structure (such as a tree, or a network. See, for example, Figure 5.12). Examples include the shortest-path problem, the minimum-spanning-tree problem, the traveling salesman problem, the knapsack, bin-packing, set-covering, and clique problems, scheduling and sequencing problems, and so on. For details on the techniques behind these micro-solution methodologies, see Refs [47,48,53], which are the classic texts by the pioneers of the integer and discrete/combinatorial methods. For network-specific algorithms and methods, see Ref. [55].

Some other specialty forms that we will not cover here exist, including dynamic programming, multiobjective or multicriteria programming, and stochastic and constraint programming.

Group II: Relationship to Macro-Methodologies

While the specific micro-methodology chosen will depend on the type of problem faced, the assumptions made by the practitioner, and the model selected, the success of the models and techniques in this group hinges on certain macro-methodology steps, particularly business understanding and problem definition (including assumptions). As mentioned earlier, the scientific method and the exploratory micro-methodologies are appropriate for fitting

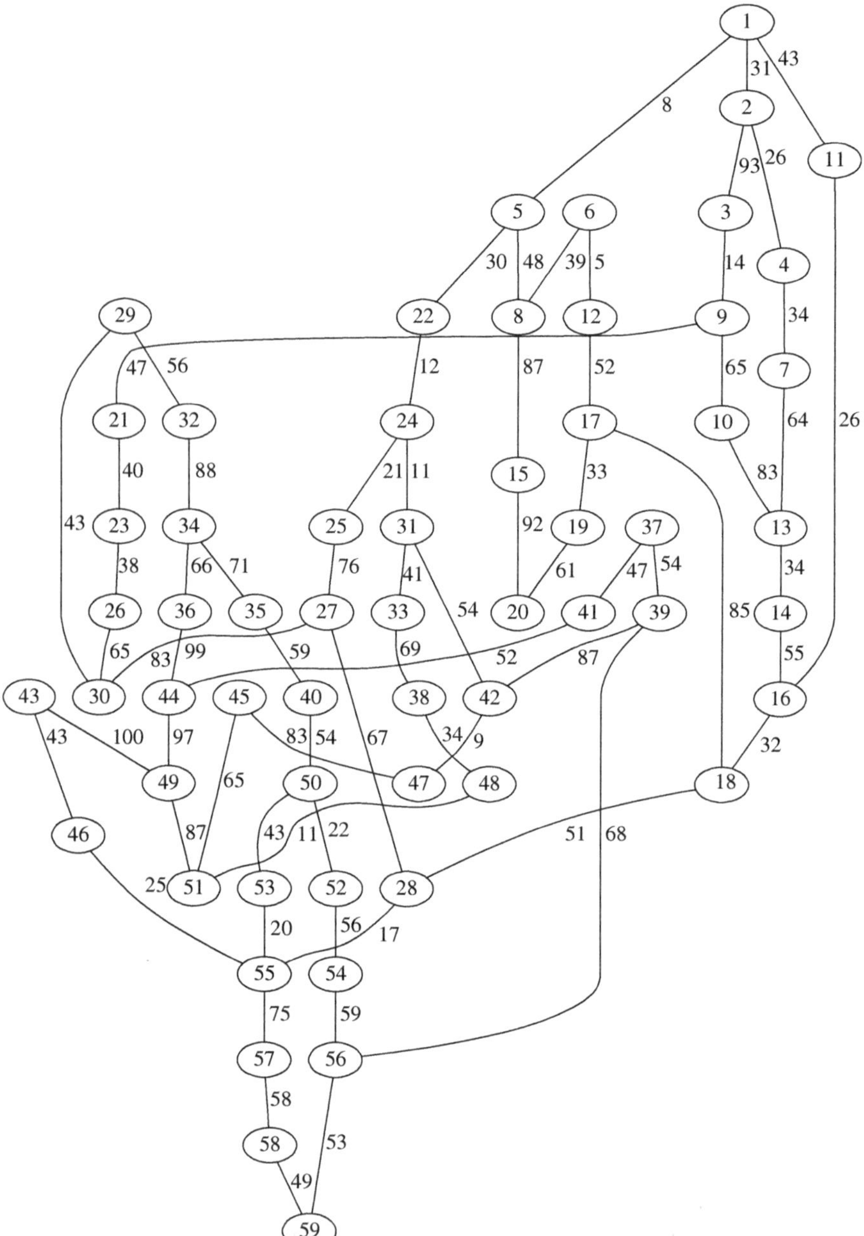

Figure 5.12 Example of an undirected network. The edge set is a configuration of the ARPANET from about 1977 [56].

model parameters and testing various assumptions (e.g., linearity, pseudo-convexity). The OR project methodology steps were designed specifically with projects using these micro-methods in this group. However, it should be noted that a few of the CRISP-DM steps can also be applicable; for example, when data are sought for parameter fitting–specifically the data understanding and data preparation steps. In some cases, more advanced transformations of data are needed in preparation for use in these modeling techniques. In fact, in some cases, the analytics we would like to introduce as parameters is derived from forecasting–that is, a special class of predictive modeling, which we turn to next.

Group II: Takeaways

Historically, the operations research discipline has been a collection of quantitative modeling methodologies that have their roots in logistics and resource planning. Over the past two decades, with the surge in data available for problem-solving, "research on operations" (i.e., operations and systems understanding), and model building, an emphasis of operations research (and management science) has shifted to embrace insights that can be derived directly from data. In this section, many of the traditional OR modeling approaches and their techniques were presented with the main message that these are largely models that have solution techniques that are independent of, but not isolated from, data.

5.3.5 Group III: Micro-Solution Methodologies Using Models Where Techniques to Find Solutions Are Dependent on Data

This section considers the final group of micro-methodologies, that is, those where the models involve solution techniques that are not possible to execute unless there are data present. In other words, they are data-dependent. Examples of solutions, in these cases, are the explanation or creation of additional system entity attributes or a prediction about a future event based on a trend that is observable in the data.

Group III: Problems of Interest

This group of micro-methods is most often used in conjunction with data mining. While these problems share the theme of exploration and discovery with Group I, the outcomes tend to be broader in nature and with fewer assumptions (e.g., normality of data). Problems relevant here include the desire to create categories of things according to common or similar features; finding patterns to explain circumstances or phenomena, that is, seeking understanding through common factors; understanding trends in processes and systems over time (and/or space); and understanding the relationships between cause and effect for the purpose of predicting some future outcome given similar circumstances.

Typical examples of problems of interest include understanding which retail items tend to be purchased together; sorting research articles into categories based on similarities in content identified through common keywords, concepts, methodology, or conclusions; determining if the fall in sales revenue is due to a trend in consumer preferences; if a pattern of behavior exists (e.g.: *Are referees more likely to give red cards to soccer players of darker skin tone?* which was studied in Ref. [2]); and others.

Group III: Relevant Models

Some of the main models used in this micro-methodology group include the following:

- *Generalized linear models* are a collection of models including traditional linear and logistic regression models. Logistic models have discrete (category) response variables.
- *Common factor and principal component models* are used to find the common denominators in groups.
- *Clustering models* are used to find groupings of things.
- *Classification models* are used to determine which set something belongs to. The main difference from clustering methods is that these are generally considered *supervised learning* (i.e., a *training set* is known and is used to guide membership), whereas clustering techniques are generally *unsupervised*. Note that supervised and unsupervised model building are described in detail in Chapter 6, Modeling Building.
- *Graph-based models* are general purpose data structures that support various models in this group.
- *Time series models*, for example, ARMA (auto-regressive-moving-average model), are used to model trends over time.
- *Neural networks* are generally used to direct inference in pattern recognition.

It should be noted that there is intersection with Groups I and II. Specifically, these methods borrow heavily from statistical analysis and even optimization (e.g., by solving an underlying total distance minimization problem).

Group III: Data Considerations

By design, this group is most distinguished in consideration of the data dependency on model building and solution techniques. Furthermore, data for this group of micro-methods are generally assumed to be abundant–for example, digital history of sales transactions, Internet sites visited, searched keywords, and so on. Data are often collected by observing digital interactions by a large number of people with systems such as Internet services and applications via browser connections or a mobile device that has passive data collection (e.g., location services) allowed, either intentionally or unintentionally.

A key distinction of these data is that they are not planned in the same way that exploratory methods, say, related to the scientific method of inquiry, may involve experimental design, observation, and data collection. In fact, for data to be considered "usable" in this group, it often must be interpreted, or derived by mining, analyzing, inferring, or applying models and techniques to create meaningful new features.

Group III: Solution Techniques

The following are some of the most common micro-solution techniques for this group:

- *Generalized Linear Model Techniques*–See Refs [26,57] for a review of techniques. Techniques include the following:
  - Imputation of missing data, for systematically replacing missing values with contants or other values.
  - The method of least squares, which finds the model parameters that minimize the sum of squared residual (i.e., distance to the fitted model) terms.
  - Statistical analysis and inference (e.g., estimation and hypothesis testing) for evaluating models.
- Factor/principal component analysis techniques include the following:
  - NIPA (noniterated principal axis method)
  - IPA (iterated principal axis method)
  - ML (maximum likelihood factor analysis method)

  All of these find common factors while using differing by underlying computation approach. See Ref. [58] for details.
- Clustering analysis uses a variety of techniques, depending on the nature of the data. Some specific techniques include the following:
  - Univariate and bivariate plots such as histograms, scatter plots, boxplots, and others may be used to visually aid the clustering process (see Figures 5.3–5.5).
  - Graph-based techniques may be used to generate additional features, such as distance and neighborhoods.
  - Hierarchical and nearest neighbor clustering, see Ref. [59].
  - Specialty methods such as *collaborative filtering, market basket association,* or *affinity analysis* may be used for specific problems, such as finding the items in a retail shopping basket that are generally purchased together (see Ref. [60]).
  - Linear models for classification, see Refs [24,26,61].
- *Classification Methods*: Some specific techniques include the following:
  - Linear classifiers, such as logistic regression
  - Support vector machines (SVMs)
  - Partitioning
  - Neural networks
  - Decision trees

  See Refs [26,61,62] for overviews and commenets on these and related methods.

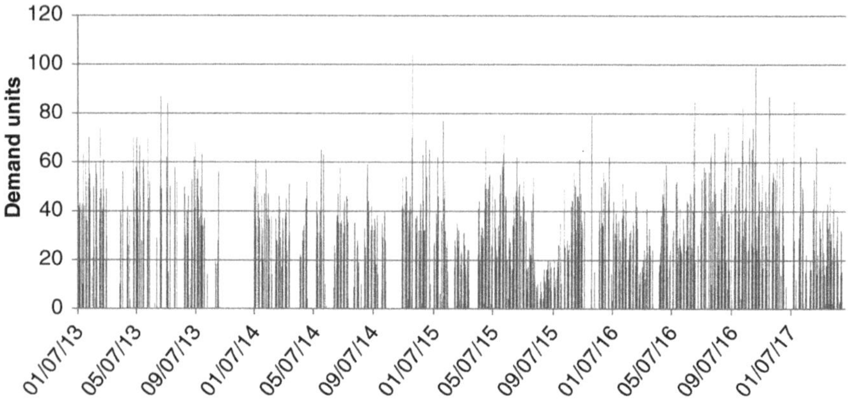

Figure 5.13 Example of raw time series data before ARMA methods are applied. Unit dispense history for a single food item at a New York City digital food pantry from January 2, 2013 to April 24, 2017.

- Graph-based modeling techniques are often used to derive features of components for other types of models. For example,
  - shortest path algorithm helps to identify nearest neighbors for clustering.
  - minimum spanning tree helps to determine connected subcomponents in a general graph.

  See Ref. [55] for a comprehensive treatment of graph models, network-based problems, and an exhaustive accounting of known algorithms. Hastie et al. [26] extend these basic graphical models for statistical machine learning techniques, including neural networks.
- Time series models, for example, ARMA (autorrgressive-moving-average model); see Ref. [63] for an exhaustive treatment of theory and techniques. See Figure 5.13 for an example of raw time series data.
- Neural networks methods are described in detail in Ref. [61].

Group III: Relationship to Macro-Methodologies

While CRISP-DM is likely the most common of the used macro-methodologies for this group, analytics projects leveraging data-dependent methods are likely to benefit from any and all macro-methodologies. In fact, because this set of methods is most closely related to evaluation and discovery of complex cause-and-effect relationships, as well as differentiation (through classification and categorization; which are sometimes prone to discrimination that can lead to inequity and unfair treatment of groups of people), practitioners should take utmost care in verifying, validating, and creating project documentation that promotes study replication.

Group III: Takeaways

While it may seem that this group of methods is *all about the data*–because they are data-dependent–that is not really true. Like all analytics solution methods, it really is still *all about the problem*. Because to have meaning, solutions must solve a problem. Also note that these analytics methods are sometimes referred to as *the advanced analytics methods*. The author would like to point out that they are, in fact, the newest, least established, and least proven in practice, of all methods in our discipline. This implies that they are the *least* advanced analytics methods and suggests that we should all be working harder to deepen their theory and rigor–which is actually what we are good at as an INFORMS community.

5.3.6 Micro-Methodology Summary

In summary of micro-methdologies, we emphasize that analytics problems encountered in practice seldom require techniques that fall into only one micro-methodology category. Techniques in one category may build on techniques from another category–for example, as noted earlier, linear regression modeling within data dependent methodologies relies on solving an underlying optimization problem. Regression modelers who use software packages to fit their data may not be aware that the least squares optimizatoin problem is being solved in the background. However, to truly understand our methods and results, it is important to be aware of the background mechanics and connections. This specific type of dependency is, in fact, common–particularly in the realm of contemporary statistical machine learning.

Projects in practice often leverage methodologies in progression as well–for example, using descriptive statistics to explore and understand a system in the early stages of a project may lead to building of an optimization model to support a specific business or operations decision. If the decision needs to be made for a scenario that will take place in the future, then forecasts may be used to specify the optimization model's input parameters. At the same time, it is important to keep in mind that there may be trade-offs to consider when combining different techniques. For instance, in this same example project requiring forecasted parameters of an optimization model, the practitioner has a choice between using a sophisticated predictive technique that yields more accurate forecast but leads to a complex, difficult-to-solve nonlinear optimization model, or using a simpler predictive approach that sacrifices some of the forecast accuracy, but leads to a simpler, linear optimization model.

The micro-solution methods available to analytics practitioners are many. However, it should be noted that making this selection is analogous to being an artist and deciding among watercolor, oil, or acrylic paint; deciding what kind of surface to paint on, for example, canvas, wood, paper, and so on; deciding how big to make the piece, and so on. But it is probably not unlike being the painter in these ways as well: You are most likely to pick the method you are most familiar with, just as the watercolor specialist is less likely to choose charcoal for a new painting of the sunset.

5.4 General Methodology-Related Considerations

5.4.1 Planning an Analytics Project

A critical success factor in technical projects, particularly where there is any element of exploration and discovery, is project planning. This is no different for analytics projects. In fact, when one adds the expectation for a usable outcome (i.e., a tested and implemented process coded in software, running on real data, complete with a user interface and full documentation, all while providing smashing insights and impactful results), the project risks and failure odds go up fast. As mentioned in the macro-methodology section, the macro-methods align nicely with project planning because they give a roadmap that equates to the high-level set of sequential actities in an analytics project. When considering macro- and micro-method planning together, skills and details of activities can be revealed, so that task estimation and dependencies are possible. In fact, one of the traditional applications of network models taught to students of operations research is the PERT (program evaluation and review technique)/CPM (critical path method)–a micro-method that practitioners can apply to macro-methodology for helping to smoothly plan and schedule a complex set of related activities (see Ref. [14]).

When there are expectations for a usable software implementation outcome, practitioners can augment their macro-methodology steps with appropriate software engineering steps. The software engineering requirement step is recommended for planning desired outcome function, as well as usability needs and assumptions. In fact, complex technical requirements, such as integration into an existing operations environment, or perhaps data traceability for regulatory compliance, are best considered early in requirements steps that compliment domain and data understanding steps.

Overall, while prototyping and rapid development often coincide with projects of more exploratory nature, which analytics projects often are, some project planning and ongoing project management is the best way to minimize risks of failure, budget overruns, and outcome disappointments.

5.4.2 Software and Tool Selection

Most if not all of our analytics projects need some computational support in the form of software and tools. Aside from DIY software, which is sometimes necessary when new methods or new extensions are developed for a project, most micro-solution methods are available in the form of commercial and/or open-source software.

Without intending to endorse any specific software package or brand, a few packages are named here to provide illustrations of appropriate packages, while leaving to the reader to decide which packages are most appropriate for their specific project needs.

For (*Group I*) exploration, discovery, and understanding methods, popular packages include R, Python, SAS, SPSS, MATLAB, MINITAB, and Microsoft EXCEL. Swain [64] provides a very recent (2017) and comprehensive survey of statistical analysis software, intended for the INFORMS audience. Most of these packages also include GLM, factoring, and clustering methods needed to cover (*Group III*) data-dependent methods, as well.

For (*Group II*), a fairly recent survey of simulation software, again by Swain [65] and a very recent linear programming software survey by Fourer [66], are resources for selecting tools to support these methods, respectively. An older but still useful nonlinear programming software survey by Nash [67] is a resource to practitioners. MATLAB, Mathematica, and Maple continue to provide extensive toolboxes for nonlinear optimization needs. For Branch and Bound, the IBM ILOG CPLEX toolbox is freely available to academic researchers and educators. COIN-OR, Gurobi, GAMS, LINDO, AMPL, SAS, MATLAB, and XPRESS all provide various toolboxes across the optimization space. More and more, open source libraries related to specific languages, such as Python, now offer tools that are ready to use–for example, StochPY is a Python library addressing stochastic modeling methods.

As a final note, practitioners using commercial or open-source software packages for analytics are encouraged to use them carefully within a macro-solution methodology. In particular, verification, that is, testing to make sure the package provides correct results, is always recommended.

5.4.3 Visualization

Visualization has always been important to problem-solving. Imagine in high school having to study analytical geometry without 3D sketches of cylinders. Similarly, operations research has a strong history of illustrating concepts through visualization. Some examples include feasible regions in optimization problems, state space diagrams in stochastic processes, linear regression models, various forms of data plots, and network shortest paths. In today's world of voluminous data, sometimes the best way to understand data is to visualize it, and sometimes the only way to explain results to an executive is to show a picture of the data and something illustrating the "solution."

Other chapters in this book cover the topic of analytics and visualization, for example, see Chapters 3 and 6. The following points regarding visualization from a solution methodology perspective are provided in order to establish a tie with the methods of this chapter:

- Analytics and OR researchers and practitioners should consider visualizations that support understanding of raw data, understanding of transformed data, enlightenment of process and method steps, and solution outcomes.
- Visualization in analytics projects has three forms, which are not always equivalent:
  1) *Exploratory*–that is, the analyst needs to create quick visualizations to support their exploration and discovery process. The visualizations may help to build intuition and give new ideas, but are not necessarily of "publish" or "presentation" quality.
  2) *Presentation*–that is, the analyst needs to create visualizations as part of a presentation of ideas, method steps, and results to sponsors, stakeholders, and users.
  3) *Publishing*–that is, the analyst wants to create figures or animations that will be published or posted and must be of suitable quality for archival purposes.

5.4.4 Fields with Related Methodologies

Many disciplines are using analytics in research and practice. As shown in the macro-methodology section summary, all macro-methodologies are derivatives of the scientific method. In fact, many of our micro-solution methodologies are shared and used across disciplines. As a community, we benefit from and have influenced shared methods with the fields of science, engineering, software development and computer science (including AI and machine learning), education, and the newly evolving discipline of data science. This cross-pollination helps macro- and micro-solution methodologies to stay relevant.

5.5 Summary and Conclusions

This chapter has presented analytics solution methodologies at both macro- and microlevels. Although this chapter makes no claim to cover all possible solution methodologies comprehensively, hopefully the reader has found the chapter to be a valuable resource and a thought-provoking reference to support the practice of an analytics and OR project. The chapter goals of enlightening the distinctions of macro- versus micro-solution methodologies, providing enough details of these solution methodologies for a practitioner to incorporate them into the design of a high-level analytics project plan according to some macro-level solution methodology, and providing some guidance for assessing and selecting appropriate micro-solution methodologies appropriate for a new analytics project should have hopefully come through in the earlier sections and sections. In addition to a few pearls scattered throughout the chapter, we conclude by stating that solution methodologies can help the analytics practitioner and can help that practitioner help our discipline at large, which can then help more practitioners. That's a scalable and iterative growth process that can be accomplished through reporting our experiences at conferences and through peer-reviewed publication, which often forces us to organize our thoughts in terms of methodology anyway, so we might as well start with it too! The main barriers for solution methodology seem to be myths. Dispelling some of the myths of analytics solution methodology is covered in these final few paragraphs.

5.5.1 "Ding Dong, the Scientific Method Is Dead!" [68]

The scientific method may be old, but it is not dead yet. By illustrating its relationship to several macro-solution methodologies in this chapter, we've shown that the scientific method is indeed alive and well. Arguments to use it literally may be futile, however, since the world of technology and analytics practice often places time and resource constraints on projects that demand quick results. Admittedly, it is quite possible that rigor and systematic methodology could lead to results that are contrary to the "desired" outcome of an analytics study. Thus, without intentionally doing so, our field of practice may be inadvertantly missing the discovery of truth and its consequences.

5.5.2 "Methodology Cramps My Analytics Style"

Imagine for a moment that analytics practitioners used systematic solution methodologies to a greater extent, particularly at the macrolevel and then publish their applied case study following an outline that detailed the steps that they had followed. Our published applied literature could then be a living source of experience and practice to emulate, not only for learning best practices and new techniques, but also for learning how to apply and perfect the old standards. More analytics projects might be done faster because they wouldn't have to "start from scratch" and reinvent a process of doing things. Suppose that analytics practitioners, in addition to putting rigor into defining their problem statements, also enumerated their research questions and hypotheses in the early phases of their project. Would we publish experiences that report rejecting a hypotheses? Does anyone know of at least one published science research paper that reports rejecting a hypothesis let alone one in the analytics and OR/MS literature?

Research articles on failed projects rarely (probably never) get published, and these could quite probably be the valuable missing links to helping practitioners and researchers in the analytics/OR field be more productive, do higher quality work, and thrive by learning from studies that show what doesn't work. When authentically applied, the scientific method should result in a failed hypothesis every once in a while, reflecting the true nature of exploration and the risks we take as researchers of operations and systems. The modern deluge of data allows us to inquire and test our hunches systematically without the limitations and scarcity of observations we faced in the past. Macro-solution methodologies, either the scientific method or any derivative of it (which is just about all of them), could relieve analytics projects cramps not only by giving us efficient and repeatable approaches but also by recognizing that projects sometimes "fail" or reject a null hypothesis–doing so within the structure of a methodology allows it to be reported in an objective, thoughtful manner that others can learn from and that can help practitioners and researchers avoid reinvention.

5.5.3 "There Is Only One Way to Solve This"

We've all heard the saying, *if all you have is a hammer, then every problem looks like a nail.* This general concept, phrased in a number of different ways since first put forward in the mid-1960s, is credited to Maslow [69], who authored the book *Psychology of Science*. In our complex world, there are usually many alternate ways to solve a problem. These choices, in analytics projects, may be listed among the micro-methodology techniques described in this chapter or elsewhere. Sometimes, there are well-established techniques that work just fine, and sometimes a new technique needs to be created. The point is that there are many ways to solve a problem, even though many of us tend to first resort to our favorite ways because those tend to align with our personal experiences and expertise. That's not a bad approach to project work, because experience usually means that we are using other knowledge and lessons learned. However, behind this is the danger of possibly using the wrong micro-solution methodology. In fact, the problem of an ill-defined problem can lead to overreliance on certain tools–often the most familiar ones. What does this mean? That in our macro-solution methodology, steps such as understanding the business and data, defining the problem, and stating hypotheses are useful in guiding us to which micro-methodologies to choose from and thus avoiding the potential pitfalls of picking the wrong micro-method or overusing a solution method.

Interview With Alan Taber

Lockheed Martin Missiles and Fire Control's System Engineer Alan Taber offered these thoughts when asked to consider how the analytics professional determines the appropriate analytic methodology for a problem:

When the analytics professional is given a problem–when she or he is called in to someone's office and told, "We would like you to solve this problem," it is incumbent on the analytics professional to ascertain from the person with the problem what level of solution is sought and when is an answer needed. Sometimes the answers to these questions are evident from the context of the problem, but it is important to always ask! Once you have the initial input of the person whose problem you are trying to solve, you might ask for an opportunity to take a couple of very quick stabs at the model and then return to discuss the merits and drawbacks of various methodologies and results–in mechanical engineering we call this rapid prototyping. You can offer a couple of initial options and determine if any are appealing. Sometimes the response will be an enthusiastic, "Yes! I really like the first methodology you showed me!," so that's the route you go. And sometimes the response will be, "I don't like either of those," but the conversation may provide you with the insight you need to develop and offer new options. Through this iterative process, you gradually–and sometimes painfully and sometimes slowly–collect contextual information that you need in order to guide you on what methodology and what approach to solving the problem you are going to take.

If you are going to be in a profession, whether it is analytics or any other profession, you must maintain your skills and knowledge base. You can accomplish this by reading journals, attending conferences, presenting posters and talks, being active on blogs and reading other people's ideas, and so on. How have others solved problems? I find these activities to be very valuable because through them I am exposed to fields and problems and solutions that I would otherwise I never see.

There is a problem-solving methodology called TRIZ that asserts most innovation results from appropriately transferring a methodology from one field to another field. For example, when industrial diamonds were first created in a lab, nobody knew how to split them appropriately. Diamonds are very brittle; they fracture easily, and you will not achieve desired shape if you do not split it just right. Then someone borrowed the idea, "Hey, we can split and deseed peppers." Sure enough, the type of compressed air jets used to split peppers could be used to successfully split diamonds. So again, methodologies for solving problems often already exist, but we have to be aware of them. The more methodologies you have rattling around in your brain, sorted however you care to sort them, the easier it will be to find and select an appropriate methodology when you have a problem. If you are familiar with only a very sparse methodology set, you are going to struggle to answer some problems. The richer the methodology set in your mind, the more likely you will have one or even several methodologies that will fit a given question.

This is an excerpt from one of a series of interviews with analytics professionals and educators commissioned by the *INFORMS Analytics Body of Knowledge* Committee.

5.5.4 Perceived Success Is More Important Than the Right Answer

In math class, school teachers might make a grading key that lists the right answer to each exam or homework problem. In practice however, there is no solutions manual or key for checking if an analytics project outcome is right or wrong. We have steps within various macro-solution methodologies, for example, verification, that help us to try to make the best case for the outcome being considered "right," but for the most part, the correctness of an analytics project outcome is generally elusive, and projects are usually judged by the perceived results of the implementation of a solution. In analytics and OR practice, there are cases where the implementation results were judged as wildly successful, for example, analytics/OR project recognized as an INFORMS Edelman award finalist for its contribution to a company's saving of over $1 billion might actually be judged as not successful because the company creating the OR solution was not able to commercialize the assets and find practitioners in its ranks to learn and deploy them and thus reproduce the solution as a profitable product (see, for example, Ref. [70]).

Documentation of reasons for analytics project failures probably exists, but it is rarely reported as such. Plausible reasons for failure (or, perhaps more accurately, "lack of perceived success") include the following ones: the solution was implemented, but there was no impact, or it was not used; a solution was developed but never implemented; a viable solution was not found; and so on. Because of the relationship between analytics projects and information technology and software, some insights can be drawn from those more general domains. Reference [71] provides an insightful essay on why IT projects fail that is loaded with examples and experiences, many with analogues and wisdom transferable back to analytics. Software project failures have been studied in the software engineering community for over two decades, with various insights; see, for example, Ref. [72]. The related area of systems engineering offers good general practices and a guide to systematic approaches: One of the most recognized for the field of industrial engineering is by Blanchard and Fabrycky in its fifth edition [73].

It is important to remember that in practice ultimate perceived success or failure of an analytics project may not mean "finding the right answer," that is, finding the right solution. By perceived success, we mean that an analytics solution was implemented to solve a real-world problem with some meaningful impact acknowledged by stakeholders. Conversely, perceived failure means that for one of a number of reasons, the project was deemed not successful by some or all of the stakeholders. Not unlike some micro-solution methodologies of classic operations research, we have necessary and sufficient conditions for achieving success in an analytics project, and they seem to be related to perception and quality. Analytics practitioners need to judge these criteria for their own projects, while perhaps keeping in mind that there have been well-meaning and not-so-well-meaning uses of data and information to create perceptions and influence. See, for example, *How to Lie with Statistics* by Darrell Huff [74] and the more contemporary writing, which is similar in concept, *How to Lie with Maps* by Mark Monmonier [75].

The book *How to Lie with Analytics* has not been written yet, but unfortunately it is likely already practiced. By practicing some form of systematic solution methodologies, macro and micro, in our analytics projects, we may help our field to form an anchoring credibility that is resilient when that book does come out.

5.6 Acknowledgments

Sincere thanks to two anonymous reviewers for critically reading the chapter and suggesting substantial improvements and clarifications; Dr. Lisa M. Dresner, Associate Professor of Writing Studies and Rhetoric at Hostra University, for proofreading, editing, and rhetoric coaching; and Dr. Joana Maria, Research Staff Member and Data Scientist at IBM Research, for inspiring technical discussions and pointers to a number of relevant articles.

References

1 Eves H (2000) Analytic geometry, in W. H. Beyer, ed., *Standard Mathematical Tables and Formulae*, 29th ed. (CRC Press, Inc., Boca Raton, FL), pp. 174–207.
2 Silberzahn R, Uhlmann EL, Martin D, Anselmi P, Aust F et al. (2015) Many analysts, one dataset: making transparent how variations in analytical choices affect results, Open Science Framework. Available at https://osf.io/gvm2z/ (accessed July 2, 2017).
3 INFORMS (2015) Constitution of the Institute for Operations Research and the Management Sciences. Technical report, Catonsville, MD.
4 SAS (2017) What is analytics? Available at https://www.sas.com/en_us/insights/analytics/what-is-analytics.html (June 7, 2017).
5 Wikipedia Scientific Method. Available at https://en.wikipedia.org/wiki/Scientific_method (accessed July 2, 2017).
6 Harris W (2008) How the scientific method works. Available at http://science.howstuffworks.com/innovation/scientific-experiments/scientific-method.htm (accessed January 14, 2008).
7 Hoefling H, Rossini A (2014) Reproducible research for large-scale data, in Stodden V, Leisch F, Peng RD, eds., *Implementing Reproducible Research*, 1st ed., The R Series (Chapman and Hall/CRC), pp. 220–240.
8 Stodden V, Leisch F, Peng RD, eds., (2014) *Implementing Reproducible Research*, 1st ed., The R Series (Chapman and Hall/CRC).

9 Foreman H (2015) What distinguishes a good manuscript from a bad one? https://www.elsevier.com/connect/get-published-what-distinguishes-a-good-manuscript-from-a-bad-one (accessed July 2, 2017).
10 Foster KR Skufca J (2016) The problem of false discovery: many scientific results can't be replicated, leading to serious questions about what's true and false in the world of research. *IEEE Pulse* 7(2): 37–40.
11 Hardt M, Ullman J (2014) Preventing false discovery in interactive data analysis is hard. Proceedings of the 2014 IEEE 55th Annual Symposium on Foundations of Computer Science (FOCS'14), Washington, DC, IEEE Computer Society, pp. 454–463.
12 Ioannidis JPA (2005) Why most published research findings are false. *PLoS Med* 2(8): 696–701.
13 Winston WL (2003) *Operations Research: Applications and Algorithms,* 4th ed. (Duxbury Press).
14 Hillier FS, Lieberman GJ (2002) *Introduction to Operations Research,* 7th ed. (McGraw-Hill).
15 Sargent RG (2007) Verification and validation of simulation models, in Henderson SG, Biller B, Hsieh MH, Shortle J, Tew JD, Barton RR, eds., Proceedings of the 2007 Winter Simulation Conference, pp 124–137.
16 Hillier FS, Lieberman GJ, Nag B, Basu P (2009) *Introduction to Operations Research,* 9th ed. (McGraw-Hill).
17 Pete C, Julian C, Randy K, Thomas K, Thomas R, Colin S, Rüdiger W (2000) CRISP-DM 1.0 Step-by-Step Data Mining Guide. Technical report, The CRISP-DM Consortium.
18 IBM (2011) IBM SPSS Modeler CRISP-DM Guide. Technical report, Catonsville, MD.
19 Wirth R (2000) CRISP-DM: towards a standard process model for data mining. Proceedings of the Fourth International Conference on the Practical Application of Knowledge Discovery and Data Mining, pp. 29–39.
20 Cynthia D, Vitaly F, Moritz H, Toniann P, Omer R, Aaron R (2015) The reusable holdout: preserving validity in adaptive data analysis. *Science* 349 (6248): 636–638.
21 Pressman RS (1997) *Software Engineering: A Practitioner's Approach,* 4th ed. (McGraw Hill).
22 Zeigler BP (1976) *Theory of Modelling and Simulation* (John Wiley & Sons, Inc., New York).
23 Law AM, Kelton WD (1999) *Simulation Modeling and Analysis,* 3rd ed. (McGraw-Hill).
24 Kutner M, Nachtsheim C, Neter J, Li W (2004) *Applied Linear Statistical Models,* 5th ed. (McGraw-Hill/Irwin).
25 Shearer C (2000) The CRISP-DM model: the new blueprint for data mining. *Journal of Data Warehousing* 5: 13–22.

26 Hastie T, Tibshirani R, Friedman J (2009) *The Elements of Statistical Learning: Data Mining, Inference, and Prediction*, 2nd ed., Springer Series in Statistics (Springer Science & Business Media).
27 Provost F, Fawcett T (2013) *Data Science for Business: What You Need to Know about Data Mining and Data-Analytic Thinking*, 1st ed. (O'Reilly Media, Sebastopol, CA).
28 Wilder CR, Ozgur CO (2015) Business analytics curriculum for undergraduate majors. *INFORMS Trans. Educ.* 15(2): 180–187.
29 Walpole RE, Myers RH (1978) *Probability and Statistics for Engineers and Scientists*, 2nd ed. (Macmillan Publishing Co., Inc).
30 Hogg RV, Craig AT (1978) *Introduction to Mathematical Statistics*, 4th ed. (Macmillian Publishing).
31 Box GEP, Hunter JS, Hunter WG (2005) *Statistics for Experimenters: Design, Innovation, and Discovery*, 2nd ed., Wiley Series in Probability and Mathematical Statistics, (Wiley-Interscience).
32 Draper NR Smith H (1981) *Applied Regression Analysis*, 2nd ed. (John Wiley & Sons, Inc., New York).
33 Nocedal J, Wright S (2006) *Numerical Optimization*, 2nd ed., Springer Series in Operations Research and Financial Engineering (Springer).
34 Ross SM (2012) *A First Course in Probability*, 9th ed. (Pearson).
35 Ross SM (1995) *Stochastic Processes*, 2nd ed. (John Wiley & Sons, Inc., New York).
36 Suri R, Diehl GWW, de Treville S, Tomsicek MJ, (1995) From CAN-Q to MPX: evolution of queuing software for manufacturing. *INFORMS Interfaces* 25(5): 128–150.
37 Little JDC (2011) OR FORUM: Little's law as viewed on its 50th anniversary. *Oper. Res.* 59(3): 536–549.
38 Little JDC, Graves SC (2008) Little's law, in Chhajed D, Lowe TJ, eds., *Building Intuition: Insights from Basic Operations Management Models and Principles* (Springer Science+Business Media, LLC, New York, NY), pp. 81–100.
39 Kleinrock L (1975) *Queueing Systems. Volume 1: Theory* (John Wiley & Sons, Inc., New York).
40 Kleinrock L (1976) *Queueing Systems. Volume 2: Computer Applications* (John Wiley & Sons, Inc., New York).
41 Keenan PT, Owen JH, Schumacher K (2017) ABOK Chapter 1: introduction to analytics, in Cochran JJ, ed., *Analytics Body of Knowledge (ABOK)* (INFORMS).
42 Nash JC (2000) The (Dantzig) simplex method for linear programming. *Comput. Sci. Eng.* 2(1): 29–31.
43 Gass SI, Assad AA (2011) Transforming research into action: history of operations research. INFORMS Tutorials in Operations Research 1–14.

44 Karmarkar NK (1984) A new polynomial-time algorithm for linear programming. *Combinatorica* 4(4): 373–395.
45 Tsuchiya T (1996) Affine scaling algorithm, in Terlaky T, ed., *Interior Point Methods of Mathematical Programming,* (Springer), pp. 35–82.
46 Adler I, Resende MGC, Veiga G, Karmarkar N (1989) An implementation of Karmarkar's algorithm for linear programming. *Math. Program.* 44(1): 297–335.
47 Papadimitriou CH, Steiglitz K (1982) *Combinatorial Optimization: Algorithms and Complexity* (Prentice Hall, Inc).
48 Parker RG, Rardin RL (1988) *Discrete Optimization of Computer Science and Scientific Computing* (Academic Press).
49 Bazaraa MS, Javis JJ, Sherali HD (2009) *Linear Programming and Network Flows,* 4th ed. (John Wiley & Sons, Inc., New York).
50 Bazaraa MS, Sherali HD, Shetty CM (2006) *Nonlinear Programming: Theory and Algorithms,* 3rd ed. (Wiley-Interscience).
51 Bertsekas D (2016) Optimization and Computation, *Nonlinear Programming,* 3rd ed. (Athena Scientific).
52 Helander ME, Zhao M, Ohlsson N (1998) Planning models for software reliability and cost. *IEEE Trans. Softw. Eng.* 24(6): 420–434.
53 Nemhauser G, Wolsey LA (1988) *Integer and Combinatorial Optimization* (John Wiley & Sons, Inc., New York).
54 Bertsimas D, Tsitsiklis JN (1997) *Introduction to Linear Optimization,* 3rd ed., Athena Scientific Series in Optimization and Neural Computation, 6 (Athena Scientific).
55 Ahuja RK, Magnanti TL, Orlin JB (1993) *Network Flows: Theory, Algorithms, and Applications* (Prentice Hall).
56 Lukasik SJ (2011) Why the ARPANET was built. *IEEE Ann. Hist. Comput.* 33 (3): 4–21.
57 West BT, Welch KB, Galecki AT (2014) *Linear Mixed Models: A Practical Guide Using Statistical Software,* 2nd ed. (Chapman and Hall/CRC).
58 Fabrigar LR, Wegener DT (2011) *Exploratory Factor Analysis of Understanding Statistics* (Oxford University Press).
59 Everitt BS, Landau S, Leese M, Stahl D (2011) *Cluster Analysis,* 5th ed., Wiley Series in Probability and Statistics (John Wiley & Sons, Inc., New York).
60 Larose DT, Larose CD (2014) *Discovering Knowledge in Data: An Introduction to Data Mining,* 2nd ed., Wiley Series on Methods and Applications in Data Mining (John Wiley & Sons, Inc., New York).
61 Bishop CM (2007) *Pattern Recognition and Machine Learning of Information Science and Statistics* (Springer).
62 Domingos P. (2012) A few useful things to know about machine learning. *Commun. ACM* 55(10): 78–87.
63 Brockwell PJ Davis RA (1991) *Time Series: Theory and Methods,* 2nd ed., Springer Series in Statistics (Springer Science+Business Media).

64 Swain JJ (2017) Statistical analysis software survey: the joys and perils of statistics. *OR/MS Today* 44(1).

65 Swain JJ (2015) Simulation software survey. *OR/MS Today* 43(5).

66 Fourer R (2017) Linear programming: software survey. *OR/MS Today* 44(3).

67 Nash SG (1998) Nonlinear programming software survey. *OR/MS Today* 26(3).

68 Harburg EY (1939) Ding-dong! the witch is dead, *The Wizard of Oz* (Metro-Goldwyn-Mayer Inc., Beverly Hills, CA).

69 Maslow AH (1966) *Psychology of Science*, 1st ed. (Joanna Cotler Books).

70 Katircioglu K, Gooby R, Helander M, Drissi Y, Chowdhary P, Johnson M, Yonezawa T (2014) Supply chain scenario modeler: a holistic executive decision support solution. *INFORMS Interfaces* 44(1): 85–104.

71 Liebowitz J (2015) Project failures: what management can learn. *IT Prof.* 17(6): 8–9.

72 Charette RN (2005) Why software fails. *IEEE Spectr.* 42(9): 42–49.

73 Blanchard BS Fabrycky WJ (2010) *Systems Engineering and Analysis*, 5th ed., Prentice Hall International Series in Industrial & Systems Engineering (Pearsons).

74 Huff D (1954) *How to Lie with Statistics* (W. W. Norton & Company) (reissue edition October 17, 1993).

75 Monmonier M (1996) *How to Lie with Maps*, 2nd ed. (University of Chicago Press).

6

Modeling

Gerald G. Brown

Operations Research Department, Naval Postgraduate School, Monterey, CA, USA

6.1 Introduction

This chapter recalls some of the most influential real-world operations research models in history. The organization is topical and introduces in context standard operations research modeling terminology. The presentation focuses on how each problem is stated, and how solutions are interpreted. Not all model solution methods are shown, but a path to access these methods is given. Neither calculus nor linear algebra is required.

Some acute modeling pitfalls are highlighted here. Most references provided herein are from widely available, approachable sources, such as Wikipedia, rather than from scholarly open literature or textbooks that a reader might not own or want to purchase. Illustrations are open source or original.

The goal here is showing, by example, how to build a model. Your model will not likely exactly match any of these examples, but will surely require necessary and reasonable simplifying assumptions.

Operations research is all about making things better, and this hopefully shows how this has been, and can be accomplished.

6.2 When Are Models Appropriate

A *model* is an abstraction that emphasizes certain aspects of reality to assess or understand the behavior of a *system* under study. The system may be physical, logical, mathematical, or some other representation of reality, such as an enterprise or some portion of one (Figures 6.1–6.4).

We concentrate on building mathematical models rather than physical ones, although much of our advice applies to those models as well. Some mathematical models can be solved *analytically*, algebraically in closed form, while most

INFORMS Analytics Body of Knowledge, First Edition. Edited by James J. Cochran.

Figure 6.1 A map is a model. A map is an abstraction of reality emphasizing entities such as road networks, infrastructure, or natural features at the expense of others. : https://en.wikipedia.org/wiki/File:World_Map_1689.JPG. Public Domain.

mathematical models can be solved on a computer, even if they are extremely complex. The value of a closed-form solution is that the entire domain of admissible inputs is accommodated and wholesale conclusions can be drawn; while a computer solution might be for a single instance of a problem and even after many such solutions with varied inputs, we only observe model behavior in the domain we have evaluated, and may need to make estimates of model behavior intermediate between those cases we have completely evaluated.

The system performs some *function*, and may be governed by a *system operator*. The system operator may be an executive or a managerial organization (e.g., a railroad), an automated control protocol (e.g., the Internet), an economic equilibrium or invisible hand[1] (e.g., traffic flows), by government regulation (e.g., income taxation), or follow scientific laws (e.g., Newton's laws). For simplicity, let's view a *modeler* as someone trying to respond to a *problem* posed by a system

1 https://en.wikipedia.org/wiki/Invisible_hand

Figure 6.2 A physical model. This full-scale model of an aircraft in a wind tunnel excludes many important properties of the real system, including engines, flight instruments, and a pilot. However, the model is extremely useful for doing what it was designed to do: determining the aerodynamic properties of the aircraft.

operator *client* who may work for *senior stakeholders* who ultimately make decisions.

At its core, a model describes the performance of a system in a particular *state*, and a set of admissible *actions* by the system operator that can transform that state.

We may also turn to models to learn how to interact with a system to achieve improved function or to avoid undesirable states.

We will look at several examples. The idea is to present a diversity of models that have historically proven innovative and effective in dealing with certain types of problems. It is rare that a new problem will be solved verbatim by any of these historical models, although a new problem often resembles one already studied. The idea is to learn how each model has been crafted to solve each

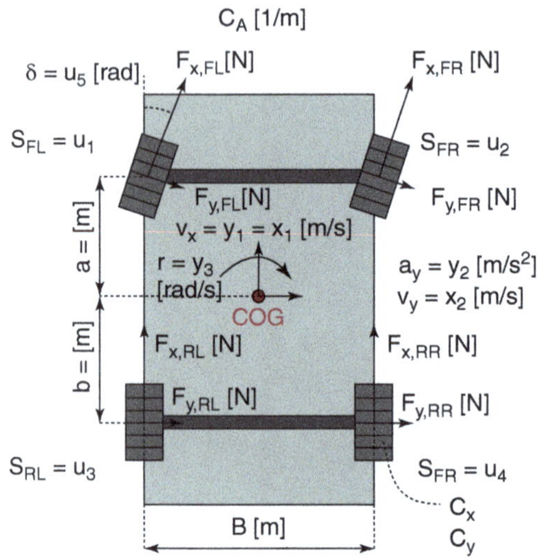

Figure 6.3 A mathematical model. This model can be used to describe the state and behavior of an automobile in mathematical terms. Here, we can see the abstraction from the physical state to the mathematical state.

Figure 6.4 Google Maps and the shortest path from Monterey to Lake Tahoe. Google Maps uses an algorithm to calculate the shortest path through a road network to minimize driving distance or time. Google invented neither maps nor the algorithm to determine the shortest path through a network (computer scientist Edsger Dijkstra conceived the method in 1956), but bundled these abstractions together with a clever graphical interface to produce something of great utility.

problem. What are the key ideas and important insights? Some of these models have had profound historical influence and many are works of peculiar genius that are still used to solve important problems and influence policy.

The notation used in these examples is as consistent as possible with the literature and should be viewed in isolation, because there is some reuse of terms between models.

Models are particularly valuable when one cannot directly interact with or influence a system, or when doing so would be prohibitively expensive and/or dangerous (operations research was born during World War II).

Before we investigate models further, please consider these five essential steps prior to building a model. We refer to the system operator in this hypothetical engagement as "the client."

6.2.1 What Is the Problem with This System?

Establishing a problem definition understood by both modeler and client is key. This may be the hardest part of any modeling project. The modeler must establish a problem definition expressed in carefully crafted language, establishing a lexicon that is, at once, precise for the modeler and understood by the client. This may involve many clarifying iterations between modeler and client. The ideal outcome is when the client declares "Yes, that describes my problem exactly."

6.2.2 Is This Problem Important?

Are we facing a problem that is, in fact, a minor annoyance, or one that is an existential threat to the system? Not all problems are worth solving with a model. Will dealing with this one be worth the cost of the modeling effort?

6.2.3 How Will This Problem Be Solved Without a New Model?

System operators have to be pretty clever, and modelers can learn a lot from the way they deal with problems. Find out how the problem is or will be solved without new modeling. And, you can bet it will be solved, with or without a modeler's help. Thumb rules, white boards, spreadsheets, and sticky notes can be very effective: The client and modeler need to collect as much of this tribal wisdom as possible. Observing anything that seems relevant but is not currently considered may reveal new insights and opportunities, and may also uncover organizational taboos.

6.2.4 What Modeling Technique Will Be Used?

For a modeler, this is the fun part. However, the modeler needs to involve the client as deeply and effectively as possible. At this point, any crippling simplifying assumption needs to be discovered. Here is when the modeler and client need to work out how and whether hypothetic model results are likely to be operationally useful.

6.2.5 How Will We Know When We Have Succeeded?

The modeler and client need to agree on objective criteria by which model success, or failure, will be judged. Nothing is more damaging than undefined and/or shifting modeling goals. Based on these criteria, the modeling effort might end up failing. The modeler and client need to be prepared for this outcome.

Interview With Jeffrey D. Camm

Jeffrey D. Camm, Associate Dean of Business Analytics and the Inmar Presidential Chair in Analytics at Wake Forest University's School of Business, shared these thoughts on the core of mathematical modeling:

Mathematical models of all types are constructed to help us better understand the entity/system being modeled. Ensemble modeling, that is, the use of two or more models to improve the accuracy of a prediction is currently in vogue. But in a sense, all analytical models, even prescriptive models, should be used in an ensemble, combining common sense, domain knowledge, and insights gleaned directly from the model to arrive at a decision that is superior to what could have been achieved with any of these three inputs alone.

The art of mathematical modeling is transforming a mess into a structured model. This is a tricky endeavor that requires strong listening skills and creativity. Managers, like patients, describe symptoms of problems rather than the actual problems. Like a good physician, a good analyst asks the right questions, listens well, and is careful not to draw premature conclusions.

This is an excerpt from one of a series of interviews with analytics professionals and educators commissioned by the *INFORMS Analytics Body of Knowledge* Committee.

Who Are the System Operator Stakeholders?

Typically, there will be *planners* who have to deal with the problem at hand, and it is these personnel who are the immediate beneficiaries of a model. The system operator is usually managed by *executives* in a variety of areas. Each of these areas presents a distinct set of concerns, and executive evaluation and compensation may be aligned with these. Executives may have conflicting objectives, and this can complicate matters for a model attempting to solve some problem to the satisfaction of all. The *client* is the executive assigned to deal at once with the planners and executives. The degree to which a model is expressed in the system operator's language can help the client guide executives to mutual agreement on actions.

6.3 Types of Models

Once we pass all prior hurdles, we embark on a new modeling project.

Models can be categorized based on what they are intended to achieve.

6.3.1 Descriptive Models

Descriptive models explain relationships between observed states. Company operating statements are descriptive models relating, usually in monetized terms, the beginning states, intermediate actions, and ending states of an enterprise. Your annual income tax return is a descriptive model of your income and other monetized activities ultimately leading to some outcome, resulting in a tax bill due. You might view a descriptive model as an explanatory reconciliation of initial state, intermediate actions, and resulting state.

Newton's Second Law (Deterministic, Descriptive)

In an inertial reference frame, the vector sum of the forces f (in Newtons) on an object is equal to the mass m (in kilograms) of that object multiplied by the acceleration a (meters per second) of the object:

$$f = m\text{a}.$$

This is a deterministic and purely descriptive model.

6.3.2 Predictive Models

Predictive models attempt to forecast future actions and resulting states, frequently employing forecast probabilities, seasonal trends, or even *subject matter expert* opinions (i.e., educated guesses). The goal is to forecast, envision, anticipate, or otherwise foretell future states. Weather forecasting models are predictive. Predictive analytics in data mining are currently fashionable.

6.3.3 Prescriptive Models

Prescriptive models seek admissible actions that, given initial state, lead to the best anticipated ending state.

6.4 Models Can Also Be Characterized by Whether They Are Deterministic or Stochastic (Random)

Deterministic models, such as the EOQ model here, are based on constant estimates of state and performance, while *stochastic (random) models* recognize the random nature of some states. Some stochastic simulation models generate states from random distributions, while others seek analytic characterizations of random processes.

Economic Order Quantity (EOQ) (Deterministic, Prescriptive)

We want to minimize the ordering cost and holding cost of a *stock keeping unit* (SKU) item. Given demand per period d items, fixed cost per replenishment order f, and holding cost per item per period h, the lowest-cost steady state is achieved by ordering

$$\mathrm{EOQ} = \sqrt{\frac{2df}{h}}.$$

This is a prescriptive model that lends insight, but ignores many other considerations, especially when the system operator manages at once many distinct types of items.

"Essentially, all models are wrong, but some are useful."

G.E.P. Box

6.5 Counting

Counting Permutations and Combinations

Frequently, we need to count the number of states that some action can cause.

Suppose we are dealing with random five-card poker hands from a 52-card deck. There is exactly one way to deal with a royal flush in spades (Ace, King, Queen, Jack, and 10) in that sequence, and the number of such ordered sequences or *permutations* is $52 \times 51 \times 50 \times 49 \times 48 = 311,875,200$. As we deal with five cards, we are *sampling without replacement* from the deck. Typically, we don't care in what order our cards are dealt, and would be quite happy to receive that spade royal flush in any sequence. Getting these five spades in any sequence can happen $5 \times 4 \times 3 \times 2 \times 1 = 120$ ways. The number of distinct five-card hands, without respect to order, or *combinations*, is $\frac{52 \times 51 \times 50 \times 49 \times 48}{5 \times 4 \times 3 \times 2 \times 1} = \frac{311,875,200}{120} = 2,598,960$. Our mathematical shorthand for this is $\binom{52}{5} = \frac{52!}{5!(52-5)!}$. Looking at this formula, we see that it counts not only the number of five-card hands but also the number of 52-5=47-card combinations remaining undealt in the deck.

6.6 Probability

A *probability*[2] is an assessment of the likelihood that a binary event (future state) will take place, numerically ranging from zero (impossibility) to one (certainty).

Independence Assumption and the Multiplication Rule (Deterministic, Predictive)

Suppose A and B each represent binary states (true–false, yes–no, on–off, win–lose, etc.) to be discovered. We refer to $P(A)$ as the probability A turns out to be true, and $P(B)$ that B does. We use $P(A,B)$ to represent the *joint probability* that both are true, and $P(A|B) = \dfrac{P(A,B)}{P(B)}$ the *conditional probability* of A given B. If $P(A|B) = P(A)$, then state A is not influenced by state B, and we refer to A and B as being *independent*, and the joint probability $P(A,B) = P(A)P(B)$. This simple *multiplication rule* is so easy to apply, it makes "the independence assumption" very attractive. *Caution*: If this assumption is not justified, your simplified model can give very misleading, even dangerous results.

Bayes Theorem (Deterministic, Predictive)

A and B are events, binary states that are not necessarily independent, that is, $P(A|B) \neq \dfrac{P(A,B)}{P(B)}$. Bayes theorem applies the definitions of conditional probabilities to derive this result: $P(B|A) = \dfrac{P(A|B)P(B)}{P(A)}$.

Suppose an employer uses a drug test to screen applicants that gives a positive or negative result. Given a drug user, the test returns a positive result 98% of the time, but also gives a positive result 10% of the time for a nonuser. If we know that 1 in 10 applicants is a user, what is the probability that an applicant with a positive test result is a user?

Let A represent the test result (positive, negative), and B the state of nature (user, nonuser). We know $P(\text{user}) = 0.1$ and $P(\text{positive}|\text{user}) = 0.98$. We also know $P(\text{positive}|\text{nonuser}) = 0.1$ (false positive result) and $P(\text{nonuser}) = 0.9$.

$$
\begin{aligned}
P(\text{user} \mid \text{positive}) &= \frac{P(\text{positive} \mid \text{user})\, P(\text{user})}{P(\text{positive})} \\
&= \frac{P(\text{positive} \mid \text{user})\, P(\text{user})}{P(\text{positive} \mid \text{user})\, P(\text{user}) + P(\text{positive} \mid \text{nonuser})\, P(\text{nonuser})} \\
&= \frac{0.98 \times 0.10}{0.98 \times 0.10 + 0.10 \times 0.90} = 0.52.
\end{aligned}
$$

Insight: With this test, if a room is full of positive testers, about half of them are innocent nonusers. Watch out for the influence of false positive and false negative results.

2 https://en.wikipedia.org/wiki/Probability

"With caution judge of probability. Things deemed unlikely, even impossible, experience oft hath proved to be true."

Shakespeare

Binomial Model of Coin Tosses (Stochastic, Descriptive)

A series of independent head-or-tail coin tosses with *p* the probability of a head on each toss is a stochastic simulation, and a computer program generating a series with state head appearing with probability *p* can simulate these tosses (Bernoulli trials) and record the outcome states.

This particular simulation can also be characterized analytically (mathematically) in closed form, for instance, yielding the probability of *h* heads in *n* tosses as

$$b(h; n, p) = \frac{n!}{h!(n-h)!}p^h(1-p)^{n-h}, \quad n! = n(n-1)(n-2)\cdots 1.$$

For example, if we toss a fair coin ($p = 1/2$) 10 times, the probability of getting exactly three heads, in any order, is $\binom{10}{3}\frac{1}{2}^3\left(1-\frac{1}{2}\right)^{10-3} = 120 \times 0.000977$ $= 0.117$.

Or, perhaps we merely want to know the mean np and variance $np(1-p)$ of the number of heads *h*, or the distribution of *h* when the number of trials *n* gets large and *p* is not too close to zero or one (by well-founded theory and experience, *h* should approach a normal distribution with this mean and variance—an important asymptotic example).

Synonyms for Probability

There are many synonyms for probability: likelihood, chance, propensity, odds, expectation, prospect, possibility, anticipation, contingency, conceivability, hazard, liability, plausibility, prayer, risk, promise, reasonableness, shot, and so on.

These are often used to imply subtle differences in meaning. *This is nonsense.*

A probability, by any name, is an assessment that a binary event will take place, ranging numerically from zero (impossible) to one (certain).

6.7 Probability Perspectives and Subject Matter Experts[3]

There are at least three basic views of probability:

6.7.1 Classical (sometimes called *a priori* or *theoretical*) probability assumes each of n possible outcomes of an event is equally likely, and assigns a probability of $1/n$ to each. This is how most people think about probability: This is appropriate if each outcome is equally likely, but generally not true and so the classic view is naïve.

6.7.2 Empirical (sometimes called *a posteriori* or *frequentist*) probability uses the observed relative frequency of outcomes of repeated experiments or experiences to estimate the probability of each future outcome. Empirical probabilities will vary with different data, but their estimate will become more reliable as data from more experiments or experiences become available.

6.7.3 Subjective (sometimes called *personal*) probability is the result of neither repeated experiments nor long-term empirical historical experience, yet is necessary, for instance, to assess the likelihood of future events with which we may have had no past experience at all. In such cases, we are forced to employ qualitative rather than quantitative experience. Speaking plainly, we may need some guesswork.

6.8 Subject Matter Experts[4]

Subject matter experts (SMEs), also called domain experts, are those with substantial experience and expert judgment in the area of interest for which we need probabilities.

A subject matter expert sometimes comes with formal credentials, such as a professional engineering certification, advanced scholarly degrees, licenses, permits, and a long, impressive resume.

However, subject matter experts may also be self-declared; before employing them, you are well advised to carefully evaluate their credentials. There is no such thing as a subject matter novice, or apprentice, so the evolution and advancement of a subject matter expert is a matter of some debate.

If the subjective probability may have results of significant consequence, we may employ more than one subject matter expert, use methods of elicitation that let us compare subjective probabilities, and attempt to reassure ourselves that the probabilities are consistent among experts, and for each expert individually.

---

3 https://en.wikipedia.org/wiki/Probability_interpretations

4 https://en.wikipedia.org/wiki/Subject-matter_expert

Discovering conflicting opinions among SMEs can be valuable for an organization.

Model results must be accompanied by documentation of the exact manner in which any subjective probabilities have been assessed. Results should also include parametric evaluations of response to a range of subjective probabilities.

In the end, subjective probabilities are guesses.

6.9 Statistics[5]

Statistics involves analysis of data. Statistics is generally descriptive or predictive. Statistics seeks relationships, perhaps hidden, between measured samples of sets of states, called data sets. Data on states are either gathered by sampling a subset of a population or recorded exhaustively from every member of the population.

6.9.1 A Random Sample

A random sample follows in Table 6.1 showing observations of height and weight for a set of American females, aged 30–39 [1].

We will assume this is a representative, random sample of all American females aged 30–39 at the time, and reuse these data in the following examples.

6.9.2 Descriptive Statistics

Descriptive statistics develops measures that can be used to characterize sets of state data. For instance, the mean, or average, observation of some state is representative of that state overall. Descriptive statistics also seeks distributions that can be used to represent data sets, either by theoretical observation or by empirical record keeping; recall that increasing numbers of trials for the binomial model above leads to a normal distribution.[6]

6.9.3 Parameter Estimation with a Confidence Interval

Parameter estimation with a confidence interval[7] uses sample data to estimate population parameters. The random sample in Table 6.1 can be used to estimate the population mean (average) weight at the time. Because the sample exhibits mean weight 62.07 kg and sample standard deviation[8] 7.05 kg, and because this mean comes from a sum of weights we assume to be independent and identically

---

5 https://en.wikipedia.org/wiki/Statistics
6 https://en.wikipedia.org/wiki/Normal_distribution
7 https://en.wikipedia.org/wiki/Confidence_interval
8 https://en.wikipedia.org/wiki/Standard_deviation

Table 6.1 Observations of height and weight for a set of American females aged 30–39.

Observation	Height (m)	Weight (kg)
1	1.47	52.11
2	1.50	53.12
3	1.52	54.48
4	1.55	55.84
5	1.57	57.20
6	1.60	58.57
7	1.63	59.93
8	1.65	61.29
9	1.68	63.11
10	1.70	64.47
11	1.73	66.28
12	1.75	68.10
13	1.78	69.92
14	1.80	72.19
15	1.83	74.46

For observation 9, 1.68 m is about 66 in. (or 5′ 6″) and 63.11 kg is about 139 lb.

distributed, the central limit theorem tells us this sample mean is normally distributed with these parameters. A normally distributed random variable falls within 1.960 standard deviations of its mean 95% of the time, and within 5.576 standard deviations 99% of the time. Thus, we can conclude from this sample with 95% confidence that the population mean lies within 48.25 and 75.89 kg.

To make this prediction more precise, we need to increase our random sample size. Unfortunately, to double our precision (i.e., halve the confidence interval width) we would expect to have to square our sample size.

6.9.4 Regression[9,10]

Regression estimates how some variable (measurement of state) is influenced by values of one or more other variables. The influenced variable is called *dependent*, and the other variables are *independent* or explanatory. Given a set of numerical observations, each with a value for the dependent variable and each

9 https://en.wikipedia.org/wiki/Ordinary_least_squares
10 https://en.wikipedia.org/wiki/Regression_analysis

independent one, and a candidate function relating the response of the dependent variable to the influence of the independent ones, coefficients for the function are estimated such that the function gives the best average prediction for the observations. Best is typically taken to mean that the squared difference between the function prediction at each observation and actual dependent variable value, summed over the observations, is minimized. Such estimation is called least squares regression.

Linear Least-Squared Error Regression (Deterministic, Descriptive)

Using data in Table 6.1, the least-squared-error function that best predicts observed height from observed weight is

$$\text{wgt} = 125.534 - 139.277\text{hgt} + 60.813\text{hgt}^2.$$

This fitted function is shown in the following graph along with the 15 observations.

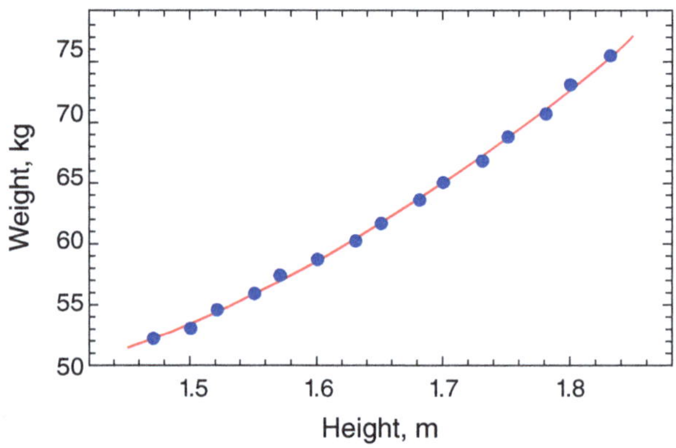

This is linear regression, because the estimated function is linear in the estimated coefficients, and the weight and height are observed data.

Given some simple assumptions about the observations, such as that they are statistically independent, that the variability of the dependent variable is about the same over the range of the observations, and that such variability follows a normal distribution, a host of statistical techniques apply to help decide if a candidate, fitted regression function, is better than some other one, whether enough variability of the dependent variable is explained by the fitted function, and so forth. In this case, the fit is very good, indeed (i.e., the probability of such a fit occurring at random is quite small).

Although regression is a statistical technique, it might be listed later with optimization examples because of the minimization of squared errors, and because in practice constraints may be imposed on the fitted function due to other considerations, making this what will be called a constrained optimization problem.

6.10 Inferential Statistics

Inferential statistics[11] assesses how much some state can be expected to vary, making statements in terms of probabilities of exceedance of a given threshold. Inferential statistics also uses probability to make decisions about whether or not some relationship exists between two types of state. A null hypothesis states there is no relationship. Based on probabilistic modeling, the null hypothesis may be accepted incorrectly, a *Type I* or *false positive error*, or the null hypothesis may be rejected incorrectly, a *Type II* or *false negative error*. Each type of error has its cost, and a test is designed to recognize these costs in setting the limits governing a decision. Reducing the risk of committing an error can be achieved by coarsening the decision rule, or gathering larger samples upon which to base statistics.

We have been told that in 2014 the average population weight of U.S. females over 20 years old was 76 kg (about 169 lb) [2]. We wonder, has this population significantly changed average weight since the 1975 random sample was collected?

Suppose the data in Table 6.1 have been hidden from us (more on this in a moment).

Statistical Hypothesis Test (Deterministic, Descriptive)

Null Hypothesis H_0*:* 2014 average weight of 76 kg is no different than it was when the 1975 sample was collected.

Alternative Hypothesis H_1*:* Average weight is different from 76 kg.

Significance level or Probability of Making a Type I Error: Rejecting H_0 even though it is true: 99% (i.e., a *critical probability* $\alpha = 0.01$).

Test Statistic: Mean of a random sample of 15 1975 observations.

Decision Rule: If sample mean is within 2.576 sample standard deviations from the hypothetic mean 76 (in the interval 43.91–80.23 kg), do not reject H_0, otherwise reject it.

Now we open the data envelope. A sample mean of 62.07 kg is within our interval, so we fail to reject H_0.

"But, wait, obviously the average population weight has increased. Why didn't we reject the null hypothesis? We would have rejected it with a critical probability of 95% and a decision interval of 48.25–75.89 kg."

11 https://en.wikipedia.org/wiki/Statistical_hypothesis_testing

We create the hypothesis test significance level[12] *before* we see the test statistic—otherwise, the test statistic is not a statistic, but a known constant, and then we can rig our significance level to conclude whatever we please.

When you read about experimental results (especially, for some reason, in medical literature), you should be suspicious when you find statements such as "this result has 95.12% significance" (as might have been claimed with our example). This is a symptom of misuse: A likely explanation is that "this is as far as we could push our realized statistic to support our preferred hypothesis test conclusion." If the scientific method is properly employed, a test is designed *before* the sample data are viewed. The potential costs of committing a Type I or Type II error are assessed, and the significance level is fixed. The decision and its potential risks of having made an error are known once the sample data are known. Otherwise, are you saying that the decision might be different if we change our prior estimates of the costs of making an error? That's an entirely different analysis. It is human nature to seek certainty, and scientists are human; there is continuing debate on the misuse of statistical methods [3].

We might also be criticized for unstated assumptions. For instance, the 1975 sample is from adult U.S. females aged 30–39, while the 2014 statistic applies to U.S. females aged 20 years and above. This may or may not be a serious problem, and should certainly be part of the documentation accompanying the statistical work.

6.11 A Stochastic Process

A stochastic process[13] is a descriptive or predictive probability model yielding a location or time sequence representing the state of a system that is subject to random variation. We may want to examine the *transient behavior*[14] of such a process, from some starting state to some limit of our interest, or from some signal state change, following system behavior afterward. Or, we may seek to examine the long-term *equilibrium*,[15] or *steady state*,[16] if such can be anticipated.

Queueing Model (Stochastic, Descriptive)

A queueing model describes behavior of a stochastic process, a system with customers arriving randomly to wait in a single queue to receive a random service time. The random state of the system is the number of customers either waiting for or receiving service. A goal is to develop mathematical predictions or

12 https://en.wikipedia.org/wiki/Statistical_significance
13 https://en.wikipedia.org/wiki/Stochastic_process
14 https://en.wikipedia.org/wiki/Transient_response
15 https://en.wikipedia.org/wiki/Equilibrium
16 https://en.wikipedia.org/wiki/Steady_state

numerical estimates of, say, the average long-term state in terms of parameters describing the random distributions of arrivals and services, so that one can see the influence of any parameter changing, or of changing the number of servers or the discipline used to assign arriving customers to servers.

Some systems have networks of queues, customers with limited patience, and a host of other realistic complications. Many such systems have been characterized analytically, in closed form, and all can be empirically evaluated by digital simulation.

Exponential, Poisson, and Memoryless Models

A continuous exponential random variable can be used to express the continuous time t between random state changes, with probability density $f(t;\lambda) = \lambda e^{-\lambda t}$, $t, \lambda \geq 0$, and parameter λ the constant rate at which state changes (events) take place over time (events/time). Sometimes the parameter λ is replaced by $\theta = 1/\lambda$ (time/event). The mean and variance of this distribution are respectively $1/\lambda$ and $1/\lambda^2$. Its cumulative distribution function is $F(t;\lambda) = 1 - e^{-\lambda t}$. This distribution exhibits a remarkable *memoryless property*: The probability that T will be at least $\tau + s$ given T is at least τ is the same as the probability that T will be at least τ:

$$\{1-(1-e^{-\lambda(\tau+s)})\}/\{1-(1-e^{-\lambda s})\} = e^{-\lambda(\tau+s)}/e^{-\lambda s} = e^{-\lambda t} = 1-(1-e^{-\lambda t}).$$

This means that *no matter how long it's been since the last event, the distribution predicting the time that will elapse before the next one stays the same.*

If times between events are exponential, then the number of events in an interval of t time units follows a Poisson distribution with probability mass function:

$$\Pr(k;t,\lambda) = e^{-\lambda t}\frac{(\lambda t)^k}{k!}, \quad k = 0, 1, \ldots.$$

These results often appear in models as initial simplifying assumptions, often merely by implication: If the words exponential, Poisson, or memoryless appear in any modeling discussion, they are intended to invoke these properties.

Markov Chains (Stochastic, Descriptive)

Let the state of a queue be the number of customers it contains. The following state-space diagram shows the transitions between adjacent states over time. Arrivals occur with rate λ (moving to the right), and service completions occur with rate μ (moving to the left) (the rates λ and μ represent the expected number of transitions per time unit):

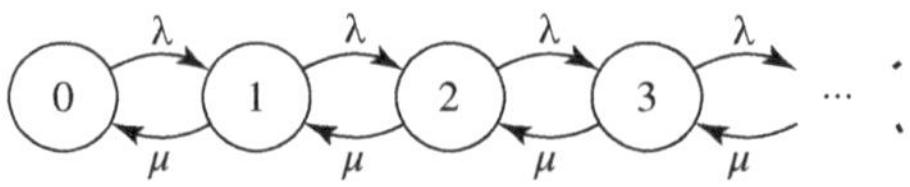

We can summarize these in a transition rate matrix:

$$Q = \begin{array}{c} \\ 0 \\ 1 \\ 2 \\ 3 \\ 4 \end{array} \begin{array}{c} \begin{array}{ccccc} 0 & 1 & 2 & 3 & 4 \end{array} \\ \begin{pmatrix} -\lambda & \lambda & & & \\ \mu & -(\mu+\lambda) & \lambda & & \\ & \mu & -(\mu+\lambda) & \lambda & \\ & & \mu & -(\mu+\lambda) & \lambda \\ & & & & \ddots \end{pmatrix} \end{array}$$

Transition rates in each row add to zero, representing mutually exclusive and exhaustive accounting for state. For example, from row state 1 to column state 0, μ represents a service completion of the sole customer in our system. The negative numbers on the diagonal are chosen so that the sum of each row is zero to account for flow balance. Let π_i represent the *stationary* probability of the system being in state *i*. Write a linear equation representing the balance of transitions for states entering and leaving row *i* (here, for $i > 0$, $\mu_{i-1}\pi_{i-1} -(\mu + \lambda)\pi_i + \lambda\pi_{i+1} = 0$). Whether the number of states is finite or infinite, simultaneously solving these equations gives stationary probabilities associated with any state, and these can be used to derive further properties, such as the *infinite-state probability* $\pi_i = \rho^i(1-\rho)$, or the *finite n-state probability* $\pi_i^n = \rho^i(1-\rho)/\sum_{s=1,n}\rho^s(1-\rho)$.

M/M/1 Queue (Stochastic, Descriptive)

Returning to our prior queueing model, suppose arrivals with exponential [sic] rate λ and service times are exponentially distributed at rate μ. Suppose service is rendered first-come, first-served, and we have infinite capacity to hold arrivals until they receive service. We can study either the *transient* behavior of this queue, following some starting state, or the *steady-state* (*stationary, equilibrium*) behavior given it is in constant operation and if the server utilization $\rho = \lambda/\mu < 1$ (otherwise, our queue will continue to grow forever). In steady state, ρ is the probability our server is busy at any given time. The expected length of the queue is $L = \rho^2/(1-\rho)$, the expected waiting time in the queue is L/λ, and the probability of *n* customers being in our system is $\rho^n(1-\rho)$. You can see this queue is well characterized analytically in closed form. Some embellishments can be accommodated such as multiple servers or finite queue capacity and still yield closed-form results. Otherwise, analysis may require simulation.

For example, if the arrival rate $\lambda = 9$ per hour and the service rate $\mu = 10$ per hour, $\rho = 9/10$, $L = (9/10)^2/(1 - 9/10) = 8.1$ and waiting time is $8.1/9 = 0.9$ h. This is a busy queue.

6.12 Digital Simulation[17]

Here is an abstract simulation model. This is akin to a computer procedure, written in primitive but unambiguous terms. This can be implemented in many computer languages.

Coin Toss Simulation (Stochastic, Descriptive)

given data

n number of coin tosses

p probability that a coin toss results in a head

variables

toss toss number

h number of heads in *toss* tosses

procedure

input n and p

output h

$toss \leftarrow 0$

$h \leftarrow 0$

while $toss < n$

    $toss \leftarrow toss + 1$

    if [0, 1] random_uniform_number $\leq p$

        $h \leftarrow h+1$

print h

When a simulation needs to represent a random event, such as the coin toss here, a pseudo-random number generator[18] (an intrinsic function in almost all general-purpose and simulation computer languages) provides a stream of random numbers. Any statistical distribution can be generated this way.[19] One handy feature of these generators is that they can be rigged to produce the same sequence of random numbers each time the simulation is run, the

17 https://en.wikipedia.org/wiki/Simulation
18 https://en.wikipedia.org/wiki/Random_number_generation
19 https://en.wikipedia.org/wiki/Pseudo-random_number_sampling#Continuous_distributions

better to be able to exactly reproduce any experiment, or isolate some curious event, or bug, in the simulation behavior.

Simulations are attractive because they directly represent system operation and states, and are designed to directly exhibit symptoms of the problem being modeled.

Simulations, especially those employing random numbers, have the disadvantage, for instance, that one needs to decide how many replications are necessary to achieve a trustworthy estimate of long-term, or *equilibrium* (*steady-state*) system operation. How much "warm-up" time or distance is needed before a simulation can be trusted to be behaving as it would, essentially, forever? When we seek to discover some anticipated but rare state, how long must the system be simulated before we conclude whether or not this event should have been encountered? These are serious issues that have received substantial attention in the scholarly literature.

6.12.1 Static versus Dynamic Simulations[20]

The static and dynamic adjectives reveal whether the behavior of a system varies over time. Dynamic simulation frequently involves describing state relationships and constraints with systems of differential or partial differential equations, and solving these with numerical methods. Some refer to this branch of modeling as the study of *system dynamics*.

Whether using a simple static simulation or a dynamic one with more mathematical detail and advanced numerical solution tools, the analysis required to properly design, use, and interpret results from a simulation can be very sophisticated. Packages animating system operation with visual icons may be entertaining and instructive, but do not relieve the modeler from responsibility to explain results carefully and correctly.

6.13 Mathematical Optimization[21]

The economic order quantity (EOQ) introduced previously constitutes the solution of a model, rather than the model. Now let's actually state the optimization model leading to this policy, and solve it analytically in closed form.

20 https://en.wikipedia.org/wiki/Dynamic_simulation
21 https://en.wikipedia.org/wiki/Mathematical_optimization

Economic Order Quantity: Optimization (Deterministic, Prescriptive)

given data (units)

d demand (SKU items/time)

h inventory holding cost (cost/time)

f fixed cost per order (cost)

variables (units)

c total cost of ordering policy (cost/time)

x economic order quantity (SKU items)

formulation

$$\min_x \quad c = h\frac{x}{2} + \frac{fd}{x}$$

solution

$$\frac{dc}{dx} = \frac{h}{2} - \frac{df}{x^2} = 0 \Rightarrow EOQ = x* = \sqrt{\frac{2df}{h}}$$

$$\frac{d^2c}{dx^2} = \frac{2fd}{x^3} > 0 | f > 0,\ d > 0 \text{ and } x > 0$$

This Economic Order Quantity model yields a stationary solution, is stated in terms of a continuous variable, and exhibits a convex objective function; so we are confident that our solution is valid and in this case unique.

Not all optimization models can be satisfactorily solved with advice from continuous variables. Some decisions are go, no-go (binary), others involve small numbers that need to be whole numbers (for instance, if our demand is just for a few items per planning period within our decision horizon). For example, our Economic Order Quantity model may present some trouble if a particular numerical solution turns out to be, say, $x* = 1.6$ (in this single-variable model, we can discover the better whole number solution by evaluating the cost when we round down to one, and up to two—but we will see this strategy won't work in general).

6.14 Measurement Units

Measurement Units

Even though metric units are the international standard[22] with all scientists and engineers, English units[23] are still used (and intuitively familiar to many clients) only in the United States, Liberia, and Myanmar. Use both, as is done in this chapter.

22 https://en.wikipedia.org/wiki/Metric_system
23 https://en.wikipedia.org/wiki/English_units

Units in Expressions

The constants and variables appearing in every algebraic expression exhibit units of some kind. Every additive operator, plus or minus, requires that the two terms have exactly the same units. If you encounter an example like this:

$$APPLES + ORANGES,$$

there must be some implicit conversion of one or the other of these terms to make the expression commensurate and sensible. If the intent here is units of "fruit," this should be made explicit.

Multiplication or division of terms converts units, thus some term with units "cost per car, or cost/car" multiplied by another with units "cars" yields a product "cost."

Probabilities are not merely unitless fractions. Expressions among probabilities generate probabilities, and those such as

$$probability \times APPLES$$

convert to an expectation of the units of *APPLES*.

Be particularly cautious of rates and their inverse. If cars require two hours each to produce, then total production during some number of hours is at the rate of ½ cars per hour:

$$CARS = 1/2 \times HOURS$$

6.15 Critical Path Method[24,25]

This is an invention from military operations research.

A *project*, from industrial to software development, consists of a number of separable *activities*, such as clearing a production site, pouring concrete foundations, or framing a new building. Completing necessary activities is required to achieve *milestone events*, such as completing framing so that finishing can begin. Each activity defines a *partial order* between adjacent events, where the activity cannot be commenced until all its preceding events have been achieved, and its succeeding events cannot be commenced until the activity has been completed. Each milestone event may have multiple predecessor activities, and cannot be achieved until the last of these has been completed.

Figure 6.5 shows a simple directed graph representing a project consisting of 13 activities (directed arcs) and 9 milestones (nodes). Each activity has an adjacent predecessor and successor milestone, and time duration required for its

24, https://en.wikipedia.org/wiki/Critical_path_method
25 https://en.wikipedia.org/wiki/Program_evaluation_and_review_technique

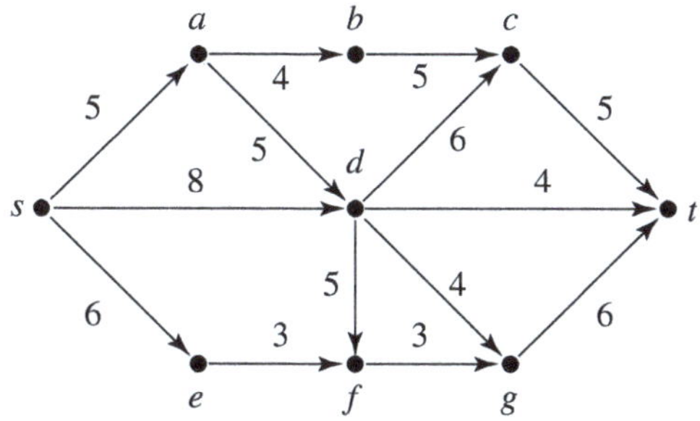

Figure 6.5 Critical path method. This is an activity-on-arc directed graph representing a project. Each node (letter) represents a milestone, each directed arc (arrow) represents an activity separating two adjacent milestones, a predecessor and a successor, and each number labeling an arc represents the number of months to complete that activity. The project starts at milestone *s* and completes at *t* when the longest additive time path from *s* terminates there. By inspection, the longest directed *s–t* path here—the critical path—is *s,a,d,f,g,t* with duration 24 months. We see that the earliest we can achieve milestone *f* is 15 months via path *s,a,d,f,* so the 9 month path *s,e,f* can be delayed by 6 months without delaying the project at all.

completion. The project starts at milestone *s*. No activity can commence until its predecessor milestone has been achieved, and that only happens when all its predecessors are achieved. Thus, we are led to follow the directed paths (alternating directed arcs and their adjacent nodes) from *s* through this network, determining the longest path from *s* to each node, and ultimately to *t*. An *s–t* path with the longest additive duration is called a *critical path*.

This problem, which is a bit tedious to solve manually, can be solved by some relatively simple and very fast *network algorithms.* It is included with optimization examples because, like with a number of simple network problems (e.g., finding connected components, cycle detection, degree assessment, shortest path, assignment, transportation, transshipment, and maximum flow) when you add realistic complications, such as consumption rates for each activity for a number of key resources, and constraints on the amount of each resource available over time, you quickly complicate the network structure with these embellishments, and need to employ numerical methods designed for these more complicated problems.

Project schedules are traditionally displayed with Gantt Charts.[26]

Gantt Chart (Deterministic, Descriptive)

A Gantt chart of a project schedule displays a heading calendar row (here across calendar days), and a following row for each activity showing its name, predecessor activity, and duration, followed by a horizontal stripe showing its (here contiguous) planned activity days.

26 Source: https://en.wikipedia.org/wiki/Gantt_chart, Used under CC BY-SA 3.0, https://creativecommons.org/licenses/by-sa/3.0/.

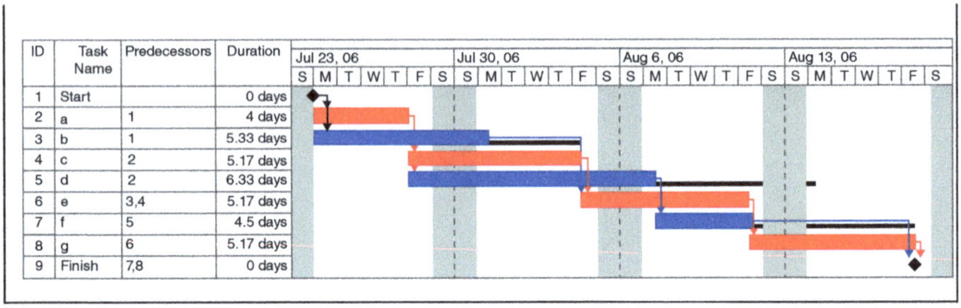

6.16 Portfolio Optimization Case Study Solved By a Variety of Methods

Suppose we want to choose items from a number of item types, each with a per-item value, weight, and area. See Table 6.2.

Unfortunately, our weight and area capacities are finite, but we still want to choose a set of items with maximum value that will fit.

6.16.1 Linear Program

We can write a mathematical optimization, a *linear program*,[27] for this problem as follows:

Numerical Optimization, Portfolio Selection (Deterministic, Prescriptive)

$$\text{Maximize}_{A,B,C,D} \quad 120A + 79B + 55C + 34D \quad \text{[C0]}$$
$$\text{subject to} \quad 12A + 8B + 6C + 3D \leq 100 \quad \text{[C1]}$$
$$22A + 16B + 11C + 10D \leq 200 \quad \text{[C2]}$$
$$A, B, C, D \text{ non-negative whole numbers} \quad \text{[C3]}$$

Table 6.2 Data for a portfolio optimization.

Item type	Value/item ($)	Weight/item (kg)	Area/item (m^2)
A	120	12	22
B	79	8	16
C	55	6	11
D	34	3	10
Capacity		100	200

100 kg is about 220 lb and 200 m^2 is about 2153 ft^2.

27 https://en.wikipedia.org/wiki/Linear_programming

The labels in square brackets identify each row of this model. This reads, in English, select the number of items *A*, *B*, *C*, and *D* that together maximize total value, computed by [C0]. Constraint [C1] limits the total weight of selected items, [C2] their total area, and [C3] reminds that we must select whole numbers of items.

Solving this model with continuous variables, we get 7.41 *A*'s and 3.70 *D*'s. This is not admissible for restriction [C3]. We need whole numbers of item selections.

You can solve this model with whole numbers of items by trial-and-error inspection (well, with some patience), using complete, exhaustive enumeration, employing a simple local search heuristic,[28] or with optimization.

6.16.2 Heuristic

Given the data in Table 6.2, we might try something simple like a greedy, nonbacktracking *heuristic*,[29] a thumb rule. Suppose we proceed by choosing the most-valuable item that will still fit until we can fit no more. This leads to choosing 8 *A*'s and 1 *D*, with a portfolio weight of 99 out of the 100 kg allowed, cube of 186 out of 200 m^2 capacity, and value 994. Not bad.

But, could we improve things by systematically adding and dropping items? There are many ways to do this, ranging from slightly more complicated heuristics to outright exhaustive enumeration.

6.16.3 Assessing Our Progress

How many portfolios are possible? Well, we can select as many as 8 *A*'s, reasoning that either weight or cube will limit our selection of all *A*'s, and this gives us the largest whole number of *A*'s: $\left\lfloor \min\left(\frac{100\text{ kg}}{12\text{ kg/item}}, \frac{200\text{ m}^2}{22\text{m}^2/\text{item}}\right)\right\rfloor$. Using this reasoning, we get upper limits on the numbers of items (*A*,*B*,*C*,*D*) of (8,12,16,20). This means, remembering that we can select none of an item, we have no more than $9 \times 13 \times 17 \times 21$ possible portfolios. 41,769 is a modest number, but we'll likely need a computer program to grind through these. However, if this portfolio problem is merely a pilot model, and real portfolios have hundreds of items, or real weight and cube and other resource limits admit hundreds of each item, we can see we are facing a combinatorial explosion.

6.16.4 Relaxations and Bounds

Well, if we are unwilling to commit to enumerating all possible portfolios, how about we develop an upper bound on how valuable these portfolios might be,

28 https://en.wikipedia.org/wiki/Local_search_(optimization)
29 https://en.wikipedia.org/wiki/Heuristic

even though we haven't evaluated all of them? If we just focus on the weight constraint, ignoring the cube one and the requirement to select whole numbers of items, we see we would maximize portfolio value by selecting 33.33 *D*'s, with a value of 1133.33. If we only consider the cube constraint, ignoring the weight limit and whole numbers, we would select 9.09 *A*'s with value 1090.91. The values of these *relaxations*[30] of our problem are each valid upper bounds on the as yet unknown true, optimal solution.

6.16.5 Are We Finished Yet?

So we have in hand a quick heuristic solution worth 994, and an upper bound on the best solution we may not have discovered yet of 1090.91. In many modeling situations, for instance, if the value units are U.S. dollars, this may be good enough. If the value units are billions of euros, we might want to do some additional analysis, as we will later in this chapter.

> Simple heuristics can be very effective, and you will find a large volume of enthusiastic literature proposing many techniques. Sadly, you will not find much on developing bounds for heuristic solutions. This is evidently not as interesting to researchers, but can you appreciate how important it is to see how much of your client's money you might be leaving on the table?

The optimal, whole-number selection is 6 *A*'s, 2 *B*'s, 1 *C*, and 2 *D*'s (this does not much resemble the inadmissible continuous variable selection). This selection has total value \$1001, uses our full 200 kg weight capacity and 195 of our 200 m^2 area capacity. We are confident the total value of our selection is as good as it can be due to the solution method we employed. Some solution methods (such as trial-and-error) may yield advice that appears to be good, but solution methods that guarantee optimality provide a warranty that there is no better selection left undiscovered. Whether or not prospective methods yield such reassurance may influence the method you choose.

This simple optimization problem resembles many that arise in portfolio selection,[31] cargo loading, target selection, satellite surveillance, capital budgeting,[32] and so on. We frequently want to select the most-valuable affordable set of items. Sometimes, the expressions, such as [C0–C3] above, may be nonlinear, or maybe not even available in closed algebraic form.[33] Sometimes, the data are not

30 https://en.wikipedia.org/wiki/Relaxation_(approximation)
31 https://en.wikipedia.org/wiki/Modern_portfolio_theory
32 https://en.wikipedia.org/wiki/Capital_budgeting
33 https://en.wikipedia.org/wiki/Response_surface_methodology

known exactly, or may vary randomly.[34] For a vast domain of models to make things better, subject to constraints, we have methods available to do so.

> "If the system exhibits a structure which can be represented by a mathematical equivalent, called a mathematical model, and if the objective can be also so quantified, then some computational method may be evolved for choosing the best schedule of actions among alternatives. Such use of mathematical models is termed mathematical programming."
>
> G. Dantzig

6.17 Game Theory[35]

Game theory prescribes actions for opponents in conflict. In the simplest *two-person, zero sum*[36] case, each of the two opponents chooses an action in secret, and when these actions are taken, their joint consequence is some payoff from one player to the other. The following example from World War II illustrates.

In late February 1943, US-allied intelligence learned Japan would convoy troops from Rabaul, New Britain, to Lae, New Guinea, a 3-day passage (see Figure 6.6). But, allies did not know whether Japan would choose a northern route where poor visibility was forecast, or the southern route with clear weather. Allies had limited reconnaissance aircraft and fuel, and could only search either north, or south, but not both. Allied planners considered both Japanese courses of action, and both allied ones, estimating the time to detect and remaining time to attack a detected convoy (see Table 6.3). We can assume the Japanese planners did the same analysis, and came to essentially the same conclusions.

Military planning considers the enemy's most damaging course of action. If each opponent here *minimizes the worst outcome*, the Japanese would sail north

Table 6.3 Payoff matrix: The asterisk shows the "saddle point," the optimal actions of both opponents, and resulting days for allied attacks.

	Convoy sails north	**Convoy sails south**	**Allied action**
Allied air searches north	2 days*	2 days	1
Allied air searches south	1 day	3 days	0
Japanese action	1	0	2 attack days

For instance, if the allies search south, and the Japanese convoy sails south, the clear weather would lead to almost immediate discovery and 3 days of allied attacks.

34 https://en.wikipedia.org/wiki/Stochastic_programming
35 https://en.wikipedia.org/wiki/Game_theory
36 https://en.wikipedia.org/wiki/Zero-sum_game

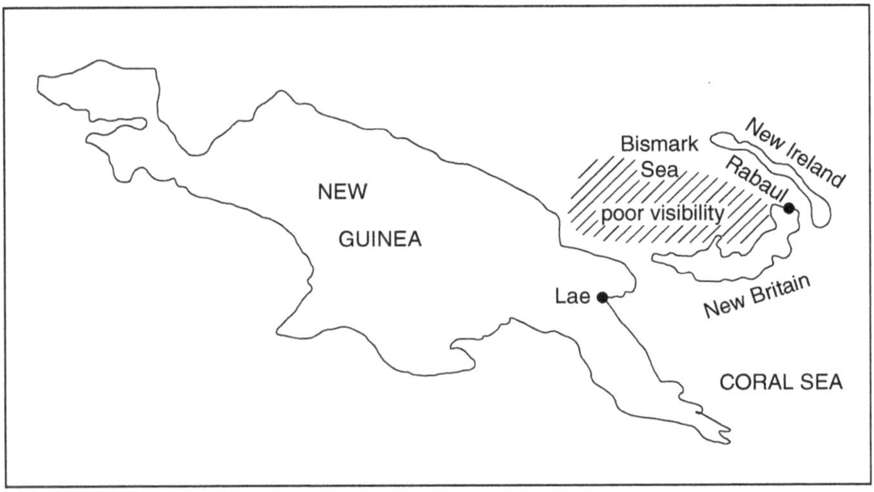

Figure 6.6 March 2, 1943, intelligence intercepts tell the US allies that Japan will convoy reinforcements from Rabaul to Lae, but not the three-day route they will use. Japan can choose to sail north or south of New Britain. With limited reconnaissance aircraft and fuel, the allies must either search north or south, but not both.

in poor visibility, facing at worst 2 days of attacks, rather than the possible 3 days if they sailed south. This is called a *minimax* strategy. The allies would search north expecting no worse than 2 days of attacks, rather than as little as 1 day if they searched south. This is called a *maximin* strategy. Because the minimax and maximin strategies identify a single, dominant action for each opponent, there is a *saddle point* in this game. And that is what happened.

The Japanese were located during their northern passage, and the Battle of the Bismark Sea,[37] March 2–4, 1943, resulted in 13 allied casualties, and about 3000 Japanese lost.

The *value* of this game is 2 days of allied attacks.

Now, suppose Japan had advanced intelligence that the allies would search south. They would convoy north. The value of this intelligence to Japan is $2-1=1$ attack day avoided. This is a symmetric measure of the value of intelligence to Japan, and the value of secrecy to the allies.

When one player must play first, and reveal his strategy before the other, this is known as *Stackelberg Game*.[38] The first player is called the *leader*, and the second the *follower*. These have become newly fashionable with defender–attacker models of infrastructure defense [4]. The leader (the defender) moves first with measures to defend, harden, add redundancy, or otherwise invest to make some infrastructure (e.g., the electric grid, highways and bridges, and

37 https://en.wikipedia.org/wiki/Battle_of_the_Bismarck_Sea
38 https://en.wikipedia.org/wiki/Stackelberg_competition

petroleum distribution) harder to damage. The follower (the attacker) observes these defensive preparations before deciding whether, where, and how to attempt to inflict maximum damage. Two-sided optimization models of this situation simultaneously minimize the defender's damage while at once maximizing the attacker's effects. The best worst-case solution is advised.

> "Mother Nature rolls dice, terrorists do not."
>
> E. Kaplan

Now let's hypothesize a different weather forecast, resulting in a slightly modified payoff matrix shown in Table 6.4. Now there is no saddle point. Analysis here leads to discovering a *mixed strategy*, where each opponent will choose an action based on a probability.

This game has an expected value of 2.6 allied attack days. There are a number of ways to arrive at this mixed strategy solution, with perhaps the easiest via optimization. We can state a simple linear program as follows:

Linear Program to Solve Convoy Game (Deterministic, Prescriptive)

Nonnegative decision variables

AN allies search north

AS allies search south

DAYS attack days

Formulation

$$\max_{AN,AS,DAYS} \text{DAYS}$$

s.t. $\text{DAYS} \leq 5\text{AN} + 2\text{AS}$ (Japan convoys north)

$\text{DAYS} \leq 1\text{AN} + 3\text{AS}$ (Japan convoys south)

$\text{AN} + \text{AS} \leq 1$ (mixed strategy probabilities)

The optimal actions for Japan can be found with a symmetric optimization (or recovered from the *dual solution*[39] of this linear program, a topic not pursued here).

Before we leave the game in Table 6.4, let's again wonder how Japan could have analyzed their prospects before deciding to convoy at all, but in anticipation of receiving intelligence before deciding to deploy. Their mixed strategy solution for the allies would have predicted a northern search with probability 0.2, and they would plan to convoy south in that case. With probability 0.8, allies could be anticipated to optimally search south, and Japan would convoy north. This expected loss of $0.2 \times 1 + 0.8 \times 2 = 1.8$ attacked days is better than 2.6 expected

39 https://en.wikipedia.org/wiki/Duality_(optimization)

Table 6.4 Payoff Matrix: A changed weather forecast slows the convoy on the northern route.

	Convoy sails north	Convoy sails south	Allied action
Allied air searches north	5 days	1 day	0.2
Allied air searches south	2 days	3 days	0.8
Japanese action	0.4	0.6	2.6 expected attack days

Now the opponents should use a mixed strategy, choosing actions as shown. Allies should search north with probability 0.2, or south with probability 0.8. The convoy should sail north with probability 0.4, or south with probability 0.6.

attacked days with no intelligence, but still might have weighed on the desperation of Japan to mount the convoy at all.

Conditioned analysis like this leads to the next class of models.

6.18 Decision Theory

Decision theory[40] offers two sorts of insight:

1) It can advise how to make optimal decisions based on probabilities of achieving particular gains or losses as a consequence.
2) It can explain why human decision makers choose other than such optimal decisions.

The amounts of potential gains or losses are typically weighed by some estimate of the probability they will occur, their *expected value*. Individuals have differing views of the *utility*, or value, of a gain or loss, and differing views of the *risk*, or probability that a decision will prove to be an incorrect one.

Home insurance has an expected payoff far less than the cost of the premium. Evidently, those buying home insurance are willing to pay more than its expected nominal monetary value because they have an even higher utility for such a loss than this nominal value. A lottery ticket has an expected payoff much less than its purchase price, but the utility of a large, if unlikely, payoff evidently overwhelms its comparably trivial cost.

Psychological study of decisions attempts to explain seemingly irrational choices by decision-makers. *Preferences* between paired alternatives may not be transitive, for instance, a shopper may prefer vehicle A over vehicle B, B over C, and C over A.

40 https://en.wikipedia.org/wiki/Decision_theory

Decision theory typically applies to an individual decision-maker, unlike game theory that applies to decisions of opponents. Decision theory admits not just human choices, but those of Mother Nature too.

Mathematical models for sequences of decisions, some choices by the decision-maker, and some purely random events chosen by Mother Nature are often expressed as *decision trees*. The *root node* of such a tree is viewed as the beginning of the decisions, and weighed with a total initial probability of 1. Nodes in a decision tree are conventionally either rectangles, for *decision nodes*, or ovals, for random *chance nodes*. From each decision node, a branching set of successor *arcs* represents alternatives that may be chosen. From each chance node, a branching set of successor arcs represents random outcomes with the probability for each, and those probabilities summing to 1. Each arc may have a gain or loss associated with its traversal. Importantly, *each probability may be conditioned on the entire preceding history of events and decisions*. The ultimate states at the culmination of all decisions are represented by *leaf nodes* that have no successor.

Any *directed path* from root to leaf is a possible outcome of successive decisions and random events, and each leaf node is a possible ultimate outcome state.

The idea is to evaluate the expected value of every leaf node. This is done by backward induction, starting with the leaf nodes and backtracking toward the root, computing and accumulating expected values for successor arcs of each node.

If one maintains the decision tree as a sequence of decisions is made, and events are experienced, the tree is conditioned to have as its new root node the latest node in this set of experiences, with total conditional probability 1. All its leaf nodes have expected values conditioned by this influence.

Figure 6.7 shows a decision tree faced by a competing figure skater who must decide whether or not to attempt a very difficult "quadruple jump" during competition.

What if the probability of success is conditioned on prior experience? For example, suppose the probability of success of the first quad attempt is 0.5, and given a first success the second quad attempt will also succeed with probability 0.6, or given a first failure, the second quad will succeed with probability 0.4 (see Figure 6.8). In this case, the skater should attempt the first quad (with expected value 0.687) and attempt second quad even if the first has already succeeded (with conditioned expected value 0.854), and not attempt a second quad if the first one failed (with conditioned expected value 0.30), but attempt a quad if there was no first attempt (with conditioned expected value 0.575).

A key disadvantage of decision trees is that the number of leaf nodes (or, equivalently, directed paths or alternate outcome states) grows exponentially with the degree of the nodes (the number of successor arcs from each) and the number of successive nodes in each directed path. One signal example [5] exhibits no less than 10,032,906,240 leaf nodes, each with directed-path-conditioned probabilities.

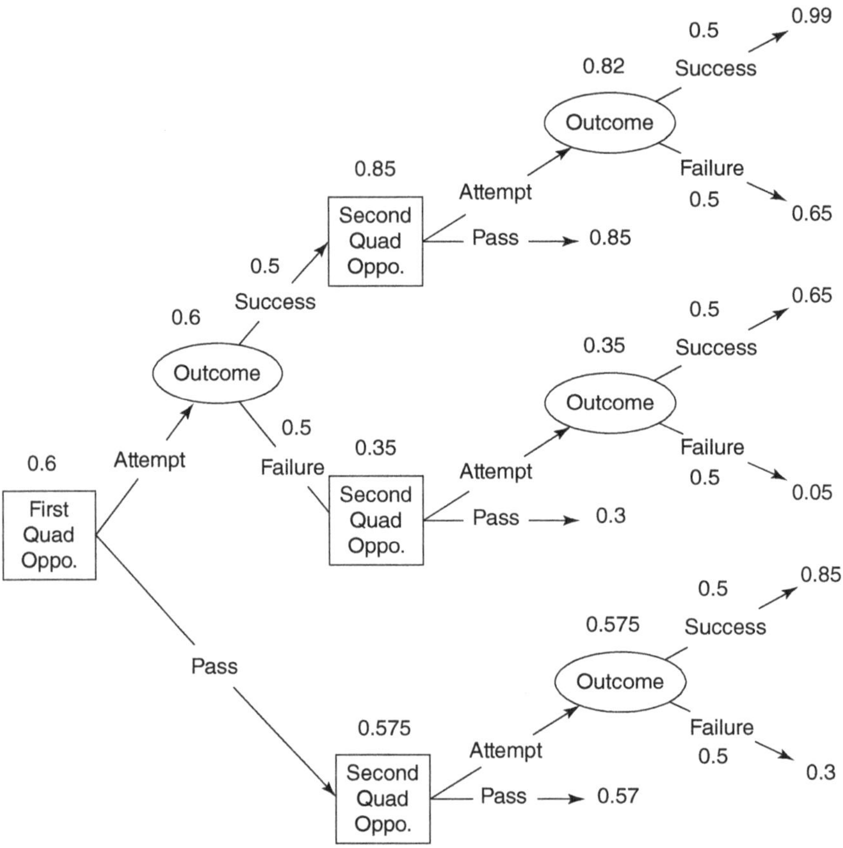

Figure 6.7 Figure skating competition. During a competitive performance, a figure skater can decide to attempt one or two extremely difficult "quadruple jump" maneuvers or none at all. The probability of success on each trial is 0.5. Square nodes represent the skater's decisions, oval nodes the probability of each attempt outcome, and the numbers at the right the final payoff in terms of the probability the skater will win the competition. Each directed path from the root node at the left to some leaf node at the right represents a sequence of decisions and outcomes of each attempt, and of the competition. The numeric labels over the nodes represent the conditional expected payoff, computed right to left, given the skater reaches that point in the competition. Upon reaching a decision node, the skater chooses the larger expected payoff. The optimal strategy is to attempt the first quad, with expected payoff 0.6, and if that succeeds, pass on the second one. If the first quad attempt fails, the best conditional strategy is to attempt the second one, with expected payoff 0.35.

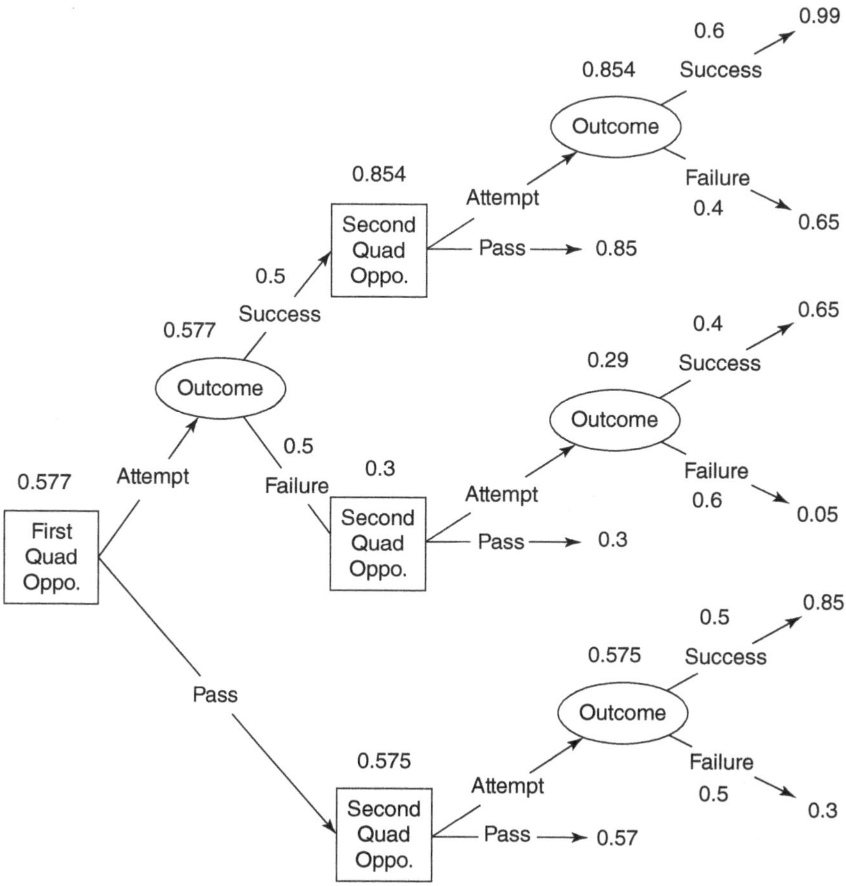

Figure 6.8 Figure skating competition with conditioned success probabilities: the first outcome influences the probability of success of a second attempt. The optimal strategy is to attempt the first quad, with expected payoff 0.577, and if that succeeds, attempt the second one too, with conditional expectation 0.854. If the first quad attempt fails, the best conditional strategy is to attempt the second one, with expected payoff 0.575.

6.19 Susceptible, Exposed, Infected, Recovered (SEIR) Epidemiology[41,42]

Recent experience with the Ebola,[43] Zika,[44] and other infectious diseases sharpens our interest in modeling the establishment and spread of infectious

41, https://en.wikipedia.org/wiki/Epidemic_model
42 https://en.wikipedia.org/wiki/Compartmental_models_in_epidemiology
43 https://en.wikipedia.org/wiki/Ebola_virus_disease
44 https://en.wikipedia.org/wiki/Zika_virus

diseases. Mathematical epidemiology has produced many models, among which SEIR is a good example. This is a compartmental model that divides a population into four states: susceptible, exposed, infectious, and recovered (or in the vernacular of this domain, "removed").

Movement of individuals between adjacent states is governed by transition probabilities.

For some starting state (S, E, I, R) population recovery is $(N, 0, 0, 0)$, if the disease abates, while a long term, *steady state* is called an endemic equilibrium.

Susceptible, Exposed, Infected, Recovered (SEIR) Epidemiology (Deterministic, Predictive)

<u>given data</u>[units]

N	population size [individuals]
μ	romoval rate [individuals/time]
β	contact rate [individuals/time]
γ	recovery rate [individuals/time]
ε	incubation (latency) rate [individuals/time]

<u>variables</u> [units]

S	susceptibles [individuals]
E	exposed [individuals]
I	infected [individuals]
R	recovered (immune or removed and assumed replaced by new births) [individuals]

<u>formulation</u>

$\frac{dS}{dt} = \mu(N - S) - \beta\frac{I}{N}S$	recovered less those susceptibles in contact with infecteds
$\frac{dE}{dt} = \beta\frac{I}{N}S - (\mu + \varepsilon)E$	susceptible contact with infecteds less incubateds and removals
$\frac{dI}{dt} = \varepsilon E - (\gamma + \mu)I$	incubated exposeds less recovereds and removals
$\frac{dR}{dt} = \gamma I - \mu R$	infecteds who recover less removals
$S + E + I + R = N$	static population size

<u>key insights</u>

$R_0 = \frac{a}{\mu + a}\frac{\beta}{\mu + \gamma}$	reproduction number
$R_0 \leq 1$	population recovers
$R_0 > 1$	population reaches endemic equilibrium

Models such as SEIR are used to evaluate quarantine policies and vaccination regimes.

6.20 Search Theory[45]

Search theory was the invention of mathematicians and physicists during World War II. Today, it is not widely taught outside military circles, but remains useful for finding lost things, whether they are submarines, tethered undersea mines, or you lost at sea.

One of the simplest, most elegant results is the following, known as Koopman's area search equation:

Area Search (Stochastic, Predictive)

given data [units]

A search area [km^2]

w sweep width of a searching sensor [km]

v velocity of the searcher [km/h]

variable [units]

t search time [h]

formulation

p instantaneous probability of success

$$p = vw/A$$

$P(t)$ cumulative probability of success after searching t hours:

$$P(t) = 1 - e^{-pt}$$

This is a conservative estimate of the probability of search success, and we can do much better if we can afford to conduct an exhaustive search. Nonetheless, this is a useful descriptive model. One corollary insight is that the instantaneous probability of success stays the same over time (that memoryless property, again), so even if you haven't succeeded so far, the expected time until success is the same. You can view this in two ways. After some time without success, a pessimist wonders: "Why have I wasted so much effort?" While hope blooms eternal for an optimist. If you are lost, hope for optimistic searchers.

6.21 Lanchester Models of Warfare[46]

This is another military model developed by the British engineer F. W. Lanchester in 1914, and published 2 years later, to describe combat exchanges

45 https://en.wikipedia.org/wiki/Search_theory

46 https://en.wikipedia.org/wiki/Lanchester's_laws

between opposing air forces. It has been more widely used to describe continued land combat between armies.

Lanchester's Aimed Fire Square Law (Deterministic, Predictive)

given data [units]

A_0	initial size of force A[combatants]
α	attrition rate [force B combatant killed/force A combatant]
B_0	initial size of force B [combatants]
β	attrition rate [force B combatant killed/force A combatant]

variables [units]

A	size of force A [combatants]
B	size of force B [combatants]

formulation

$\frac{dA}{dt} = -\beta B$ rate of attrition inflicted on force A by aimed fire from force B

$\frac{dB}{dt} = -\alpha A$ rate of attrition inflicted on force B by aimed fire from force A

solution

$$\alpha(A^2 - A_0^2) = \beta(B^2 - B_0^2)$$

annihiliation prediction :

$$\text{if } \alpha A_0^2 \geq \beta B_0^2 \text{ then } B \to 0 \text{ and } A \to \sqrt{A_0^2 - \frac{\beta}{\alpha} B_0^2}$$

$$\text{if } \alpha A_0^2 \leq \beta B_0^2 \text{ then } A \to 0 \text{ and } B \to \sqrt{B_0^2 - \frac{\alpha}{\beta} A_0^2}$$

This is for what is called "aimed fire": inflicted attrition is a function of the number of shooters and their accuracy, it is assumed every shooter has a target. This applies, for instance, to opposing infantry forces.

There is a parallel set of results for unaimed, "area fire" leading to Lanchester's linear law. In this case, every shooter has targets spread uniformly over a target area, so attrition is the product of the lethality of each shot, the number of shooters, and the number of dispersed targets. This applies to artillery fire.

Lanchester's Area Fire Linear Law (Deterministic, Predictive)

given data [units]

A_0	initial size of force A [combatants]
α	attrition rate [force B combatant killed/force A combatant]
B_0	initial size of force B [combatants]
β	attrition rate[force B combatant killed/force A combatant]

variables [units]

A	size of force A [combatants]
B	size of force B [combatants]

formulation

$\frac{dA}{dt} = -\beta AB$ rate of attrition inflicted on force A by unaimed fire from force B

$\frac{dB}{dt} = -\alpha AB$ rate of attrition inflicted on force B by unaimed fire from force A

solution

$\alpha(A - A_0) = \beta(B - B_0)$

annihilation prediction :

if $\alpha A_0 \geq \beta B_0$ then $B \to 0$ and $A \to A_0 - \frac{\beta}{\alpha} B_0$

if $\alpha A_0 \leq \beta B_0$ then $A \to 0$ and $B \to B_0 - \frac{\alpha}{\beta} A_0$

These are descriptive models, but you can deduce it is good to have either a superior initial force or one with more lethal aiming soldiers and/or area-firing artillery.

These are also models of *transient* behavior rather than steady state. Annihilation of either opponent terminates combat.

You may wonder what use this has for other than military planners. There are many applications among competitors, biologic predator–prey models, and other systems whose operation involves continuous exchange rates of some sort, where the exchanges harm mutual competitors.

These closed-form, analytic solutions are derived by solving ordinary differential equations. Make modifications to the model, such as adding more constraints, and you may or may not be able to solve the result analytically. However, you can always evaluate the exchanges with simulation.

Simulation of Lanchester's Square Law

input

B_0, β, R_0, α	as specified
T	time horizon [h]

variables

t, B, R, B_hold

procedure

$B \leftarrow B_0$

$R \leftarrow R_0$

$t \leftarrow 0$

```
while t < T and R > 0 and R > 0
        t ← t + 1
        B_hold ← B
        B ← B − αR
        R ← R − βB_hold
print t, B, R
```

This deterministic, discrete time-increment simulation will approximate the Lanchester exchanges. To be more accurate, we can increase *model fidelity* by decreasing the time increment with the exchange rates α and β reduced proportionately to apply to these shorter epochs of combat. While a simulation like this can lend insight, quite a few replications with varied inputs would be required before you begin to suspect whether or not something like the square law still holds.

To simulate Lanchester's linear law, replace the attrition terms $-\alpha R$ and $-\beta B\_hold$ by $-\alpha BR$ and $-\beta B\_hold\,R$, respectively.

6.22 Hughes' Salvo Model of Combat[47]

The classic Lanchester models are aimed [sic] at large-scale sustained combat involving large numbers of combatants, many shots fired by each, and continuous warfare. In fact, thousands of shots may be required to achieve a single kill.

In contrast, we anticipate modern naval missile exchanges to involve a single, signal engagement between a small number of adversaries, perhaps capital ships, with a small number of attacking missiles that are extremely accurate, highly lethal, and perhaps vulnerable to defensive missiles trying to nullify each attack [6]. This leads us to explicitly represent the lethality of each attacking shot, the effectiveness of each defending shot, and the number of "leakers" (attacking missiles not nullified) required to kill a discrete combatant unit.

Hughes' Salvo Equations (Deterministic, Predictive)

given data [units]

A_0 initial size of force A [units]

B_0 initial size of force B [units]

α number of well-aimed missiles fired by each A unit [missiles]

β number of well-aimed missiles fired by each B unit [missiles]

a_1 number of hits by B's attacking missiles needed to put one A out of action [missiles]

47 https://en.wikipedia.org/wiki/Salvo_combat_model

b_1 number of hits by A's attacking missiles needed to put one B out of action [missiles]
a_3 number of well-aimed attacking missiles nullified by each A [missiles]
b_3 number of well-aimed attacking missiles nullified by each B [missiles]

variables [units]

B size of surviving blue force [units]
ΔA number of A units put our of action by B's salvo [units]
ΔB number of B units put our of action by A's salvo [units]

formulation

$$\Delta B = \frac{\alpha A - b_3 B}{b_1}, \quad \Delta A = \frac{\beta B - a_3 A}{a_1}$$

winner prediction

if $a_1\ \alpha A^2 - a_1 A b_3 > b_1 \beta B^2 - b_1 B a_3 A$, then A wins the salvo exchange; or
if $b_1\ \beta B^2 - b_1 B a_3 > a_1 \alpha A^2 - a_1 A b_3 B$, then B wins the salvo exchange.

If in either inequality the second term is larger than the first, the respective defense is so strong no damage is done by the attacker, and zero (not negative) loss results.

These results are expressed as discrete difference equations, rather than differential equations, and can be solved with elementary algebra.

After a missile exchange, we can make a number of assessments analogous to the continuous Lanchester ones.

A charm of the Hughes Salvo equations is they admit embellishments such as scouting to locate targets and improve aim, decoys, evasion, distraction, surprise, and so on. The resulting difference equations are still trivial to solve numerically.

6.23 Single-Use Models

Single-use models can respond to an exigent question quickly and effectively, and don't necessarily need to employ advanced mathematics.

Compound Interest and Net Present Value (Deterministic, Descriptive)

Suppose we have an investment that pays r % interest at the end of every year, and we decide to reinvest interest at the end of each year. What is the value of our investment after y years? Following year-by-year, our value increases by a factor of $(1 + r/100), (1 + \mathrm{r}/100)^2, \ldots, (1 + r/100)^y$.

Now, suppose we have an alternative investment that pays r % interest continuously (imagine computing the interest payment second-by-second, instead of once per year). Our value increases continuously at the rate $e^{(r/100)y}$.

For example, an investment returning 5% at the end of each year returns 105% after 1 year, 110.25% after 2 years, and 162.89% after 10 years. A continuous investment at the same 5% rate returns 105.13% after 1 year, 110.52% after 2 years, and 164.87% after 10 years.

Conversely, suppose you save some cash under your mattress for some time, and wonder what its buying power will be if prices of the things you want to purchase inflate at the rate 5% per year? The continuous deflation of the value of your cash cache amounts to a factor of 95.12% after a year, 90.48% after 2 years, and 60.65% after 10 years.

Net present value arises in all long-term planning models when the value proposition is monetized.

"Compound interest is the eighth wonder of the world.

He who understands it, earns it . . . he who doesn't . . . pays it."

A. Einstein

The following algebraic model has been used to illustrate to navy fleet commanders the impact of their policy for naval combatants to always maintain a certain level of fuel as safety stock against running out in case of unanticipated demand. Maintaining a safety stock, expressed as a percentage of fuel capacity, is a wise policy, but making minor adjustments to this percentage responding to changes in your estimate of threat level—the likelihood you will need to use a lot of unanticipated fuel right away—can have significant influence on the policy cost, born by the supply ships that must sortie to supply fuel to the combatant customers at sea.

Cost to Maintain Safety Stock

US Navy Combat Logistics Force (CLF) ships refuel and resupply US and coalition customer ships while they are underway, which enables these customers to operate for extended periods at sea. Navy policy is that no ship should ever have less than some stated amount of fuel onboard. Maintaining this so-called *safety stock* is a major demand signal for CLF to travel to a customer ship for an underway replenishment.

Suppose for some customer s is the safety stock, expressed as a percentage of its fuel capacity, meaning that $1-s/100$ is the useable fraction of fuel capacity. This dictates that the "hit rate" at which CLF must visit this customer is proportional to $h(s)=1/(1-s/100)$. Changing the safety stock from s to s' has the following percentage influence on this hit rate: $100[h(s')-h(s)]/h(s)$. For example, if the safety stock requirement $s=60\%$ is relaxed to $s'=50\%$, this reduces the CLF hit rate by 20%. Regardless of customer fuel capacity or rate of fuel use, if the customer operations are independent of those of the CLF ships providing fuel, this 20% reduction applies directly to reduction of CLF fuel consumption.

This simple algebraic example illustrates the influence safety stock has on CLF operations and costs, which is good for fleet commanders to understand when they set safety stock policy. Millions of dollars of CLF fuel consumption are in play.

6.24 The Principle of Optimality and Dynamic Programming

The Principle of Optimality

An optimal policy has the property that whatever the initial state and initial decision are, the remaining decisions must constitute an optimal policy with regard to the state resulting from the first decision.

This terse statement by Richard Bellman has stood the challenge of time, and to this day nobody has found a way to expand, contract, or clarify this seminal version. What a compliment to this genius.

Rather than restating this, let's try to understand it, and exploit its implications.

Returning to our portfolio example, let's view solving this from Bellman's perspective. Let's start by choosing some number of A items, given we might have any amount of remaining weight and area capacity to fill. We would greedily and correctly (in fact, optimally) pack as many A's as we could fit. If we create a schedule of the number of A's that will fit in any particular remaining weight and area capacity, we could record this is a matrix for remaining weight $w = 0, 1, 2, \ldots, W$, and remaining area $a = 0, 1, 2, \ldots, A$.

Now, suppose we choose some number of B items, also for any remaining weight and area capacity. Now an optimal choice would account for the immediate value of the B items chosen, plus the value of A items for which we leave unused weight and area capacity. So, for example, if we have remaining (weight, area) capacity (21,40) and chose one B item with value 79, this would leave capacity ($21-8=13$, $40-16=24$), and by lookup in the A item table, we could see that we still have capacity to choose one A item with value 120. We can continue similarly with C and then D items.

This works because the weight and area requirements for items are whole numbers, so there are a finite number of (weight, area) capacities in the *state space* of this problem. We refer to each sequential item-by-item enumeration of what will immediately fit in any available remaining capacity the *stage* of selection.

Expressing this more compactly, represent item types by index $i = 1, 2, \ldots, I$ and the remaining weight capacity $w = 0, 1, \ldots, W$ and area $a = 0, 1, \ldots, A$. Let $f_i(w, a)$ be the value of $X_i(w, a)$ items selected through stage i.

$$\begin{aligned}&\text{for} \quad i = 1, 2, \ldots, I:\\&\qquad f_i(w,a) = \max_{X_i}\\&\qquad \{\text{value}_i X_i(w,a) + f_{i-1}(w - \text{weight}_i\, X_i(w,a), a - \text{area}_i\, X_i(w,a))|_{i>1}\}.\end{aligned}$$

Verbally, this tells us that for given available capacity (w, a) at stage i, we choose X_i that maximizes the immediate item i value $value_i X_i$ in addition to the optimal prior selections, if any, through stage i–1 of the capacity, our choice of $X_i(w, a)$ would leave $f_{i-1}(w - \text{weight}_i X_i(w,a), a - \text{area}_i X_i(w,a))$. In more detail, we would carefully control the maximization at each stage:

$$X_i(w,a) \leq \lfloor \min\{w/\text{weight}_i, a/\text{area}_i\} \rfloor.$$

That is, we only search over whole numbers of items that can fit in remaining capacity (w,a).

We read out the final, optimal solution by tracing back through our stage matrices. $f_I^*(W, A)$ gives us the optimal portfolio value. This portfolio includes $X_I^*(W, A)$ items of type *I*. $X_{I-1}^*(W - \text{weight}_I, A - \text{area}_I)$ gives us the number of items *i–1*, and so forth.

For our numeric example and stage sequence A, B, C, D we get

$$\begin{aligned}&X_D^*(100, 200) = 2, && f_D^*(100, 200100, 200) = 1001,\\&X_C^*(94, 180) = 1, && f_C^*(94, 180) = 933,\\&X_B^*(88, 169) = 2, && f_B^*(88, 169) = 878, \text{and}\\&X_A^*(72, 137) = 6, && f_A^*(72, 137) = 720.\end{aligned}$$

The optimal solution leaves five area units unused, which is revealed by $f_D^*(100, 200) = f_D^*(100, 199) = \cdots = f_D^*(100, 195)$.

This systematic enumeration is called *dynamic programming* [7],[48] and although it works just fine for our example, the total number of evaluations inside the maximization for all item stages and all states is 467,814, so we would have been better off enumerating by brute force the 41,769 item permutations. This is evidence of what is called *the curse of dimensionality*.[49] However, for many problems, dynamic programming offers dramatic improvements in enumeration.

48 https://en.wikipedia.org/wiki/Bellman_equation
49 https://en.wikipedia.org/wiki/Curse_of_dimensionality

6.25 Stack-Based Enumeration

Directed graphs and networks (graphs with data attributes) frequently arise in modeling, and we end up seeking ways to maneuver through these with some goal in mind.

In order to enumerate all directed paths (alternating sequences of nodes and arcs) from a given source s to a given sink t in a directed graph $G = (N,A)$, we must use an algorithm and data structures that ensure (1) completeness of the enumeration (we don't miss any paths) and (2) uniqueness (we don't want to report any path twice).

The basic concept behind path enumeration is the construction of a *partial* s–t path, and the enumeration of all possible *completions* of that path. A partial s–t path is a directed path between s and some other node, k (see Figure 6.9). The set of all completions of this partial path is the set of all k–t paths that do not use any nodes already in the partial path from s to k.

This is just a recursive description, and it seems circular, but note that the k–t path enumeration subproblem is solved on a *smaller* network than G. In fact, when the current partial path includes every node but t, there is at most one single completion of the current path: jump directly to t, if there's an arc.

To fully define an algorithm, then, we need a way of defining unambiguously a current partial path, and we need to know how to build all completions of the current partial path (avoiding nodes already on the path). This is accomplished with a mechanism to extend a current partial path by adding one arc.

The best way to understand a procedure like this is to define an instance precisely.

6.25.1 Data Structures

1) Stack[50] PATH (also known as a last in, first out, or LIFO queue) records the nodes visited, in an order, on the current partial path. The top of the stack PATH is the tip of the current partial path, k. PATH has n positions, and top gives current location of top of stack: PATH(top) = k.
2) Array[51] onPath(i) records the position of each node i on the current partial path, or is zero if a node is not on the current partial path.

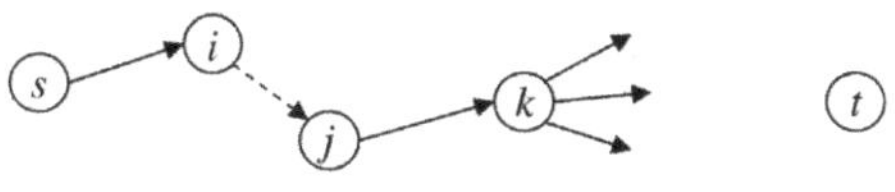

Figure 6.9 A partial k–t path in a directed s–t graph.

50 https://en.wikipedia.org/wiki/Stack_(abstract_data_type)
51 https://en.wikipedia.org/wiki/Array_data_structure

3) Adjacency list[52] A(*i*) of traversable edges out of node *i*, represented in forward-star structure by point(*i*) and head(*e*) arrays.
4) Current-arc structure next_arc(*i*) records the next arc to examine in A(*i*). Reset to first arc in A(*i*) every time *i* appears at tip of current partial path.

Let's assume a directed graph has been created in adjacency list form. In particular, all the head nodes `j` adjacent to tail node `i` can be accessed by `j=head[point[i]], ...,j=head[point[i+1]-1]`.

Stack-Based Enumeration

```
<![CDATA[
  top = 0;
  for i = 1 to n
    onPath(i) = 0;
  top = top + 1;
  PATH[top] = s;
  onPath[s] = top;
  next_arc[s] = point[s]
  while top > 0
    i = PATH[top];
    while next_arc[i]<point[i+1]
      j = head[next_arc[i]];
      next_arc [i] = next_arc[i]+1;
      if (onPath[j] == 0 and OK_to_add(j))
        top = top + 1;
        PATH[top] = j;
        onPath[j] = top;
        next_arc[j] = point[j];
        if j == t
             print(PATH[]);
             onPath[t] = 0;
             top = top - 1;
        end
        i = PATH[top];
      end
    end
    onPath[PATH[top]] = 0;
    top = top - 1;
  end
]]>
```

52 https://en.wikipedia.org/wiki/Adjacency_list

6.25.2 Discussion

The *s–t* path enumeration algorithm enumerates all directed paths from *s* to *t*; this can be proved by induction on the length of the partial path ($n-1$ is the first case, then $n-2$, etc.).

The only way that a path can be listed twice is if some partial path appears on the stack twice (that is, a node appearing on the stack recreates a partial path that has already appeared). Again, this can be proved by (forward) induction on the length of the partial path: no partial path with two nodes appears twice, because `next_arc(s)` is monotone increasing, and when *s* leaves the stack, we're done enumerating. If no partial path with *p* nodes appears twice, then clearly no partial path with $p+1$ nodes appears twice.

Therefore, the algorithm provides a complete enumeration, and does not repeat any *s–t* path.

There are more elegant recursive statements of pathfinding procedures, but these do not run as quickly on a computer, and speed is important, because there are a lot of paths to enumerate in directed graphs of interest—in fact there are an exponential number of these. To prove this to yourself, consider a dense directed graph with every node adjacent to all others. We can start a path from any node, and there are *n* of these. We can continue this path to the $n-1$ unvisited adjacent nodes, so now we have $n(n-1)$ partial paths, and so forth, eventually enumerating $n(n-1) \ldots 1 = n!$ paths.

6.25.3 Generating Permutations and Combinations

Many problems arise involving sequences, such as the order in which successive operations are carried out, cities are visited, and tests are conducted. How do you generate all permutations[53] of a set of objects? Suppose we have a set of *n* objects $S = \{a, b, c, \ldots\}$ and we are interested in looking at all permuted sequences of *k* of these at a time. The number of such permutations can be denoted $P(n, k) = n(n-1)\ldots(n-k+1) = n!/(n-k)!$

One of the most straightforward ways to generate permutation sequences on a computer is, surprisingly, by designing a directed graph with the *n* objects used as node labels and the subset size *k* as the number of echelons (Figure 6.10).

For the directed graph in Figure 6.10, there is a path for every subset of labels (*a*,*b*,*c*), lexicographically ranging from *s*,*a*1,*a*2,*a*3,*t* to *s*,*c*1,*c*2,*c*3,*t*.

To restrict our program to produce permutations of the labels *a*, *b*, and *c*, we arrange the test `OK_to_add(j)` in our stack-based enumeration procedure to be false if a label *j* already appears in the stack. This is as simple as associating with each node *j* a single bit representing stack occupancy.

---

53 https://en.wikipedia.org/wiki/Permutation

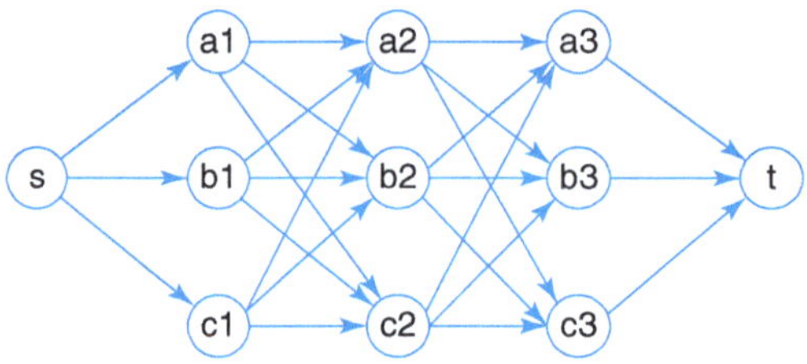

Figure 6.10 A directed *s*–*t* subset graph. Each node has an object (letter) label and an echelon number. There is a row of nodes for each of $n = 3$ objects, and each row has $k = 3$ echelons. Nodes adjacent to node b2 include *a*3, *b*3, and *c*3. One *s*–*t* path is *s*,*b*1,*a*2,*c*3,*t*. Every *s*–*t* path in the graph is a distinct subset of the nodes, and there is a path for every such subset.

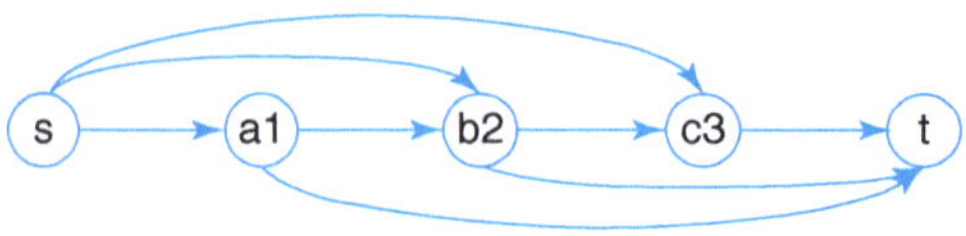

Figure 6.11 A directed *s*–*t* combination graph. Each *s*–*t* path includes a combination of the labels *a*, *b*, and *c* in alphabetic order.

If we want permutations of all subsets of size *k*, we just eliminate graph echelons beyond the *k*th one, and run the same procedure. We can also condition the permutations and filter out undesirable ones by either editing the directed graph to eliminate unwanted adjacencies in any permutation, or adjusting the test `OK_to_add(j)` to sense any undesirable permutation as soon as the `top-1` contents of the stack appear so. This admits adding numerical attributes to the nodes and/or adjacency arcs, creating from our graph a network, and computing fitness of partial orders in the logic of `OK_to_add(j)`.

We can also alter the directed graph to produce other results with path enumeration. For instance, the directed paths in Figure 6.11 select all combinations[54] (unordered) subsets of the labels a, b and c, in alphabetic order. If we again employ our handy filter `OK_to_add(j)` to be false when `top` equals $k-1$, we will enumerate all $C(n,k) = \binom{n}{k} = \frac{n!}{k!(n-k)!}$ combinations of *k* out of *n*.

6.26 Traveling Salesman Problem: Another Case Study in Alternate Solution Methods

The traveling salesman problem[55] has been known since antiquity by, well, traveling salesmen. How do you start from your home city and visit $n-1$ other

54 https://en.wikipedia.org/wiki/Combination

55 https://en.wikipedia.org/wiki/Travelling_salesman_problem (*Caution*: This otherwise excellent condensed review has an error in the last constraint of the "Integer Linear Programming Formulation", which should read $1 < i \neq j \leq n$)

cities exactly once and return home while having traveled the minimum possible distance? Such a *tour* is called *Hamiltonian*[56] (after a nineteenth century Irish mathematician who first stated it carefully), or traceable path (because you can connect dots on a map on such a tour without touching a dot twice or lifting your stylus before returning to your home city).

Table 6.5 shows a 10-city example that is symmetric and fully dense.

Getting an admissible solution to this problem is as easy as selecting n cells in this table, one per row (even though these distances are symmetric, let's say a row is an origin), and one per column (destination). Although any such selection would work, the one with minimum total travel time has length 5726 km. This selection is shown in Table 6.6.

This relaxed solution consists of subtours, each separated in the table. The value of this relaxation provides a lower bound on the value of an as-yet unknown optimal solution.

Now just visit these cities in alphabetic order, which in this full-dense problem will surely be admissible as a tour (see Table 6.7).

This is a Hamiltonian tour with total distance 13,125 km. This is an admissible solution with no subtour, so its value provides an upper bound on an as-yet unknown optimal solution.

There are $(n-1)! = 9! = 362,880$ possible tours in this full-dense problem. Why not enumerate with our directed graph enumeration procedure? We could do so here. But with a full-dense problem with 60 cities, there are on the order of 10^{80} admissible solutions. This is more than the cosmologists' estimate of the number of atoms in the universe.

There is a huge literature on how to get better solutions faster for this problem, and more realistic ones with side constraints on, for instance, time windows when we can visit each city.

Nevertheless, even simple heuristics and some elementary computer procedures can have beneficial effect. For instance, here we start with the alphabetic tour of our full-dense problem and select a random city in this tour. We evaluate moving it from its present position in the tour to some other random one. If this decreases our total tour length, we make the change. Otherwise, we randomly try again. Figure 6.12 shows the progress of our Monte Carlo simulation.[57]

In this relatively simple instance, an optimal solution with length 7,486 km is shown in Table 6.8.

Now, let's move to a just slightly larger problem, with all of the 24 European cities in our reference database. Now we have 23! solutions or somewhat less than 10^{23} of these. (The 23 coincidence is just that.) A minimal length selection with each city an origin once and a destination once has length

56 https://en.wikipedia.org/wiki/Hamiltonian_path
57 https://en.wikipedia.org/wiki/Monte_Carlo_method

Table 6.5 Distances (km) between some major European cities.[191]

	Barcelona	Belgrade	Berlin	Brussels	Bucharest	Budapest	Copenhagen	Dublin	Hamburg	Istanbul
Barcelona	0	1528	1498	1063	1968	1499	1758	1469	1472	2230
Belgrade	1528	0	999	1373	447	316	1327	2145	1230	809
Berlin	1498	999	0	652	1293	689	354	1315	255	1735
Brussels	1063	1373	652	0	1770	1132	767	773	490	2179
Bucharest	1968	447	1293	1770	0	640	1572	2535	1544	446
Budapest	1499	316	689	1132	640	0	1011	1895	928	1065
Copenhagen	1758	1327	354	767	1572	1011	0	1238	288	2017
Dublin	1469	2145	1315	773	2535	1895	1238	0	1073	2950
Hamburg	1472	1230	255	490	1544	928	288	1073	0	1984
Istanbul	2230	809	1735	2179	446	1065	2017	2950	1984	0

This matrix is symmetric and fully dense (that is, each city is adjacent to all others).

Table 6.6 An optimal selection of travel legs with each city appearing just once as an origin and once as a destination.

	Subtour 1	
Barcelona	Dublin	691
Dublin	Brussels	773
Brussels	Barcelona	1063
	Subtour 2	
Belgrade	Budapest	316
Budapest	Belgrade	316
	Subtour 3	
Berlin	Hamburg	255
Hamburg	Copenhagen	288
Copenhagen	Berlin	354
	Subtour 4	
Bucharest	Istanbul	446
Istanbul	Bucharest	446

This results in four subtours and a total travel length of 5726 km. Because this is a relaxation of the traveling salesman problem, it provides a lower bound on the minimum travel distance for a single tour of all cities.

Table 6.7 An alphabetical tour of 10 European cities with tour length 13,125 km, an upper bound on an optimal traveling salesman tour.

Barcelona	Belgrade	1528
Belgrade	Berlin	999
Berlin	Brussels	652
Brussels	Bucharest	1770
Bucharest	Budapest	640
Budapest	Copenhagen	1011
Copenhagen	Dublin	1238
Dublin	Hamburg	1073
Hamburg	Istanbul	1984
Istanbul	Barcelona	2230

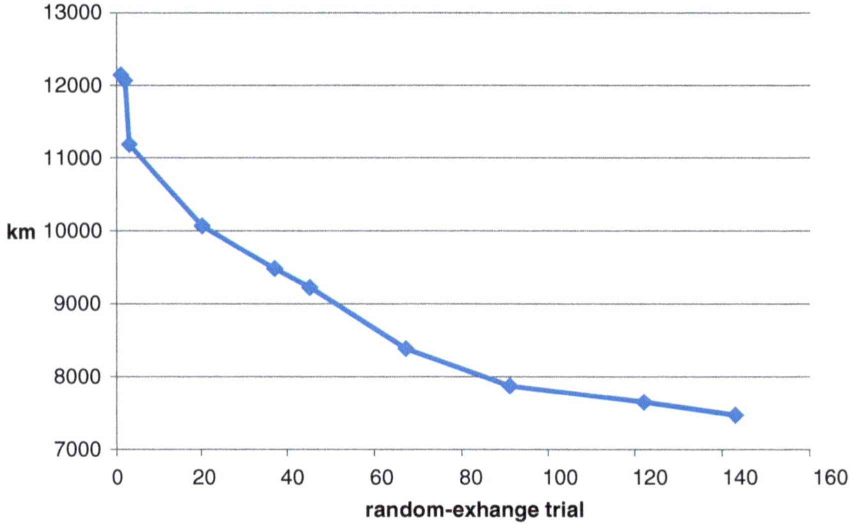

Figure 6.12 A Monte Carlo sequence of random-exchange trials. Starting with a tour in alphabetic order by city name, each random-exchange trial chooses a city to remove from a candidate sequence and randomly inserts it elsewhere in the sequence. If the tour distance is decreased by this, we record a new incumbent sequence. After less than 150 trials, we have discovered an optimal solution, although we would not know this without knowledge of the true global optimal tour distance.

Table 6.8 An optimal traveling salesman tour of 10 European cities with length 7486 km.

Barcelona	Dublin	1469
Dublin	Brussels	773
Brussels	Hamburg	490
Hamburg	Copenhagen	288
Copenhagen	Berlin	354
Berlin	Budapest	689
Budapest	Bucharest	640
Bucharest	Istanbul	446
Istanbul	Belgrade	809
Belgrade	Barcelona	1528

Table 6.9 An optimal traveling salesman tour of 24 European cities with length 12,288 km.

Barcelona	Rome	857
Rome	Milan	476
Milan	Munich	349
Munich	Prague	300
Prague	Berlin	280
Berlin	Warsaw	516
Warsaw	Vienna	557
Vienna	Budapest	217
Budapest	Belgrade	316
Belgrade	Sofia	329
Sofia	Istanbul	503
Istanbul	Bucharest	446
Bucharest	Kiev	744
Kiev	Moscow	757
Moscow	Saint Petersburg	633
Saint Petersburg	Stockholm	688
Stockholm	Copenhagen	522
Copenhagen	Hamburg	288
Hamburg	Brussels	490
Brussels	Paris	261
Paris	London	341
London	Dublin	463
Dublin	Madrid	1450
Madrid	Barcelona	505

10,194 km (a lower bound on the solution we seek, an alphabetic tour has length 33,117 km (an upper bound), and an optimal tour has length 12,288 km (see Table 6.9).

Starting with the alphabetic tour, our Monte Carlo method eventually discovers a tour with length 12,965 km at random-exchange trial 1545, *and discovers no further improvement over an additional ten million random-exchange trials* (see Figure 6.13).

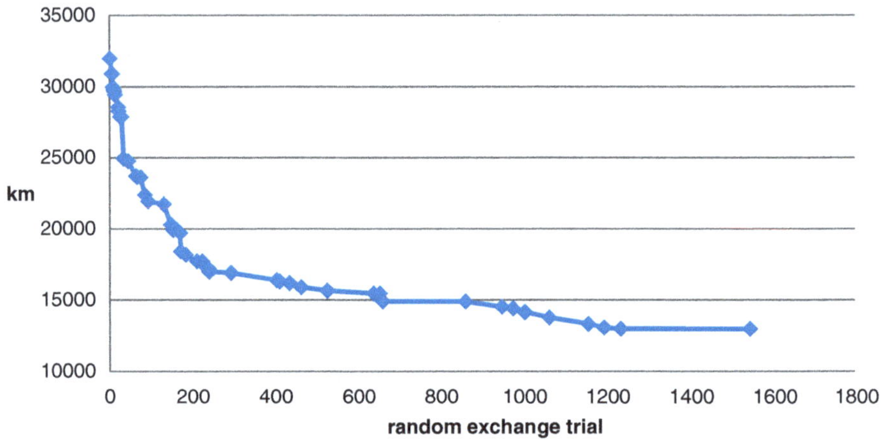

Figure 6.13 A Monte Carlo sequence of random-exchange trials. Each random trial chooses a city to remove from a candidate sequence and randomly inserts it elsewhere in the sequence. If the tour distance is decreased by this, we record a new incumbent. Trial 1545 discovers an improved tour with length 12,965 km. After 10,000,000 trials, we discover no further improvement.

6.27 Model Documentation, Management, and Performance

A carefully crafted model formulation can help assess model performance as the model is scaled up to include either more elements, or more detail on each element.

6.27.1 Model Formulation

Model formulation deserves close attention, and no model is complete without such documentation. A careful formulation can substitute for long verbal descriptions exhibiting much less specificity.

Standard Model Formulation

1) Define *index sets* (these are the subscripts you intend to use to represent the dimensions of data and variables representing states and actions). With each definition, give the approximate cardinality. These definitions include tuples of index sets, and the approximate cardinality of these.
2) Define *data elements* using the index sets already defined. Give the units of each data element. These elements and their indices and units provide the specification of a "data call" to instantiate any model example.

3) Define state and action *variables* using index sets already defined. Give the units of each of these.
4) State the *model* using only notation already defined in the previous three steps.
5) Give a *plain-language discussion* of the model, feature-by-feature.

The overarching principles here are "define before use," and "give precise specifics before plain-language discussion." We want to see the technical details before reading what is intended, the better to reconcile the two.

This form of model definition makes it a bit easier to predict, perhaps based on experiments with pilot models, how model performance (in particular, runtime and space requirements) will change with scale. We generally track in terms of the index set cardinalities, the anticipated volume of data, the time to process it, the storage to save it, the time and space to perform computations, and the volume of outputs.

6.27.2 Choice of Implementation Language

Choice of implementation language can have dramatic effect on model speed. Interpreted languages, such as Visual Basic for Applications[58] or Python,[59] run about two orders of magnitude slower than compiled, optimized languages such as C[60] or FORTRAN.[61] Before committing to use of special-purpose languages such as R,[62] Pyomo,[63] or a variety of simulation languages,[64] a modeler needs to assure that model performance at full scale is affordable or even tolerable.

6.27.3 Supervised versus Automated Models

Some models are used as needed and *supervised* by a modeler, while others are completely unsupervised or *automated*.

The Transmission Control Protocol (TCP) and the Internet Protocol (IP) models[65] are designed for continuous, unsupervised automated application.

Completely automated models require careful design to sense and interpret erroneous state data, as well as to choose constructive actions when such data

58 https://en.wikipedia.org/wiki/Visual_Basic_for_Applications
59 https://en.wikipedia.org/wiki/Python_(programming_language)
60 https://en.wikipedia.org/wiki/C_(programming_language)
61 https://en.wikipedia.org/wiki/Fortran
62 https://en.wikipedia.org/wiki/R_(programming_language)
63 https://en.wikipedia.org/wiki/Pyomo
64 https://en.wikipedia.org/wiki/Simulation_language
65 https://en.wikipedia.org/wiki/Internet_protocol_suite

are suspected. Sometimes this requires building a supervisory shell data model to detect and alarm suspected anomalies.

Automated power flow models are solved continuously by the independent system operator controlling an electric grid.[66] These have alarms for over-capacitated generators, transformers, and conductors.

6.27.4 Model Fidelity

Model fidelity is a key choice of a modeler and is perhaps the most important decision in modeling. It is often assumed by lay people that additional fidelity is always good. This couldn't be further from the truth.

A model is not made better by including excess complexity, but this addition makes the model more complex. Excess complexity can be used as an elusive smoke screen, increasing the apparent validity of results without foundation.

> "Everything should be made as simple as possible, but no simpler."
>
> A. Einstein

For instance, we may use an EOQ model to set ordering policies, but when a variety of items can be bought and shipped in a batch, or when we have limited inventory space, or when we must pay possession tax on the value of items in stock on particular dates, we are led to building more accurate models with higher fidelity for individual item types, time periods, facility locations, and so on.

Fidelity too coarse may miss essential details, and too fine may require state data not available or trustworthy.

Appropriate fidelity may be suggested by how often states are assessed in an existing system, and how often actions are required to operate the system. A telltale of excessive fidelity is when existing state data must itself be subjected to modeling to produce data with synthetically enhanced fidelity. Asking for state data with fidelity exceeding than already existing may be a modeling mistake, or a valuable discovery of a flaw in system operation. Excessive fidelity can slow responsiveness of a model while adding no additional insight.

Conversely, a model may change the suggested actions of the system operator in a qualitative way that necessitates changing how the system's states are evaluated.

For instance, a seminal supply chain design model may turn out to be considerably easier to state and solve than estimating origin–destination shipping costs per hundred-weight (i.e., hundred pounds, or about 45 kg), where the configuration of the supply chain and the dispatching rules for shipments may be changed by the model.

66 https://en.wikipedia.org/wiki/Power-flow_study

The first model you build of any textbook supply chain will almost certainly lead to another one, likely much more complicated, for estimating consolidated freight shipment costs or forecast demands of multiple items.

Interview With Cole Smith

When asked why a complex model that most accurately captures the problem being modeled is not always the best model to use in practice, Cole Smith, Professor and Chair of Clemson University's Department of Industrial Engineering, responded

George Edward Pelham Box, one of the great statisticians in the long history of the discipline, once stated, "All models are wrong, but some are useful." This is an incredibly important insight for the mathematical modeler.

Particularly in the field of optimization, many strong assumptions are often made in modeling a real-world problem. Linear programs require data to be known with certainty, assume that unknown parameters can take any continuous value as allowed by problem constraints, and assume all functions that score the value of the solution and constrain the problem variables must be linear. Because these conditions rarely hold in practice, one rarely devises a model that is correct, or even particularly close. This fact begs the question of why linear programming models, and more generally, other simplified models, might be used in the first place.

The two common reasons for the use of simplified models are that reality can be very difficult to model and that a very realistic model can be wholly intractable. With regard to the first issue, stochastic models that account for uncertain information are becoming much more common in research and application. Still, these models often assume distributions of unknown data, or that stochastic processes are stationary, and so on. Furthermore, models typically assume that there are no unknown unknowns. Moreover, models that involve human decision-makers invariably need to model actions made by humans in the loop. This fascinating field of research is rich and complex because human decision-making behavior is notoriously difficult to model accurately.

As for the latter issue regarding tractability, linear programs are convex optimization problems that can be solved in space and effort bounded by a polynomial function of the input size. Generally speaking, however, relaxing any assumption of linear programs requires the use of exponential time algorithms. Several of these models can still be solved in practical settings, even when they are of an impressively large scale. However, there are limits, and given the choice of a flawed but roughly accurate model that can be optimized or an accurate model that cannot be examined in any useful sense due to complexity, the flawed model wins in practice.

There are other ancillary benefits to using simpler models. First, simpler models are easier to explain to decision-makers, easier to implement, and easier to debug and experiment with. Second, simpler models tend to generate solutions that have certain patterns or features that others may find intuitively appealing. This is especially the case, for instance, in scheduling and routing problems. These intuitive features make the process of getting buy-in from managers and clients far easier. Third, these simpler models yield solution insights such as sensitivity analyses—answers to what-if questions—that one would usually not obtain from complex model analysis. Fourth, many state-of-the-art algorithms intentionally simplify models first to obtain a solution, and then iteratively refine the solution until it meets the important but more nuanced features of the true system being modeled. In that sense, simpler models form the basis for addressing more complex systems. Finally, a realistic model may take so long to develop that it ultimately will not be timely. If a problem must be resolved before it is possible to construct an accurate model, a relatively quick and flawed but roughly accurate model will certainly be preferable. And if the problem is dynamic and quickly evolving, a relatively quick and flawed but roughly accurate model may again be preferable.

Ultimately, it is important to state any recommendations that one makes based on mathematical models with full regard of the model's limitations. Put simply: It is vital to understand that the resulting solution might be brilliant and effective, or it may be completely useless because of flaws in the model's accuracy. As with any scientific discipline, a healthy skepticism of the validity of one's own work is paramount to success.

This is an excerpt from one of a series of interviews with analytics professionals and educators commissioned by the *INFORMS Analytics Body of Knowledge* Committee.

6.27.5 Sensitivity Analysis

Sensitivity analysis is essential in any model. The goal is assessing how changes to state data might influence model results. Some models are remarkably stable and robust. For example, the *Central Limit Theorem*[67] states that the sum of independent numbers drawn from a distribution with finite mean and variance becomes normally distributed, regardless of the distribution. The means that an arithmetic sample mean (a sum divided by the number of its terms) coming from a sample of any suitable distribution will be normally distributed, so this sample mean offers a host of standard statistical results based on what we know about the normal distribution.

Conversely, some models exhibit extreme sensitivity, seemingly amplifying small changes to input state data into wholesale revisions of suggested actions and/or output states. Such a model may need additional features to calm it down [8].

67 https://en.wikipedia.org/wiki/Central_limit_theorem

For many models, we can observe a starting state (initial condition). If a model of a system has no means to represent a starting state, or if this state is inadmissible in the model, or if the model quickly and qualitatively diverges from what you know to be an admissible state, close inspection is a necessity.

6.27.6 With Different Methods

Descriptive and predictive models are typically easier to plan, formulate, and implement as simulations. Whether deterministic or stochastic, simulations define entities to represent states, and procedures to represent actions.

To lend some concreteness to this, consider a simple queueing simulation. The state of the system might be the number in the queue. An action would be an arrival to the queue, or a service completion and subsequent departure from it. Concurrent states might assess resident times in the queue.

For prescriptive models, there is some debate about *mathematical optimization*[68] versus *simulation optimization*.[69]

Mathematical optimization requires that the initial and final states be specified, and actions be represented by decision variables governed by algebraic constraints on admissible actions. These actions influence intermediate and final states. The goal is to optimize some objective stated in terms of the achieved states and actions.

Both the *starting and ending states* must be represented and present an ambiguity. If the purpose of our model is to prescribe actions, how can we predict the ending state before we solve the model? We have encountered production–inventory models over some planning horizon that, lacking any ending state, simply advise an optimal policy to empty the system. This necessitates study of the *end effects* of a model applied over space or time, and specification of ending state can be a bit of an art form. What would be ideal? What is achievable? Resolving these problems may require a number of model runs and substantial analysis.

Models require us to make *simplifying assumptions*, and perhaps ignore some details that have consequence on our solutions. Model *restrictions*, such as adding constraints on admissible states to keep results reasonable, or to make the model easier to solve, will never improve the value of our model advice. Model *relaxations*, such as assuming continuity of actions that are, in fact, discrete, will never degrade the value of our model solutions, but may make the model advice unachievably optimistic.

Returning to our Bernoulli trial model, as the number of coin flips increases, our closed-form analytic solution becomes uncomputable, with a huge combinatoric

68 https://en.wikipedia.org/wiki/Mathematical_optimization

69 https://en.wikipedia.org/wiki/Simulation-based_optimization

number of outcomes,[70] all counted to have some given number of heads, and each outcome with an infinitesimally small probability of occurring exactly. We have mentioned using a continuous normal distribution approximation with the same mean and variance. But, when do you make the transition from discrete to continuous modeling of actions and states? With this example, you are well advised to apply a continuous approximation at about 15 trials (15! is about a 12 decimal digit integer). Similar continuous approximations may be advisable in other models of large numbers of discrete actions, such as building automobiles or buying items.

If the constraints are linear functions of actions in terms of state assessments, we have a linear program. If some or all actions need to be discrete (e.g., yes or no), we may formulate a linear integer program. If the constraints are necessarily nonlinear, and/or some decisions need to be discrete, we have a complicated optimization that may or may not yield to conventional solution methods in reasonable computing time.

Simulation optimization fixes an initial state, and then randomly draws from admissible actions to induce a random set of subsequent states. Each random action and subsequent state offers a candidate solution. By performing many such random evaluations and keeping track of the best candidate, we discover the best incumbent set of actions. The complication here is with assessing how much better the objective could have been with even more experiments. The topology of objective functions may be benign, with better solutions being obvious, or pernicious, with optimal (vice optimum) solutions being hard to discover by random, even systematically random probes.

6.27.7 With Different Variables

A variable is any number, quantity, or characteristic that can be measured or counted. In a statistical model, variables may be data items. Statistics frequently seeks hidden relationships between variables, sometimes concluding that some dependent variables are influenced by other, independent ones. In optimization, decision variables represent actions by the operator that influence system state. In either case, we must choose the fidelity of our variables in consideration of the relationships or actions we want to discover. Sometimes, certain statistical variables may turn out to have substantial influence on others, and we may seek different sets of variables to isolate the strongest influence. In optimization, we might express a model in terms of variables giving the quantities of activities to pursue, then decide to change to variables expressing the time to achieve certain quantities. We fix some variables, essentially making them states, and release

70 https://en.wikipedia.org/wiki/Orders_of_magnitude_(numbers)

others. In any sort of model, we seek variables that effectively lead us to an insightful solution.

6.27.8 Stability

Some models and solution methods exhibit intrinsically unstable behavior. For instance, heuristics may reliably produce subjectively good-quality solutions, but without some bound on how good an undiscovered solution might be, caution is called for. We might test such a heuristic by trying alternate starting solutions, even random ones, to detect any unstable performance. Other models, such as nonlinear optimizations or systems of differential equations, depend on numerical solution techniques that may not always converge to an acceptable solution, a solution with nearly the best possible value, or a solution at all. You will read advice to "tune" such methods, as well as applying alternate starting solutions. Generally, you should be able to conclude with a little research whether or not a particular solution method is stable. If your candidate method is known to misbehave, you have to decide whether its simplicity or effectiveness makes it worth the risk.

6.27.9 Reliability

Some models are intrinsically unreliable and just cannot be trusted to behave reasonably for a number of reasons. For instance, a time-discretized set of difference equations used to represent continuous-time differential ones may require a fair amount of fiddling and tuning to produce acceptable approximate results. A *piecewise linearization*[71] (an *inner approximation* always underestimating a function, or *outer approximation* overestimating) of a nonlinear model function may suffice, but may not solve reliably if the approximated nonlinear function is not approximated well, or the function has perverse structure. Why would you even risk an unreliable model? Perhaps because you have in hand a familiar solution method (say, linear numerical optimization) and wish to avoid tangling with more complex nonlinear numerical optimization.

6.27.10 Scalability

Generally, if your model is expressed in standard form, it should be straightforward to assess the impact of changing the cardinality of indices. It is frequently possible by analogy to estimate with some reliability the impact this will have on computation time or success.

---

71 https://en.wikipedia.org/wiki/Piecewise_linear_function

6.27.11 Extensibility

Extensibility applies to adding new detail, functionality, or bolting together models into a unified federation. This is usually possible if the various components are of the same model class, and more of a challenge when they are not. For instance, combining an existing numerical optimization with another is often feasible, especially if this possibility has had influence on the model designs, but extending an optimization with a simulation, or a closed-form solution with a numerical one, may not be easy. Perhaps the most vexing problem with extending models is establishing some verifying certification that the unified solution addresses at once the concerns of all model components.

> "A model should be able to produce an answer while we still remember the question, and care about the answer."
>
> G. Brown

Interview With James J. Cochran

James J. Cochran, the Rogers-Spivey Faculty Fellow and Associate Dean for Research at the University of Alabama's Culverhouse College of Business Administration, comments on the extrinsic benefits of modeling a problem.

Modeling is, of course, intrinsically valuable because if it is done properly, it can lead to a potential resolution of a problem or improvement of a system. But modeling a problem can also yield many other important benefits.

Through the process of modeling a problem, the analyst can become aware of or better understand aspects of the problem that were not originally apparent. The job of an operations research analyst is often much like that of a computer programmer; just as the programmer has to work closely with the end user of the program through several iterations to ensure that the final program will perform the desired tasks, an operations research analyst will often iterate through several versions of a model and incorporate feedback from the client at each iteration to enhance the model. In mathematical programming, for example, this process may lead to the identification of a constraint that was originally inaccurate or had not been considered. In simulation, the iterative process of modeling may lead to a better understanding of some characteristic of a process. Through these insights, a clearer, more accurate, and more comprehensive understanding of the problem may emerge.

In other instances, modeling may lead to the realization that some factor was not adequately considered in the design of a process or system. Modeling can be used to identify a design flaw before a process or system is put into use to improve the performance of the process or system from its implementation, or it can be used after a process or system is put into use to find causes of poor performance.

Modeling can also be used to convince a client of the value of operations research. For example, suppose you have a mathematical programming model coded in a spreadsheet. You can easily substitute the current decision variable values into the model to find the associated value of the objective function. You can then incrementally change the values of these decision variables in a search for superior solutions. For some problems, it is actually relatively easy to use this approach to identify solutions that are superior to the current solution before applying an optimization algorithm.

This is an excerpt from one of a series of interviews with analytics professionals and educators commissioned by the *INFORMS Analytics Body of Knowledge* Committee.

6.28 Rules for Data Use

"What gets measured, gets managed."

P. Drucker

The number and diversity of sources of data necessary to support a real-world model is surprising. A simple model of shipment planning can quickly call for road networks, freight rates from commercial sources, postal rates, state regulations, labor agreements, carrier policies, system operator policies, restrictions on mixed commodities, operating costs, and so on. Not only does a model need a lot of data from many sources, it will depend on the currency and accuracy of this data over all its life.

The following is a representative sample of common types of data sources, and the rules that govern data from each.

6.28.1 Proprietary Data

Proprietary data are routinely involved in modeling studies. It is best to establish model protocols early for storage, indexing, governance, and use of any data source. Remember that proprietary data are viewed by its owner as "property," and that use of such property must be with permission of the owner in exactly and only the way the owner permits.

6.28.2 Licensed Data

Licensed data from commercial sources usually comes with restrictions on how it can be used. For example, Graphical Information System (GIS) data on roads, rivers, railroads, undersea cables, and so on come with explicit limits on the permitted application domain. Ironically, a modeler may spend a lot of time filtering out what the data provider views as the most valuable data elements, such as appearance and shapes of features, because the modeler merely needs something as simple as the adjacent distance between one feature such as a road junction and another.

6.28.3 Personally Identifiable Information

Personally identifiable information (PII) contains, is as it sounds, data that may enable or ease identification of individuals, and this cannot be permitted without explicit, knowledgeable permission of the individuals involved. Many organizations additionally restrict dissemination and use of such data, and require special protocols for its storage and use [9].

6.28.4 Protected Critical Infrastructure Information System (PCIIMS)

Protected Critical Infrastructure Information System (PCIIMS) has been created by Department of Homeland Security (DHS) to house and control access to information on our national critical infrastructure gathered from its owners (85% of this infrastructure is not owned by the government) and from many studies since 9/11 seeking to understand the function of these infrastructures, probing for weaknesses and opportunities to improve resilience [10].

6.28.5 Institutional Review Board (IRB)

Institutional review board (IRB) human subject research documentation requirements require submission and review of special study protocols to insure safety (e.g., Ref. [11]).

6.28.6 Department of Defense and Department of Energy Classification

Categories such as For Official Use Only (FOUO), Secret, Top Secret, etc., require personnel clearances, special storage facilities, and extensive rules for how such data can be used, and how results based on such data must be marked and treated.[72]

6.28.7 Law Enforcement Data

Law enforcement data are governed by its own set of rules for access and use [12].

6.28.8 Copyright and Trademark

Copyright and trademark laws may apply to data not otherwise encumbered [13].

---

72 https://en.wikipedia.org/wiki/Classified_information_in_the_United_States

6.28.9 Paraphrased and Plagiarized

Paraphrased and plagiarized data can present a vexing problem. Data, even obtained from otherwise open sources, may carry restrictions.[73]

6.28.10 Displays of Model Outputs

Displays of model outputs need to carry with them explicit notice if they are based on data governed by any of the above restrictions. Some call this "derivative classification," and as a model is developed, each new display needs to be classified in some way. This classification may derive from how you used the data, rather than on its restricted nature.

6.28.11 Data Integrity

Data integrity can be enhanced by adding a digital signature to data fields, especially those that are not expected to change often or at all. A simple hash signature can reveal when a change may have occurred, necessitating refreshment.[74]

6.28.12 Multiple Data Evolutions

Multiple data evolutions result from model development based, for instance, on alternate predicted futures. This invites data set indexing and naming conventions that track the provenance of each of these forecasts, so planners can sort out whose ideas are expressed in each evolution, and what and how data have been applied.

6.29 Data Interpolation and Extrapolation

Data interpolation and extrapolation[75] are predictions used to fill in gaps, especially in temporal or spatial data series where we have some observations of a dependent state value for some, but not enough associated values of independent states (see Figure 6.14). As the names imply, interpolation applies inside the range of observed independent state values, and extrapolation elsewhere.

A modeler needs to assure there are no reasons to suspect systematic misbehavior of the prediction function within independent state values, and

73 https://en.wikipedia.org/wiki/Wikipedia:Plagiarism

74 https://en.wikipedia.org/wiki/Hash_function

75 https://wiki.engageeducation.org.au/further-maths/data-analysis/extrapolation-and-interpolation/

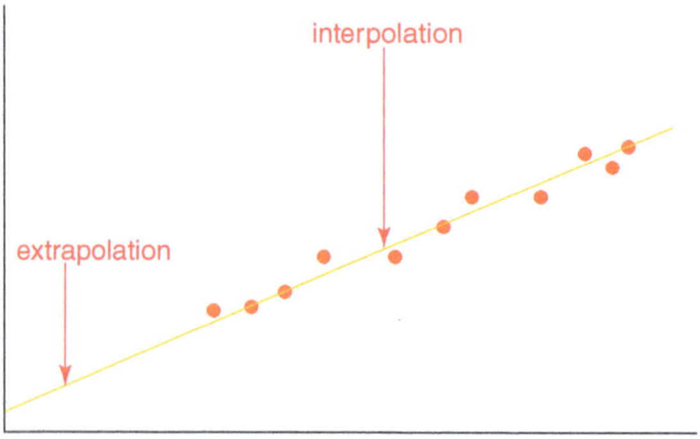

Figure 6.14 Extrapolation and interpolation of a data series. The horizontal axis is in units of some independent state variable, and the vertical one in units of associated values of a dependent state. There is an assumed functional relationship here that is used to predict behavior of the dependent state. If the prediction is within independent state observations, we are interpolating, otherwise we are extrapolating.

especially outside them. Extrapolation requires more faith that the validity of the prediction function extends beyond the range of our observations, and for this we need as much supporting evidence as possible.

Referring back to our regression model of weight as a function of height for a group of American females aged 30–39, interpolation might be reasonable within the range of observed heights from 1.47 to 1.83 m (about 4′ 10″ to 6′). But an average newborn female has height 0.51 m (about 20″) and weighs 3.6 kg (about 7.7 lbs), and extrapolation yields 70 kg (about 154 lbs). Similarly, extrapolating to heights greater than the observed 1.83 m will soon yield results beyond those to be expected for human females.

6.30 Model Verification and Validation

> "To get a large model to work you must start with a small model that works, not a large model that doesn't work,"
>
> D. Knuth

Model verification and validation have long been a source of debate.[76,77,78]

76 https://en.wikipedia.org/wiki/Verification_and_validation_of_computer_simulation_models
77 https://en.wikipedia.org/wiki/Regression_validation
78 https://en.wikipedia.org/wiki/Model_checking

6.30.1 Verifying

Verifying a model consists of providing a reasonable, representative set of starting states, and observing a responding set of actions consistent with known practice, or merely common sense, leading to a set of resulting states that can be reconciled with actions. Perhaps "debugging" a model is a better term, making sure it works as intended.

6.30.2 Validating

Validating a model establishes that it is in good alignment with reality. *This is seldom possible*. An experienced modeler will advise that to assert that a model is validated is, well, foolhardy. The goal is assurance that we have a reasonable representation of reality.

But, some models do lend themselves to validation.

For example, the physics of electricity and the performance of generators, conductors, and transformers is well enough known and we feel confident, to first order, that our power flow models validate well. Conversely, the effects of cascading failures in an electric grid are not at all well understood.

For example, water flow through systems of dams, reservoirs, channels, pumps, and pipes is physically well modeled, and assessments of leakage and evaporation are reasonably reliable.

Models of traffic flow seem to validate at large scale, and completely break down at fine scale: There is no perfect model explanation of traffic congestion and individual or group behavior of commuters.

While, say, an economic or military model may have been subjected to rigorous validation exercises, the result is inevitably that the model seems to be a good enough representation of reality to be useful.

6.30.3 Comparing Models

When comparing competing models, perhaps a legacy one with a new candidate, *Occam's razor*[79] provides good, very general advice. Everything else being equal, the model based on fewer assumptions is probably superior.

Ohm's Law? (Deterministic, Descriptive)

Ohm's law states that a potential difference (voltage) across an ideal conductor is proportional to the current through it. The constant of proportionality is called the "resistance," r. Ohm's law is given by $v = i\,r$, where i (amps) and resistance r (Ohms) are related to the potential difference v (voltage).

79 https://en.wikipedia.org/wiki/Occam's_razor

Ohm's law is unfortunately named because it is not a fundamental law of nature, but only an approximation. It breaks down for any of a host of reasons, for instance, when the resistance leads to energy loss due to waste heat.

Nonetheless, Ohm's law is a good model, under ideal conditions, but not under all conditions, and should probably be called Ohm's approximation.

6.30.4 Sample Data

Sample data for model verification is best derived with as much realism as possible. *Randomly generated sample data*[80] are notorious for being non-representative, slowing model operation, and obscuring insights.

6.30.5 Data Diagnostics

Data diagnostics are vital defenses of a model, especially if data sources are partially automated and not necessarily in control of the planner(s) using the model. There is no shame in testing for impossible conditions and issuing well-considered diagnostics.

6.30.6 Data Vintage and Provenance

Data vintage and provenance should be established for every instance a model uses, and should be displayed prominently and widely with model outputs.

6.31 Communicate with Stakeholders

"The purpose of computing is insight, not numbers."

R. Hamming

This is a continuous requirement from first client meeting to presentation of intermediate or final results. Model development inevitably encounters surprises. Anticipated, required data may not be available when needed, or trustworthy, or expressed in immediately useful form. The stakeholders may include not just the client, but also representatives of other interested groups (for instance consultants, domain experts, regulating agencies, and others). If the modeler has prepared well, successful presentations based on concrete model outcomes are the ideal outcome. There will likely be presentations for planners who deal with the problem and others for the executives who pay the bills. For

80 https://en.wikipedia.org/wiki/Test_data_generation

each stakeholder, one must carefully adhere to the lexicon worked out in the five-step preparation (see prior definition of such) for the modeling project.

> "Almost all senior analysts have similar stories about how they learned not to brief a senior executive."
>
> J. Kline

Managing relationships with multiple stakeholders, and following their multiple, possibly conflicting, objectives is beyond the scope of this introduction. But suffice to say, if you have scrupulously followed the five-step preparation above, you are as well situated as possible.

Most contemporary models are used via a *graphical user interface* (GUI), and in many cases the GUI consumes more development effort than the model. This is to be expected. However, the GUI developer(s) and modeler(s) are not necessarily the same individuals, and this calls for constant, careful coordination. There are wonderful models with poor GUI's, or none at all, and fantastic GUI interfaces to terrible models. Better to sort out which is which. Some of the most successful GUIs have appeared (though is has some vexing bugs traversing the International Dateline) for mobile applications on portable devices. Google Earth[81] is a superb example.

Although there are a host of commercial GUI developer kits available, some particularly successful applications have been developed quickly, and on the cheap, by co-opting, for instance, the Microsoft Office suite of applications (Excel, Access, Project, etc.) These packages represent hundreds of millions of dollars of development, all aimed at the sort of system operator the modeler is likely to encounter. Despite all the criticisms of any particular software suite, such as this one in particular, it presents a huge opportunity to develop a model, supporting data base(s), and GUI, with the reasonable expectation that the planners will already be quite familiar with the tools employed.

6.31.1 Training

Training for model use can involve elaborate, formal courses, including not merely model options and controls, but material on any underlying theory, and interpretation of model behavior. Training materials can range from manuals to pop-up windows on screen displays. If the model is supported by an elaborate graphical user interface, videos with voice-over instruction can be very effective.

6.31.2 Report Writers

Report writers are designed not just to convey the "what" of a solution, but also to lead to recognition of the "whys." These are not as attractive as a well-designed

81 https://en.wikipedia.org/wiki/Google_Earth

GUI, but offer much richer detail. They create detailed accounts, frequently in a tabular format. Alternatively, report writers can populate a database inviting *ad hoc* queries and planner-designed custom reports and graphics. Report writers typically get more sophisticated with model use, as successive questions arise inviting additional solution analysis and diagnosis. It is not uncommon for report writers to consume much more development time than the model they support.

One effective means to communicate strategic business results is by generating a set of forecast *operating statements*.[82,83] Nothing grabs a senior executive's attention more than details that follow all the way to projected influence on shareholders' equity. Gordon Bradley [14] and Art Geoffrion [15] did this for General Telephone and Electric (GTE) Corporation in the early 1980s.

Sometimes, a model represents a problem well, but not the way the decision must be made.

6.31.3 Standard Form Model Statement

Returning to our optimization example, let's restate it in our standard form:

Standard Formulation of Portfolio Selection Model (Deterministic, Prescriptive)

index use [~cardinality]

$i \in I$	item i in set items I [~100]

given data [units]

$value_i$	value per item [$]
$weight_i$	weight per item [kg]
$\max\_weight$	maximum selected weight [kg]
$area_i$	area per item [m^2]
$\max\_area$	maximum selected area [m^2]

decision variables [units]

X_i	number of items of type i to select

formulation

$$\max_X \sum_{i \in I} value_i X_i \qquad [C0]$$

$$s.t. \quad \sum_{i \in I} weight_i X_i \le \max\_weight \qquad [C1]$$

$$\sum_{i \in I} area_i X_i \le \max\_area \qquad [C2]$$

$$X_i \in \{0, 1, 2, \ldots\} \quad \forall i \in I \qquad [C3]$$

82 https://en.wikipedia.org/wiki/Accounting
83 https://en.wikipedia.org/wiki/Income_statement

This is the exact portfolio selection model introduced before, but in a standard form that scales up and is amenable to algebraic modeling languages (not the topic here).

6.31.4 Persistence and Monotonicity: Examples of Realistic Model Restrictions

Now suppose we have briefed our solution, and the client admits that the maximum weight budget was an estimate, that there may be some variation in the true maximum weight, and asks us for a *parametric sensitivity analysis* of possible maximum weights ranging from 95 to 105 kg. The client wants to be ready to brief for these contingencies.

The 11 rows following "optimal" in Table 6.10 show optimal selections as we vary maximum weight from 95 to 105 kg. *A reasonable client will hate these optimal results.* How do you explain a selection portfolio that exhibits so much chaotic turbulence as one simple parameter, maximum weight, is

Table 6.10 Parametric solutions to the numerical optimization portfolio selection model varying maximum weight budget.

max weight	Sel wgt	Sel area	A	B	C	D	Total value	Base portfolio
Continuous								
100	100	200	7.41	0.00	0.00	3.70	1014.8	
Optimal								
95	95	198	6	1	0	5	969	
96	96	194	7	0	0	4	976	
97	97	194	6	2	0	3	980	
98	98	199	6	1	1	4	990	
99	99	195	7	0	1	3	997	
100	99	*195*	*6*	*2*	*1*	*2*	*1001*	*base*
101	101	200	7	1	0	3	1021	
102	102	196	8	0	0	2	1028	
103	103	196	7	2	0	1	1032	
104	104	192	8	1	0	0	1039	
105	105	197	8	0	1	1	1049	

The Continuous row shows the relaxed solution when we do not require whole numbers of items. Anticipating a possible change to the weight budget, we parametrically evaluate Optimal whole number solutions from 95 to 105 kg. The Optimal portfolios bear little resemblance to their adjacent neighbors. This would be a difficult set of solutions to brief. (Note that rounding the Continuous relaxation to nearest whole numbers would not discover the Optimal base portfolio solution.)

gradually, systematically increased? Well, declaring "this is optimal" will not likely suffice.

Clients expect parametric transitions that are intuitive. Here, the client might prefer a *base portfolio*, say the one we produced with maximum weight 100 kg. Then, each time we reduce the weight budget, we only permit deletion of a single item from this base portfolio when we must, retaining the rest. Conversely, as we increase weight budget from base, we only allow a new item to be selected in addition to those already in the portfolio. This is intuitive. Less weight budget, fewer item selections, more weight budget, more item selections, while always preserving all but one item in the portfolio. This is concise. These are called *monotonic* parametric solutions. In Table 6.11, you will find "monotonic" results for the base portfolio with maximum weight 100 kg.

The client may be disappointed that from 100–105 kg maximum weight budget, no new monotonic item is selected. This is because the area constraint requires that we make room for a new selection by deleting some existing ones, and the client told us not to do this. The monotonic solution is a restriction of the optimal solutions that did make such substitutions (with excess gusto) and with respect to the base portfolio, this restriction reduces our maximum total value selected.

A reasonable client might then agree, "OK, you can make room for a new item selection by deleting an existing one, but never more than one of each to keep things simple to explain." This is an example of *persistent* parametric solutions, here limiting the changes to at most two per adjustment of maximum weight budget. These results are shown in the "persistent" section of Table 6.11.

These persistent results are a restriction of optimal ones, but not as restricted as monotonic selections. Note the maximum total value selected is no better than optimal, and no worse than monotonic.

A reasonable client may ask for more variations like these, seeking insight, but more important seeking some way to understand results in order to persuasively convince others to change policy and follow our advice. Figure 6.15 shows portfolio values as we vary our maximum weight budget.

> "A manager would rather live with a problem he cannot solve than accept a solution he cannot understand,"
>
> G. Woolsey

6.31.5 Model Solutions Require a Lot of Polish and Refinement Before They Can Directly Influence Policy

Solutions must have the following attributes:

- Understandable. Is it clear what our item selection advice means?
- Actionable. Do we have authority to select these items in these numbers?
- Legal. Are we allowed to choose this portfolio of items?

Table 6.11 Monotonic and persistent solutions to the numerical optimization portfolio selection model.

max weight	Sel wgt	Sel area	A	B	C	D	Total value	Base portfolio
Monotonic								
95	94	175	6	2	1	0	933	
96	94	175	6	2	1	0	933	
97	97	185	6	2	1	1	967	
98	97	185	6	2	1	1	967	
99	97	185	6	2	1	1	967	
100	*99*	*195*	*6*	*2*	*1*	*2*	*1,001*	*base*
101	99	195	6	2	1	2	1001	
102	99	195	6	2	1	2	1001	
103	99	195	6	2	1	2	1001	
104	99	195	6	2	1	2	1001	
105	99	195	6	2	1	2	1001	
Persistent								
95	95	189	6	1	1	3	956	
96	95	189	6	1	1	3	956	
97	97	194	6	2	0	3	980	
98	97	194	6	2	0	3	980	
99	97	194	6	2	0	3	980	
100	*99*	*195*	*6*	*2*	*1*	*2*	*1001*	*base*
101	99	195	6	2	1	2	1001	
102	102	200	6	3	0	2	1025	
103	102	200	6	3	0	2	1025	
104	102	200	6	3	0	2	1025	
105	102	200	6	3	0	2	1025	

The Monotonic solutions allow at most one item deletion per max weight reduction from the base portfolio, and at most one item addition for each budget increase from the base portfolio. These do not use all the weight budget, and may be too restrictive. The persistent solutions at the bottom allow at most one deletion and at most one addition. This intermediate restriction may be acceptable.

- Monotonic. See prior.
- Persistent. See prior.
- Robust. How good is our solution if our assumptions are wrong?
- Resilient. How good is our solution if some selection is thwarted?

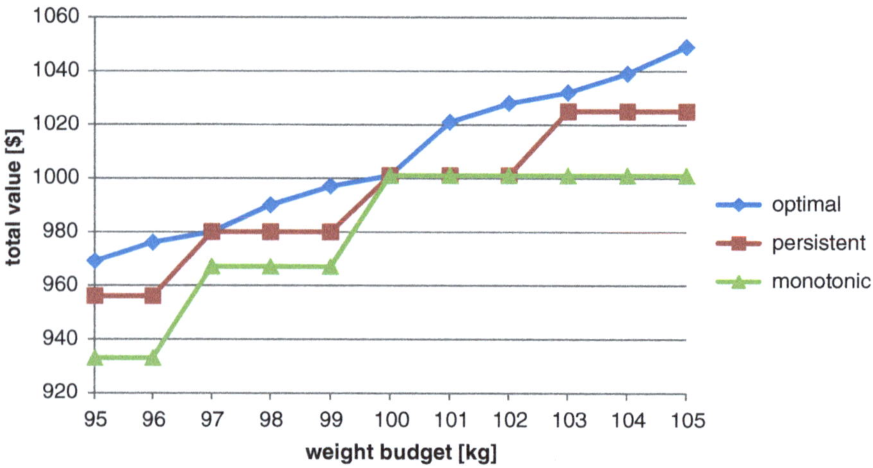

Figure 6.15 The base portfolio here is for a weight budget of 100 kg. This shows the trajectory of optimal portfolio values compared with persistent and monotonic ones as weight budget is decreased, or increased from this base portfolio. (The connecting lines serve merely to highlight each discrete series of solutions.) Persistent solutions here permit at most one item to be deleted for each one added to the base portfolio as we decrease or increase weight budget by 1 kg. Monotonic solutions permit only one item to be added or dropped from the base portfolio as we, respectively, increase or decrease weight budget by 1 kg. These three trajectories diverge from the base portfolio weight budget as we decrease or increase this budget and eventually meet as the weight budget decreases so much that no item can be selected, or increases so much that all items are selected.

We don't have space here to develop all these points in depth, but each of them arises constantly in real-world practice. Be reassured they can and have been addressed successfully in many commercial, government, and military venues. Just be ready for surprises from your client, listen well, evaluate, and brief with clarity not just the restricted advice, but the costs these restrictions have inflicted. Some restrictions are laws of physics, others "email from God" policies, but many are flexible preferences, thumb rules, tribal wisdom, or conveniences that may inflict punitive penalties. With diplomatic caution, try to expose these penalties.

When you encounter policy advising "prioritize your items and select them in decreasing priority order until you exhaust your budget," you may be dealing with a suboptimal policy. Possibly, a very suboptimal policy.

6.31.6 Model Obsolescence and Model-Advised Thumb Rules

It is time to *retire a model* when the problem it addresses is solved by other means or replaced by other concerns. However, it is premature to retire a model after it has lent so much insight to planners that they can solve the problem without model help. In such cases, continued model use can alarm when some condition has changed that the planners have misdiagnosed. Even for very successful models, looking back five years, you will see few of these are still in use. Creative destruction is a fact. Model obsolescence is the rule, not an exception.

A most impressive recent example of technical obsolescence has been replacement of legacy enterprise resource planning (ERP) systems paid for by license fees based on possession and number of users, with equivalent functionality of applications residing and used in "the cloud," some with merely per-use license fees. This is a modeling revolution that has ERP industry giants scrambling for leverage and dominance, new entrants suddenly thriving, and some of the best-known legacy ERP providers trying to buy contemporary technology and keep up.

> "The lowest level of understanding is when you convince yourself you know the answer;
>
> the next level is when you convince a colleague; and
>
> the highest level is when you convince a computer."
>
> R. Hamming

6.32 Software

An analyst generally needs familiarity with and access to a number of software tools: text editor, presentation slide maker, spreadsheet, graphics, statistics, simulation, optimization, general-purpose programming, and geographic information system. If you are affiliated with an educational institution, you likely have free access to a wide variety of software packages that would normally cost a lot more than your portable workstation. This provides a great opportunity to try a variety of packages.

Even if you are not affiliated with an educational institution, there are a number of excellent, inexpensive, or free software packages. For instance, the Microsoft Office Suite[84] includes the *text editor* Word, the *presentation slide maker* Powerpoint, and *spreadsheet* Excel, which includes *graphics* features. The *statistics* package R[85] is free and well documented. Many *simulation packages*[86] are freely available, offering libraries of random statistical generators and live animations. Some *optimization* software[87] is available in Excel, and a large

84 https://en.wikipedia.org/wiki/Microsoft_Office
85 https://en.wikipedia.org/wiki/R_(programming_language)
86 https://en.wikipedia.org/wiki/List_of_computer_simulation_software
87 https://en.wikipedia.org/wiki/List_of_optimization_software

variety of other packages are available. *General-purpose programming* languages such as Python[88] are available and well documented. Google Earth[89] provides a globe, map, and a *geographic information system*; a user can mark points, draw objects, and create animations on the Earth's depicted surface, political and geographic features can be depicted as the viewer's perspective moves over, toward, or away from the surface, and these displays are portable between computers via simple email attachments. Some *graphics* software[90] is embedded in the preceding suggestions, but more general utilities can be used to create still images and movies.

When choosing software, make sure each package can be added to your modeling federation as a compatible component. Look for examples of, for instance, a spreadsheet that can invoke a simulation as a subroutine. If you suspect you'll need some help, check for online blogs and a users' group that supports a package. Also, be mindful that the more existing users there are, the less likely the package is to exhibit annoying bugs. It is said that Microsoft Excel has more than a billion users worldwide: that's a lot of experienced potential users for anything you create with it.

6.33 Where to Go from Here

The INFORMS journal *Interfaces* is aggressively edited for clarity of exposition and includes many modeling examples explained with care; refreshingly, some recounting of false starts and lessons learned also appear. The INFORMS newsletter *OR/MS Today* features some articles about modeling, including the shared experiences of clients and modelers, some analysis puzzles, and entertaining features about the operations research (and modeling) craft. To access more of our huge open literature of articles and textbooks, some mathematical preparation will be necessary, including at least algebra, probability, statistics, and elementary modeling. An introductory modeling course with a lot of homework drills does wonders: *the best way to learn how to model is to try modeling.*

> "Anyone who has never made a mistake has never tried anything new."
> A. Einstein

6.34 Acknowledgments

The author is grateful for helpful comments (and suggested examples) from colleagues Dave Alderson (game theory), Matt Carlyle (enumeration), Rob Dell

---

88 https://en.wikipedia.org/wiki/Python_(programming_language)
89 https://en.wikipedia.org/wiki/Google_Earth
90 https://en.wikipedia.org/wiki/Graphics_software

(palatable solutions), Jeff Kline (Hughes equations), Bob Koyak (regression), Kyle Lin (figure skating decision theory), Connor McLemore (example physical and mathematical models), and Al Washburn (EOQ); two anonymous reviewers (misuse of *p*-values, data analytics); and the editor (characterizing probability).

References

1 Delury G (1975) The World Almanac and Book of Facts. Sacramento Bee.
2 Centers for Disease Control and Prevention. (2017) Body Measurements. Available at https://www.cdc.gov/nchs/fastats/body-measurements.htm.
3 Baker M (2016) Statisticians issue warning on P values. *Nature* 531(7593): 151.
4 Brown G, Carlyle M, Salmerón J, Wood K (2006) Defending critical infrastructure. *Interfaces* 36: 530–544.
5 National Research Council (2008) Department of Homeland Security Bioterrorism Risk Assessment: A Call for Change. Available at https://www.nap.edu/catalog/12206.
6 Hughes W (1995) A Salvo model of warships in missile combat used to evaluate their staying power. *Naval Res. Logist.* 42: 267–289.
7 Bellman R (1954) The theory of dynamic programming. *Bull. Am. Math. Soc.* 60(6): 503–516.
8 Brown G, Dell R, Wood R (1997) Optimization and persistence. *Interfaces* 27: 15–37.
9 General Services Administration. (2017) Rules and Policies – Protecting PII – Privacy Act. Available at http://www.gsa.gov/portal/content/104256.
10 Department of Homeland Security. (2017) The Protected Critical Infrastructure Information Management System (PCIIMS). Available at https://www.dhs.gov/pcii-management-system-pciims.
11 Office Of Naval Research. (2017) Human Subject Research Documentation Requirements. Available at https://www.onr.navy.mil/About-ONR/compliance-protections/Research-Protections/Human-Subject-Research.aspx.
12 Federal Bureau of Investigation. (2017) Security Clearances for Law Enforcement. Available at https://www.fbi.gov/resources/law-enforcement.
13 United States Patent and Trademark Office. (2017) Basic Facts Breakdown – Trademarks, Patents and Copyrights. Available at https://www.uspto.gov/trademarks-getting-started/trademark-basics/trademark-patent-or-copyright.
14 Bradley G (1986) Optimization of capital portfolios. *Proc. Natl. Commun. Forum* 40(1): 11–17.
15 Geoffrion A (1986) Capital portfolio optimization: a managerial overview. *Proc. Natl. Commun. Forum* 40(1): 6–10.

7

Machine Learning

Samuel H. Huddleston and Gerald G. Brown

Operations Research Department, Naval Postgraduate School, Monterey, CA, USA

7.1 Introduction

This chapter provides an overview of *machine learning*, that is, using automated algorithms to discover descriptive, predictive, and prescriptive results from data. Machine learning is closely related to *artificial intelligence* (in computer science) and *statistical learning* (in statistics), incorporating algorithms and procedures developed in both fields. Due to this amalgam of multiple fields, the lexicon is very broad, with many synonyms used to describe the same algorithms, procedures, and model parameters. Throughout this chapter, we seek to capture the lexicon employed by the various tribes within this emerging discipline and provide a map to terms formalized in the operations research and statistical literature.

The defining characteristic of machine learning is the focus on using algorithmic methods to improve descriptive, predictive, and prescriptive performance in real-world contexts. An older, but perhaps more accurate, synonym for this approach from the statistical literature is *algorithmic modeling* [1]. This algorithmic approach to problem solving often entails sacrificing the interpretability of the resulting models. Therefore, machine learning is best applied when this trade-off makes business sense, but is not appropriate for situations such as public policy decision-making, where the requirement to explain how one is making decisions about public resources is often essential.

There are many nonparametric and heuristic approaches employed in machine learning for which full statistical and theoretic interpretations have not yet been developed. These methods are often widely employed for years with a high degree of success in practice before formal statistical explanations for their performance are developed and validated. Machine learning practitioners balance the risk of using these emerging techniques by conducting a formal evaluation (competition) of the many available algorithms that might apply to a

INFORMS Analytics Body of Knowledge, First Edition. Edited by James J. Cochran.

problem in a way designed to identify and validate the performance of the best algorithm for the problem in a real-world context. Much of this chapter is devoted to summarizing this procedure and its underlying motivation.

Machine learning is a very large and rapidly expanding field, and a full description of the discipline would require much more space than is available in this chapter. Here we describe the major paradigms of machine learning, review the modeling in very general terms, and briefly describe some of the most popular machine learning algorithms. The reader is encouraged to conduct further investigation of the many terms and ideas summarized here.

Throughout this chapter, we have italicized terms for which Wikipedia provides more in-depth descriptions and frequently provide links to these pages in footnotes. For more in-depth technical treatments of most of the discussed algorithms, we recommend *An Introduction to Statistical Learning* [2] and *The Elements of Statistical Learning* [3], which are freely available online.[1] Additional references for specific techniques and applications are provided in the following section.

7.2 Supervised, Unsupervised, and Reinforcement Learning

There are three broad classes of machine learning problems, with classes defined based on the data provided for algorithmically "training" the models: *supervised learning* problems,[2] *unsupervised learning* problems,[3] and *reinforcement learning* problems.[4] These broad classes are not mutually exclusive, and learning methods designed for different classes of machine learning problems are often combined to solve real-world analytics challenges.

Supervised learning applies when the data set used to train models includes a labeled response variable. The goal in supervised learning is to use available observations to predict the values of the response variable associated with new observations (where the responses will not be known beforehand). Sometimes the term *semisupervised learning* is used to distinguish situations where the labeled response is available for some, but not all, observations in the training data set. Supervised learning problems are further subcategorized into *regression* problems and *classification* problems.[2]

Regression problems have response variables that take quantitative values. One classic example of a regression problem is predicting the future sale price of

1 Books published online at http://www-bcf.usc.edu/~gareth/ISL/ and https://web.stanford.edu/~hastie/ElemStatLearn/.
2 https://en.wikipedia.org/wiki/Supervised_learning
3 https://en.wikipedia.org/wiki/Unsupervised_learning
4 https://en.wikipedia.org/wiki/Reinforcement_learning

a home based on its age, zip code, square footage, and so on. Another example would be to predict the sales volume of a product in the next quarter, given historical sales records (an example of a time series forecast). The response variable in statistical regression is often referred to as the dependent variable, and the features used to predict the response variable are referred to as independent or predictor variables. Note that the term regression describes both statistical regression (an algorithm) and a subclass of supervised learning problems. However, statistical regression is not the only machine learning algorithm available to address regression problems.

Classification problems have a response variable that is categorical and unordered. The goal in a classification problem is to use the data at hand to predict the class or category for new observations (or in some cases, those for the next time period) where the class or category is not known. One classic example of a classification problem is predicting the probability of someone having (or developing) a particular disease based on risk factors (age, body mass index, cholesterol, family disease history, etc.). This is an example of binary or binomial classification. While the output is a probability (a continuous numerical quantity), this is a classification rather than a regression because the goal is to predict the probability of being in the "disease class." Multiclass or multinomial classification problems classify observations into three or more classes or groups–for example, predicting which of three products a customer is most likely to buy (and not buying any product could make up a fourth class). Most classification methods will provide a predicted probability estimate for each class for each observation (with ties possible).

Unsupervised learning seeks to identify latent (underlying or hidden) structures in a data set. Therefore, unsupervised learning requires no labeled response variable. There can be many latent structures in a data set, and there are several unsupervised learning tasks that are quite common. *Density methods* (from statistics) construct an estimate for an underlying probability distribution for the data based on the observations in the data set.[5] *Clustering methods* partition a data set into groups or clusters where each cluster (group) is a set of observations in the data that are more similar to each other in some way than they are to the other observations.[6] *Dimension reduction* seeks to provide a more compact (i.e., lower dimensional) representation of the data in a data set.[7] Dimension reduction is frequently used to provide a more compact feature set (i.e., the set of independent variables in statistical regression, also often described as the predictors in machine learning) to use for supervised learning.

Reinforcement learning (Figure 7.1) addresses a very broad category of problems in which scoring or response functions are applied iteratively over

5 https://en.wikipedia.org/wiki/Kernel_density_estimation
6 https://en.wikipedia.org/wiki/Cluster_analysis
7 https://en.wikipedia.org/wiki/Dimensionality_reduction

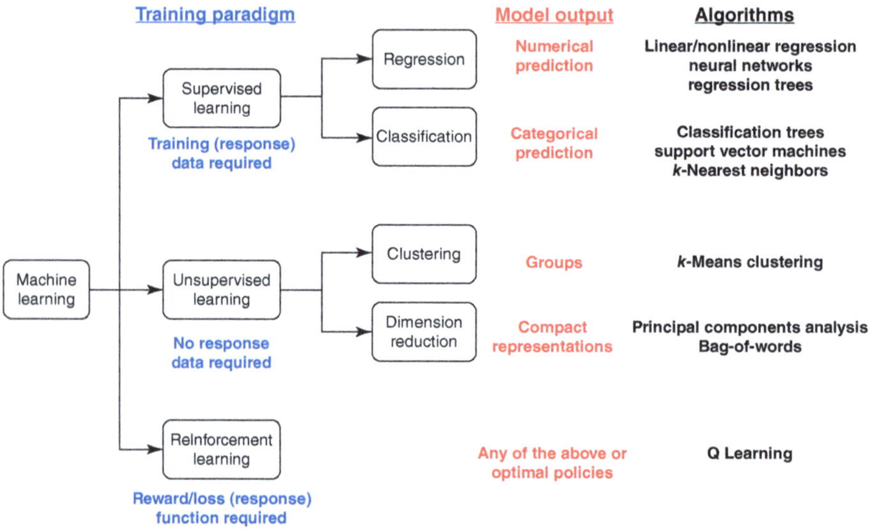

Figure 7.1 Overview of machine learning paradigms: supervised, unsupervised, and reinforcement learning. This illustration provides a summary of the data needed to train the model, the format of the resulting model output, and a list of algorithms often used for this class of problem. Algorithms listed are addressed in this chapter, except Q Learning, which is omitted due to space limitations.

time to improve the performance of models or policies. Often, reinforcement learning is applied in situations in which there is no labeled response data available now, but over time, feedback about the model outputs will provide the opportunity to improve performance. For example, one might automatically cluster documents together using topic modeling (an unsupervised approach) and then ask humans to randomly inspect the clusters made, providing feedback on which clusters are well grouped (winners) and which clusters are less useful (losers). This feedback (scoring function) provides the opportunity to apply supervised learning and then repeat the sampling and feedback (iteratively applying the scoring function). Another application is the highly generalizable multiarmed bandit problem in which the goal is to determine which slot machines (in a row of slot machines) should be played in what order to maximize profit (or minimize loss). The win–lose results (scoring function) provide the needed feedback for developing an optimal policy over time.

Artificial Intelligence (AI) systems are often trained using reinforcement learning to iteratively improve the ability of the AI system to accomplish the desired tasks (e.g., find cats in Internet images or avoid collisions while driving). One classic example of reinforcement learning is the training of an AI system (machine) to play strategy games such as Chess or Go. At the onset, the machine will make random or uninformed moves, but games or points won and lost will provide a scoring function that will allow for algorithmic evaluation of policies or decisions made. Reinforcement learning requires balancing exploitation of already-known information and exploration of new policies. In the case of the recent, and largely unexpected, success in training an AI system to defeat a Go grandmaster, the AI system competed against itself for millions of games to provide the opportunity to explore the vast decision-making space.[8]

Machine learning is an emerging field, with new algorithms regularly developed and fielded. Specific machine learning techniques are usually designed to address one of the three types of machine learning problems introduced here, but in practice several methods are often combined for real-world application. As a general rule, unsupervised learning methods are designed to be *descriptive*, supervised learning methods are designed to be *predictive*, and reinforcement learning methods are designed to be *prescriptive*. Because reinforcement learning is a more advanced machine learning topic, it will not be further expanded upon in this short overview chapter.

7.3 Model Development, Selection, and Deployment for Supervised Learning

While machine learning relies heavily on automated algorithms for model development, this does not mean that machine learning is completely automated. Rather, the development of machine learning models is an involved and iterative process that requires considerable engagement and judgment on the part of an analyst. The first step, while obvious, is often overlooked: understand the intended use for the model in the context of its "business use case" (see Chapter 6 for further discussion). Only then should you proceed with developing a model designed to meet this business need.

7.3.1 Goals and Guiding Principles in Machine Learning

In contrast to traditional approaches to modeling, machine learning does not strive to produce specific model with static coefficients. Machine learning strives to produce an algorithmic procedure with demonstrated predictive power in the business context in which it will be applied. In fact, in many machine learning applications, such as time series forecasting, we expect that the coefficients of the model(s) will change with every time period in which we apply the algorithm because we will refit the model with any new data that have arrived in the interim and use this new information to improve our forecast for the next time period. The goal is to have a useful *modeling algorithm* (i.e., sequence of steps that

8 More information available at https://en.wikipedia.org/wiki/AlphaGo and https://www.wired.com/2017/05/revamped-alphago-wins-first-game-chinese-go-grandmaster/

produce a model) to solve the real-word problem. Readers are referred to Ref. [1] for an in-depth discussion about the difference in goals between algorithmic and more traditional statistical modeling.

This difference in modeling focus yields the following list of guiding principles for algorithmic modeling, some of which are based on Breiman's influential article:

- There are likely to be many models with demonstrated predictive power.
- Analysts should investigate and compete as many models as possible.
- Analysts should measure the performance of models on out-of-sample test data sets using a procedure that mimics the real-world situation in which the model will be applied.
- Predictive accuracy on out-of-sample test data, not goodness of fit on training data, is the primary criterion for how good a model is.
- Predictive power is not the only criteria upon which model selection is made; we routinely also consider model interpretability, speed, deployability, and parsimony.

7.3.2 Algorithmic Modeling Overview

The following steps describe a standard workflow for developing an algorithmic solution for prediction in the context of a supervised learning problem:

- Data acquisition and cleaning
- Feature engineering and scaling
- Model fitting (training) and feature selection
- Model selection
- Model performance assessment
- Model implementation

7.3.3 Data Acquisition and Cleaning

Algorithmic modeling begins with data acquisition and cleaning. This is often the most time-consuming step. An oft-cited rule of thumb is that this step will consume as much as 80% of an analyst's time (with the punch line that subsequent modeling consumes the other 50%). This step requires considerable domain expertise within the area of application to ensure that data for any and all features with predictive power are collected and included in the analysis. For this reason, analysts should work closely with domain experts and practitioners in the application area to identify all possible data sources that are relevant to the problem.

Real-world data are often messy, with missing values, varied formatting, duplicated records, and so on. Analysts can expect to spend considerable time reformatting and standardizing any data acquired and will often require domain expert advice for dealing with data problems. For most supervised learning problems, data acquisition and cleaning will end with the data stored in a flat table where the rows of the table represent observations and the columns represent the features of those observations. For supervised learning problems, at least one column must represent the response variable or class (the item that will be predicted). This response variable is often referred to as the dependent variable in statistical regression and as the target variable in artificial intelligence.

7.3.4 Feature Engineering

The next step is *feature engineering*.[9] A feature in machine learning is an attribute describing an observation that all observations share (although the values of the attribute vary across observations). These features are also variously described as independent variables, predictor variables, or explanatory variables. The premise is that we can use the differences in attributes between observations to better classify or predict the response (target variable) for new observations. Feature engineering uses both algorithms and domain knowledge to generate features that provide predictive power for a machine learning algorithm. Because feature engineering usually has more impact on predictive performance than the algorithmic procedure you decide to implement, it is widely considered the most important step in generating predictive models. Feature engineering is often divided in the lexicon between *feature construction* (human-conducted) and *feature learning* (machine-automated) tasks.[9]

While feature engineering tasks overlap with data acquisition and cleaning, feature engineering also often involves the construction of new data features from existing ones. One simple example of feature engineering is constructing interaction variables by multiplying features with each other: Consider the case where the length (l), width (w), and height (h) of a product are not correlated with shipping cost (sc), but the volume estimated by the multiplication of these variables is highly predictive, that is,

$$al + bw + ch \neq sc \quad but \quad dlwh \approx sc$$

Another simple example of feature engineering might be to raise a feature to a power (e.g., square it) to model a nonlinear (e.g., quadratic) relationship between the predictor variable and the response (e.g., see the regression example provided in Chapter 6).

One example of more involved feature engineering is the process often used in crime prediction [4]. Each grid point in a city is tagged with an indication of whether or not a crime has occurred (this is usually done separately for each type of crime), and this binary variable acts as the response variable. Demographic information from sources such as the census (each point is given all of the

9 https://en.wikipedia.org/wiki/Feature_engineering

demographic features of its census block) is then added to each of the points. Finally, distances between each point and geographic features that either are associated with crime (called crime attractors or generators) or repel crime (called crime detractors) are calculated. The premise is that crimes such as assault and battery are more likely to occur in areas near crime attractors such as bar districts and are less likely to occur near geographic features such as police stations. Feature engineering would include the identification of any and all demographic and geographic features that might have an influence on the crime type an analyst is trying to predict. These feature data sets routinely involve hundreds of *potential* predictor variables. Huddleston and Brown [5] provide a detailed example of feature engineering conducted to predict gang crime.

Another common task in feature engineering is data *scaling*. While some machine learning algorithms are scale-invariant, the performance of others can be adversely affected when the scale of the predictor (i.e., independent) variables varies a great deal. Feature scaling, also called data normalization, involves transforming all features to a standard scale such as unit length.

7.3.5 Modeling Overview

Once an appropriate feature set has been constructed, analysts can begin algorithmically "fitting" predictive models. As already suggested, one of the principles of algorithmic modeling is that *analysts should investigate and compete as many models as possible.* This has two implications. First, for any particular modeling method (i.e., linear regression), there are many models that can be built from the same feature set by using different combinations of predictive features. Second, we should consider many different modeling methods (algorithms) for the task (i.e., *linear regression, regularized regression, K-nearest neighbors*, etc.), so we now need to compare the performance of these different modeling methods against each other.

This involves a multistep "contest" for choosing a model from the many (perhaps thousands when considering all the different combinations of features) available models:

- *Model Fitting:* We fit and perform *feature selection* and parameter optimization for each of the modeling methods (algorithms) under consideration on a *training data set.* The output of this step is a list of "best of breed" models that we will compete against each other in the next step.
- *Model (Algorithm) Selection:* We compete the "best of breed" models against each other on an out-of-sample *validation data set.* The best performing algorithm (on a range of criteria) is chosen for implementation.
- *Model Performance Assessment:* We assess the performance of our selected approach on an out-of-sample *test data set.* This gives us an unbiased estimate for how well the algorithm will perform in practice.

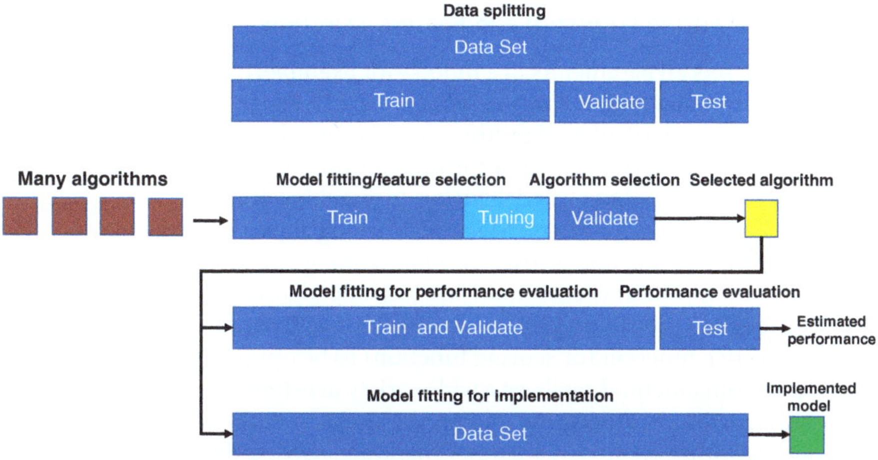

Figure 7.2 Illustration of the machine learning process as a successive algorithmic competition. The data set is partitioned into training, validation, and test data sets. Many models are trained (and tuned if necessary) on the training data set and a competition is held to predict the observations in the validation data set. The winning algorithm is trained on the combined training and validation data sets and then predictive performance on the test data set provides an estimate for real-world performance. The selected algorithm is applied to the full data set (so that the model deployed leverages all available information) and deployed.

- *Model (Algorithm) Implementation:* The selected algorithm is applied to the full data set (i.e., training, validation, and test data sets, now combined) so that the model we deploy uses all available information.

Figure 7.2 provides an illustration for of how data are partitioned into training, validation, and test data sets and then used in the context of this "contest."[10] Note that it is absolutely critical to ensure that the validation and test data sets are representative samples of the data set so that they provide realistic performance evaluation. As a general rule, the training data set is also designed to be a representative data set, but there are some exceptions. One example concerns model fitting for classification problems where the percentage of "positive" cases is very small (e.g., rare disease prediction, or flight screening for weapons and explosives). In these cases, one might build a training data set with a much higher percentage of "positive" cases than one would expect to find in practice in order to leverage classification algorithms that struggle to fit models when the number of positive versus negative cases is highly unbalanced. Some algorithms, such as artificial neural networks, also require the use of a tuning data set; in this case, the tuning data set should be subset from the training data. The remainder of this section elaborates on each of the steps already described.

10 https://en.wikipedia.org/wiki/Training,_test,_and_validation_sets

7.3.6 Model Fitting (Training) and Feature Selection

Model fitting sets all parameters of a model–for example, the model coefficients in a linear regression model. *Feature selection* selects the best set (combination) of features for a particular algorithm for a particular problem. These two processes are necessarily conflated for model fitting in machine learning.

One way to think about model fitting in machine learning is as a highly nonlinear optimization problem. Machine learning algorithms provide a working framework (via the model structure) for defining and solving this optimization problem. Specifically, each machine learning algorithm provides a model structure, defined procedures for feature selection and parameter optimization, and an objective function (or scoring function) to be optimized. Any particular machine learning method can have a wide variety of extensions and variations to the basic algorithmic procedure, and so the decision space to be explored for even one algorithm is often very large. Therefore, solving this problem is often very computationally expensive. The "learning" activity performed in machine learning is the highly iterative process of having a machine find the "optimal" solution (or perhaps a near-optimal solution) to the problem it has been given.

One simple example of how feature selection might be conducted for linear regression is known as *stepwise feature selection.*[11] In stepwise feature selection, features are iteratively added (forward stepwise procedure) or dropped (backward stepwise procedure) based on some predefined criteria that measure model quality. After a feature is added or dropped, the model parameters are fit and model goodness of fit is assessed using statistics such as the *Aikake Information Criterion (* AIC*)* or *Bayesian Information Criterion (* BIC*).*[12] These statistics are used as the objective function in the optimization, and we choose the feature set (and model) that maximizes the chosen objective function (in this case an information criterion).

However, in fitting the model for a given feature set, you also need to estimate all of the parameters for that model. In linear regression, we can use closed form linear algebra to solve for and define the parameters of a model, but many machine learning algorithms require a full numeric optimization procedure such as *stochastic gradient descent* to fit the parameters at this stage.[13] Thus, fitting machine learning models is a highly iterative and computationally intensive activity.

Feature selection and parameter estimation are performed iteratively on a training data set for each machine learning algorithm under consideration (linear regression, K-nearest neighbors, regularized regression). Thus, if you are comparing three different algorithms for a regression problem, you would

---

11 https://en.wikipedia.org/wiki/Feature_selection
12 https://en.wikipedia.org/wiki/Information_criterion
13 https://en.wikipedia.org/wiki/Stochastic_gradient_descent

conduct a highly iterative model fitting and feature selection three times, each time using the model structure, objective function, and optimization procedures defined by a particular machine learning algorithm. The result of the model fitting stage is a set of "best of breed" models that will be entered into a competition in the model selection stage of the competition.

7.3.7 Model (Algorithm) Selection

In this step, the goal is to decide which model fitting algorithm provides the best predictive performance in an out-of-sample (i.e., blind) test. We apply the "best of breed" models developed in the previous step to predict the response variable in the validation data set (the data set we have set aside as the "blind" for this purpose). Performance is measured by comparing the predicted values from each modeling algorithm with the known response variable. It is critically important to use the correct performance function to choose the winning model (and there are many of them). The performance function should measure performance in a way that is meaningful for the true application of the data. This performance measure often differs from the statistic used as the objective function for model fitting.

If one model or algorithm is better than the others in predictive performance (i.e., more accurately classifies or predicts the response variable), this does not imply that it will be the model or algorithm selected for implementation. Predictive performance describes the "payoff" for employing a model, but there may also be a figurative (for example, "political") or literal (in the case of computational resources required) cost. Some algorithms such as regression or *decision trees* are highly interpretable (meaning one can describe how the model arrives at its predicted value).[14] Others, such as *artificial neural networks*, function as "black box" oracles.[15] In many public policy applications, for example, political requirements may dictate that only interpretable models are employed so that the public can be reasonably informed about how decisions are being made.

Similarly, there may be large differences in the computational resources needed to regularly fit and deploy a model; some algorithms can be much more expensive than others (since much of this computation is performed on cloud architecture, it can cost substantially more money to train and deploy one algorithm versus another). Some models can be trained quickly, while others may require time to execute. In short, just because a model predicts well does not mean that it is the "best" for every application.

---

14 https://en.wikipedia.org/wiki/Decision_tree_learning
15 https://en.wikipedia.org/wiki/Artificial_neural_network

The final output from this step is an algorithm that provides the best performance on the range of criteria considered. This could include the decision to use many machine learning models together in a federation or ensemble.

7.3.8 Model Performance Assessment

The final step before model implementation is to assess the predictive performance of the selected algorithm(s) on an out-of-sample test data set. The selected algorithm is applied to the combined train and validation data set to build a new predictive model. Then, this model is used to predict outcomes on the out-of-sample test data set. This testing provides estimated performance for how the modeling procedure will perform when fielded. Note that predictive performance for an algorithm on the test data set is usually worse than that observed on the training data set due to some level of overfitting (a more detailed discussion of this issue will be provided later in this chapter), but test data set performance should be similar to that seen on the validation data set. A substantial drop in predictive performance between the validation and test data sets may indicate a problem with the model or algorithm.

7.3.9 Model Implementation

The final modeling step consists of combining all of the data available and the selected algorithm(s) to fit the model that will be deployed. Note that the model coefficients (e.g., for a linear regression model) may change from those that were fit in the training, model selection, and model performance evaluation steps. This is because we are seeking the algorithmic procedure that is best able to use the data available to predict future observations. The model we deploy is an implementation of the winning algorithmic approach and often a completely new model. In applications where there is a plethora of data available for training, it may not be necessary to train a new model by combining the training, validation, and test data sets together, because adding the information from the test data set will have little impact on the resulting model. In this case, the model used for performance assessment is directly fielded.

Note that model implementation often requires much more than simply identifying the model for implementation. Model implementation often requires the development of interactive visualizations, automated reports, or the identification of procedures so that the model can be employed for its intended purpose. In addition, it is wise to set up monitoring procedures so that the "online" performance of the model can be evaluated as it is used. The model performance evaluation step above provides an estimate of how well the model should perform in practice. If model performance begins to diverge from the expected performance, this can indicate that something has changed to diminish the effectiveness of the current model.

7.4 Model Fitting, Model Error, and the Bias-Variance Trade-Off

We seek to minimize predictive error–the difference between predicted and observed values–when applied to new cases. This section describes the procedures used to minimize predictive modeling error in practice.

7.4.1 Components of (Regression) Model Error

In the context of regression, model error is based on the difference between predicted values and observed values. Figure 7.3 shows three different models fit to some sample data. In this figure, the model error is the differences between the observed responses and the fitted line (the predicted values). The points in black represent data known at the time the model is fit (i.e., the training data set), while the points in gray depict points that "arrive" after the model is fit (i.e., during the model implementation phase). It is clear that the model in the middle provides the best overall performance because it minimizes the error of the gray points (i.e., it minimizes the prediction error in practice). The models depicted in the left and right panels do not fit as well. We say that the model depicted in the left panel is *biased* (or *underfit*), while the model depicted in the right panel is *overfitted* and suffers from high *variance*.[16]

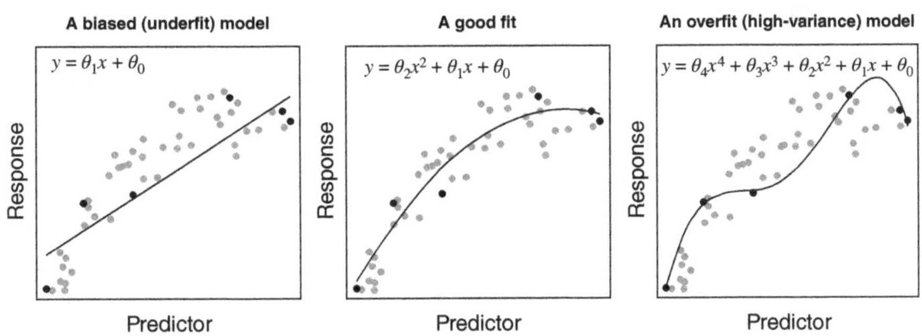

Figure 7.3 An illustration of model under- and overfitting. The model in the left panel is biased (underfit) because it misrepresents the relationship between the predictor and the response variable. The model in the middle provides a good model fit because it represents the true quadratic relationship between the predictor and the response. The model in the right panel is overfitted to the training data because it used transformations of the predictor variable to minimize the error in the training data set, resulting in a model that does not generalize.

16 https://en.wikipedia.org/wiki/Bias–variance_tradeoff

Regression model error consists of bias, variance, and noise components according to the following formula:

$$\text{Model Error} = \text{Bias}^2 + \text{Variance} + \text{Noise}$$

The noise component of error describes the inherent uncertainty and randomness of the data and is also called *irreducible error,* because it is never possible to reduce modeling error below the threshold of the inherent randomness that exists in the real world for that data.

The bias component of error is due to incorrectly representing the relationship between predictor variables and the response variable. For example, in Figure 7.3, the left panel depicts a linear model fitted to nonlinear data, producing a biased estimate for many of the data points. The variance component of error is due to oversensitivity to noise or randomness represented in the data used to train the model. The model depicted in the right panel suffers from high variance. It is easy to observe that this model provides a superb fit to the training data (depicted in black) but poor performance on data it has not seen before. This occurs because the model is too complex–in trying to minimize the bias observed in the training data, the model now represents noise (inherent randomness in the data) observed in the training data with high-order terms that do not reflect the true underlying relationship. This model has adapted to idiosyncrasies in the training data and so is overfitted (too specific) to the training data.

The variance effect of model overfitting is best explained by illustrating the effect of fitting a new model on a different set of training points, as shown in Figure 7.4 that illustrates how overfitting produces high-variance predictions; different models, and as a result different predictions, can be generated by the same algorithm. This figure updates Figure 7.3 to illustrate a new model

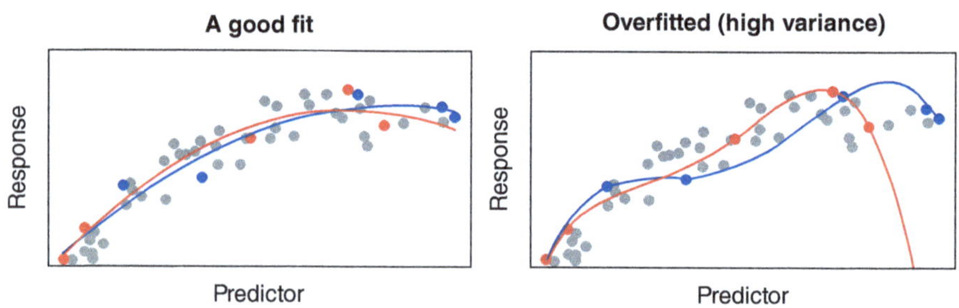

Figure 7.4 An illustration of the variance effect when using different training data sets. Using a different training data set has little effect on the fitted model (and therefore the predictions) in the panel on the left. In the panel on the right, the model fit to the data (and therefore the predictions) is very different when a different training data set is used. The models (and therefore the predictions) in the panel on the right exhibit greater variance than those on the left.

produced by using the same feature selection and algorithm but with the substitution of a different training data set. In the panel on the left, you can see that the effect of a different set of points for the training data is very small. In the panel on the right, in both cases the "fit" to the training points is better than in the left panel, but the prediction for the out-of-sample points (plotted in gray) is much worse. This effect is produced by not properly penalizing the addition of polynomial terms of the predictor variable in the model.

7.4.2 Model Fitting: Balancing Bias and Variance

Machine learning seeks to find the optimal trade-off between bias and variance as depicted in Figure 7.5, which provides a simple sketch depicting this trade-off. Note that it isn't possible to completely eliminate error, but there exists a theoretical level of model complexity that balances bias and variance in a way that minimizes the total prediction error. Brighton and Gigerenzer [6] provide an excellent in-depth discussion of this trade-off, while Huddleston et al. [7] provide a real-world example of this effect in the development of models for forecasting crime in a major U.S. city.

The modeling procedure illustrated in Figure 7.2, which provides an overview of the algorithmic modeling "competition," is designed to help identify the algorithm for model fitting that best identifies this "sweet spot" in the bias variance trade-off. By splitting data sets into the training data set to fit each model and then a validation data set to evaluate each model, we are able to see which modeling procedures best "generalize." Choosing our models in this way provides us a picture of modeling performance similar to that presented in Figure 7.3.

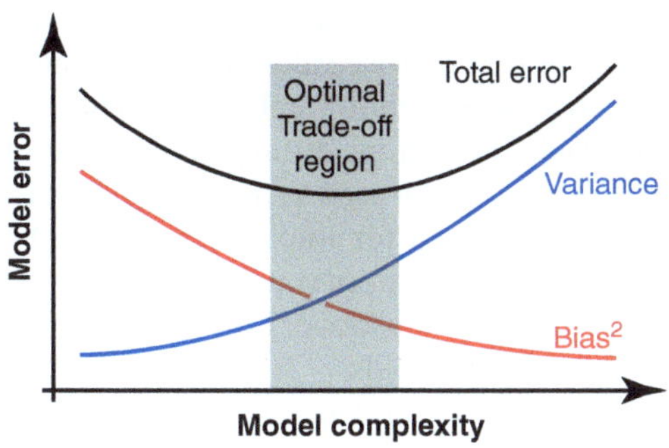

Figure 7.5 The bias–variance trade-off illustrated. This sketch illustrates how adding model complexity (by adding features etc.) reduces model bias (on the training data set) but increases variance (i.e., reduces the generalization of the model fit). There exists a theoretical "sweet spot" that balances bias and variance in a way that reduces overall error. The machine learning process in Figure 7.2 and machine learning techniques such as cross-validation and regularization are designed to find the optimal amount of complexity–the amount that reduces overall prediction error when the model is deployed.

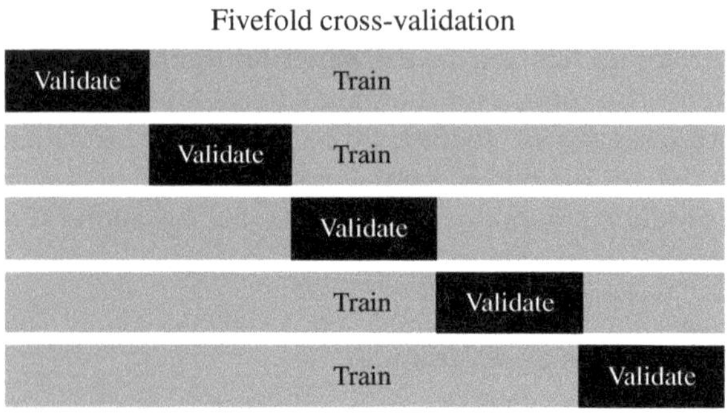

Figure 7.6 Illustration of *k*-fold cross-validation. Data observations are randomly ordered and then split into *k* sections (in this case, five sections), with each section used as a validation data set in sequence. Each model-fitting algorithm is applied *k* times, with 1/*k* of the data held out as the validation data set and the rest used as the training data set. This provides *k* independent estimates of performance for each algorithm competed. These *k* estimates are combined (usually averaged) to provide an overall estimate for each algorithm's performance.

Using a test data set gives us a "second look" at how the model is likely to perform on data that was not used to either train the model or choose the model. This protects against the situation in which the specific sample chosen for the validation set favors a particular model by random chance. The final performance observed on the test data set during performance evaluation should be very close to that observed on the validation data set. When this is not the case, it is an indication that something has gone wrong, and further investigation is needed.

There are two additional techniques commonly applied in the model-fitting stage to address the bias-variance trade-off: *k-fold cross-validation* and *regularization*. *k*-Fold cross-validation extends the practice of splitting data into training, validation, and test data sets by changing the way training and validation data sets are built.[17] To perform *k*-fold cross-validation, you repeatedly (i.e., *k* times) perform a sampling procedure (without replacement) to build a validation data set out of the data put aside for training, tuning, and validation (i.e., all of the data that is not put "into the vault" to be used as the test data set for performance evaluation on the algorithm selected for implementation). One practical way to do this is to randomly order the records in your data set and then evenly split your data into the *k* folds. Figure 7.6 illustrates how this sampling procedure structures the data into $k = 5$ different folds.

Models are fit on each fold's training data set and model performance is estimated for that fold based on predictions made on that fold's validation data set. In this fivefold case, each of the algorithms applied would have five different validation performance estimates, which are then combined (usually averaged)

17 https://en.wikipedia.org/wiki/Cross-validation_(statistics)

to provide an overall cross-validation performance summary for that algorithm. The winning algorithm provides the best comprehensive performance over all k folds. This procedure improves on the basic modeling procedure illustrated in Figure 7.2, but it requires considerably more computation and so is not always feasible.

Regularization attempts to prevent overfitting by imposing a cost (or a penalty) on the complexity of a model.[18] Returning to the models compared in Figure 7.3, you can see that as one progresses from the left panel toward the right panel, the model complexity (i.e., number of predictor variables formed by various transformations on the feature x) increases. At the time when the model is fit, only the information provided in the black dots is known and so a good model-fitting algorithm will try to minimize the error between the fitted line and the known points.

In this case, the model error for this regression problem is represented using a statistic such as *sum of squared errors* (SSE), where an error as depicted in Figure 7.3 is the vertical distance between a known black dot and the fitted line. In this case, the SSE serves as the objective function for the model selection and optimization procedure. Without some kind of penalty on the complexity of the model, any algorithm designed to reduce the error between fitted line and the observed training data will continue to reward the introduction of increasingly complex coefficients (e.g., higher-order transformations of the predictor variables as in the right panel of Figure 7.3) into the model in order to reduce the distance between fitted line and each known point. The result is a badly overfitted model. Regularization techniques improve upon basic error measures such as SSE by incorporating penalty functions for adding complexity that help to balance bias and variance more appropriately.

While there are many methods for regularization, the most commonly applied methods use an *information criterion* such as the AIC or the BIC as the target function for optimizing model fit.[19] All approaches for regularization operate in the same manner: They modify the objective function for optimizing model fit by including both some measure of goodness of fit (such as SSE) and a component that penalizes either the addition of more features (and parameters) into the model or the magnitude of the coefficients fit for those parameters.

7.5 Predictive Performance Evaluation

The objective functions used for model fitting and feature selection (e.g., AIC and BIC) will often differ from those used to conduct model selection (on the validation data set) and evaluate model performance (on the test data set). The

18 https://en.wikipedia.org/wiki/Regularization_(mathematics)
19 https://en.wikipedia.org/wiki/Information_criterion

statistics used for model fitting and feature selection are designed to find an optimal trade-off point in the bias-variance trade-off space as already discussed. They perform regularization.

The statistics used for final model selection and performance evaluation provide meaningful information about the real-world predictive performance of the fitted models when applied in a "business" context. It is critical to test models and algorithms in the same way the model or algorithm will be used in practice. Considerations for model performance evaluation for regression problems, classification problems, and problems with time dependency are briefly discussed in the following sections.

In addition, analysts must understand that any measure of performance calculated on the training data set is not a valid estimate of *predictive* performance; performance measures calculated in this way usually (and often significantly) overestimate true predictive performance. Once again, Figure 7.3 provides an illustration of this phenomenon. The training error (the difference between the black points and the fitted line) for the model in the right panel is very small, while the true predictive performance (the difference between the gray points and the fitted line) is much worse. Model selection and performance evaluation always require the use of out-of-sample data sets (i.e., data not used to fit the model). Additional considerations for model performance evaluation for regression problems, classification problems, and problems with time dependency are briefly discussed in the following sections.

7.5.1 Regression Performance Evaluation

Regression problems have response data that are numeric, either continuous or integer valued, and therefore predictive performance for regression problems is based on the numerical difference between a real-world observed response (in the validation or test set) and the predicted value for that observation. The numerical difference between model prediction and observed value (in the validation or test set) is known as a *model residual* or a *model error.* The term residual is preferred as the difference between an observation and a predicted value could be due to noise rather than a "modeling error." However, the term model error is widely used in practice, to include in the name of statistics that are often used to summarize model performance such as the *mean squared error* (MSE):

$$\text{MSE} = \frac{1}{n}\sum_{i=1}^{n}(\text{Observed}_i - \text{Predicted}_i)^2;$$

root mean squared error (RMSE):

$$\text{RMSE} = \sqrt{\text{MSE}};$$

or the *mean absolute deviation* (MAD), also known as the *mean absolute error* (MAE):

$$\text{MAD} = \frac{1}{n}\sum_{i=1}^{n} |\text{Observed}_i - \text{Predicted}_i|.$$

Note that these measures of performance all use the residuals, but each measure may favor a different model. MSE and RMSE penalize large residuals more heavily than does MAD. If the business use case for a model is a situation in which a few large prediction errors would be devastating, RMSE or MSE should be chosen over MAD for model selection and performance evaluation.

However, the best practice is to convert statistical measures of error (such as regression residuals) into real-world costs, profits, or other measures of real-world performance and then evaluate models in the context of real-world business effect. The model that you implement should be the model that provides the best business case, not the model that provides the best performance of a statistical measure. Structuring model performance in real-world terms facilitates model adoption and implementation because it contextualizes the models in a way that allows for informed decision-making about the effect of employing them.

7.5.2 Classification Performance Evaluation

The goal in a classification problem is to accurately predict the category or class of new observations given their observable features. As previously discussed, a classic example of a classification problem is predicting whether or not a person has (or will develop) a disease based on medical diagnostics. Another is predicting future loan defaults based on credit history. These are binary classification problems. Binary classifiers will usually return the probability of the "positive" case (i.e., the probability that the person develops the disease or defaults on the loan), which is usually denoted with the numerical value 1 for the response variable. A prediction of 51% would imply that the classifier believes it is more likely than not that the observation will present the positive (1) class (i.e., disease or default occurs). However, the threshold used to assert a classification can be varied and the threshold used for binary classifiers in practice can and often does vary from 50%, especially for problems with very low numbers of "positives" such as drug tests and fraud detection. If one infant in 250,000 develops isovaleric academia, a model for predicting this genetic disorder that is based on a 50% threshold would perform far worse than simply guessing that no child would develop the genetic disorder (which would be correct in 249,999 of every 250,000 infants or in 99.9996% of infants).

Binary classification performance measures are based around the *confusion matrix*, also known as a *truth table*, which is a table that records the counts of

		Actual class	
		Positive (1)	Negative (0)
Predicted class	Positive (1)	True positives	False positives
	Negative (0)	False negatives	True negatives

Figure 7.7 Confusion matrix (truth table) for binary classification. (Data from https://en.wikipedia.org/wiki/Confusion_matrix, public domain.)

occurrence for each of the four outcomes for prediction in a classification problem at a given threshold as shown in Figure 7.7.[20]

As shown in Figure 7.7, there are two ways to make a correct classification in a binary classification problem (denoted in green) and two ways to make an error (denoted in red). False positive errors are also known as *Type I Errors,* and false negative errors are also known as *Type II Errors* (see Chapter 6 for more discussion).[21] There are a wide variety of performance statistics derived from the confusion matrix, but the two most widely used are *sensitivity* and *specificity.* Sensitivity measures how well a classifier does in predicting the actual presence of disease (or predicting default) when it actually occurs. Classifier sensitivity is also known as *recall, true positive rate (*TPR*),* or *detection probability* and is calculated from the confusion matrix for a given classification threshold as follows:

$$\text{Sensitivity} = \text{TPR} = \frac{\text{True positives}}{\text{True positives} + \text{false negatives}}.$$

Classifier specificity measures how well a classifier does in predicting cases where the disease or default does actually not occur. Specificity is also known as the *true negative rate* (TNR) and is calculated as

$$\text{Specificity} = \text{TNR} = \frac{\text{True negatives}}{\text{True negatives} + \text{false positives}}.$$

Good classifiers perform well on both measures (high scores are desirable in both cases). However, as a classifier's threshold changes, the performance statistics of the classifier change as well. This creates a trade-off space that can be explored for a given classifier.

Receiver operating characteristics (ROC*)* curves are another tool often used to evaluate the performance of binary classifiers because they allow investigation of sensitivity and specificity over the full range of possible thresholds. ROC curves plot a classifier's TPR (i.e., sensitivity or recall) against its *false positive rate* (FPR) as the classification threshold is adjusted. The FPR is calculated as

$$\text{FPR} = 1 - \text{Specificity} = \frac{\text{False positives}}{\text{False positives} + \text{true negatives}}.$$

20 https://en.wikipedia.org/wiki/Confusion_matrix

21 https://en.wikipedia.org/wiki/Type_I_and_type_II_errors

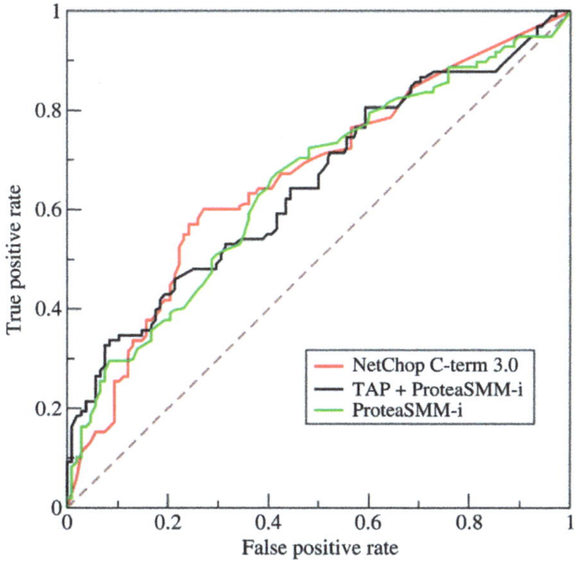

Figure 7.8 Example ROC curve. This ROC curve provides a visual summary of the available TPR and FPR trade-offs at different classification thresholds for three different models. In this case, no one model provides the best performance for all classification thresholds. For example, at FPRs below 0.15, the model depicted in black provides the best performance, while for FPRs between 0.25 and 0.4, the model depicted in red provides the best performance. Because all three lines plot above the dashed line, which illustrates the predictive performance of random guessing, all three models exhibit predictive power. (*Source*: BOR at English Wikipedia, https://en.wikipedia.org/wiki/Receiver_operating_characteristic. Used under CC BY-SA 3.0, https://creativecommons.org/licenses/by-sa/3.0/deed.en.)

Figure 7.8 provides an illustration of ROC curves plotted for three different classifiers applied to a problem. The optimal point of performance on an ROC curve is the extreme top left position at coordinate (0,1), which indicates perfect prediction (i.e., a 100% TPR and a 0% FPR). Since this level of prediction is rarely (if ever) achieved, you must choose one of the available trade-off points between TPR and FPR performance (i.e., you must select a classification threshold that best balances these objectives).

In Figure 7.8, the trade-off point where you could get a 60% TPR for a 30% FPR might be ideal (and you would select the classifier depicted in red). For a different application, where false positives are more of a concern, you might select the black classifier at the threshold that provides a TPR of 35% with an FPR of 10%. The dashed line illustrates the performance of random guessing, and any classifier whose ROC curve lies above the dashed line outperforms random guessing (i.e., has some predictive power).

Multinomial confusion matrix

		Actual class		
		Class 1	Class 2	Class 2
Predicted class	Class 1	10	3	2
	Class 2	5	5	5
	Class 3	2	4	7

Binomial confusion matrix

		Actual class	
		Class 1	Not class 1
Predicted class	Class 1	10	5
	Not class 1	7	21

Figure 7.9 Collapse of a multinomial confusion matrix (truth table) to a binomial confusion matrix (truth table). Class 2 and Class 3 are collapsed into the "Not Class 1" case in order to build a binomial truth table that summarizes classification performance for Class 1.

Performance for multinomial classification problems (i.e., when there are three or more classes considered) generalizes in a straightforward manner from the binary case. A binary confusion matrix can be built for each of the classes by consolidating all other classes as shown in Figure 7.9. More frequently, statistics such as overall *classification accuracy* are used to summarize classifier performance. Classification accuracy is calculated as

$$\text{Accuracy} = \frac{\text{Correct classificiations}}{\text{All classifications}}.$$

The overall classification accuracy for the multinomial case depicted in Figure 7.9 can be calculated directly from the number of correct (numbers on diagonal in green) and incorrect (numbers off-diagonal in red) classifications:

$$\text{Accuracy} = \frac{\text{Correct classificiations}}{\text{All classifications}} = \frac{10+5+7}{43} \sim 51\%.$$

In practice, the cost of misclassification is often asymmetric. It may cost considerably more money or resources to commit a false positive error than a false negative error (or vice versa) in a binary classification or to misclassify one particular case in a multinomial classification. For example, when evaluating loan applicants, the cost for falsely predicting a default, and therefore losing the potential profit from the loan due to denying the application, might be much smaller than the cost of failing to predict the default and incurring the much larger costs of processing a defaulted loan. In such cases, it should be possible to develop a "business case" statistic for evaluating classifier performance, translating classification performance into real-world effect. You could evaluate performance by applying the classifiers against a representative test data set of previous long-term customers (some of whom defaulted and others who did not) and calculate the profit or loss generated by different classifiers. This "real-world" assessment is likely to provide a more useful performance evaluation for models than a "statistical" measure of performance.

7.5.3 Performance Evaluation for Time-Dependent Data

Another critical consideration for performance evaluation is time-dependency. There are many real-world applications of machine learning where ignoring the effects of time on the analysis will result in very poor performance for predictive models in practice. When the factors affecting a prediction are likely to change over time, or there are other processes taking place that affect the response and exhibit time dependency, then the design of your model-fitting procedure and performance evaluation must change to account for these time dependencies. While this consideration applies obviously and directly to time series forecasting problems (such as predicting sales in future quarters based on previous history), it is also frequently important in many other regression or classification contexts.

One classic example is in the development of crime prediction models (i.e., *crime hot-spot maps* that predict where future crimes are likely to take place).[22] In this application, the question is whether or not crime hot-spots for future years can be predicted based upon features (or locations) associated with crime in past years. It may or may not be the case (depending on the city) that the locations and features associated with crime change over time (there are many factors such as urban development that can affect the spatial distribution of crime over time). Prior to conducting model-fitting and performance evaluation, it is critical to check for time dependency in your data so that specific procedures can be taken into account for this dependency in your model fitting, selection, and performance evaluation process.

A *rolling-horizons design* should be employed for model fitting and performance evaluation when there is time-dependency. In rolling-horizons design, rather than employing cross-validation procedures or splitting data sets between training, validation, and testing subsets, algorithms are evaluated based on their ability to use information from the training period to predict outcomes in a future validation or testing period. This is repeated many times, thus providing multiple estimates for how well a particular algorithm learns from previous data to predict future outcomes. Figure 7.10 illustrates a rolling-horizons design. Performance would be estimated in this case by averaging performance over periods four through seven.

22 https://en.wikipedia.org/wiki/Crime_mapping

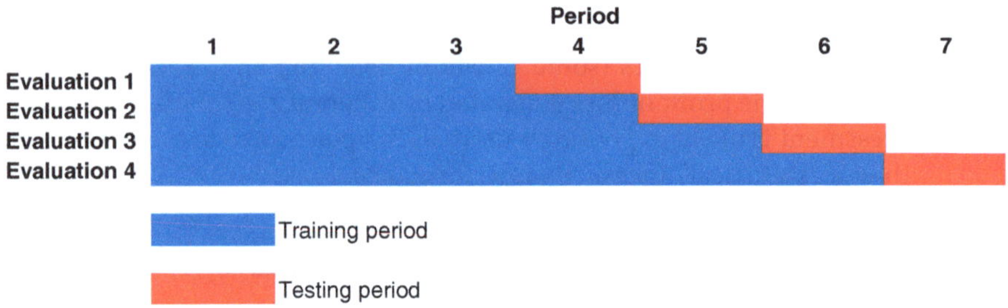

Figure 7.10 Illustration of a rolling-horizons design. Model performance is estimated by averaging predictive performance over periods four through seven.

It is not always necessary to implement a full rolling-horizons design in order to include consideration for time dependency. For many applications, one might simply reserve the most recent year (or month) of results as the test set and use previous years (or months) for training, tuning, and validating models. The key consideration is that model selection decisions and model performance evaluation should not be made on the basis of only one observation. So, in time series forecasting applications (e.g., predicting sales volume for the next time period based on previous periods), it is necessary to use a rolling-horizons design to evaluate the performance over many time periods because each time period contains only one prediction that can be used to evaluate model performance. In the case of the spatial crime prediction example, we could use the most recent time period (the last month or year of data) as the test period provided we had a large enough sample of observations (i.e., crime event locations) that occurred during that test period to develop a reasonable estimate for model predictive performance. One rule of thumb is to include both a minimum of 15 observations and at least 10% of your available data (not applicable in a rolling-horizons design) in your test data set.

7.6 An Overview of Supervised Learning Algorithms

Now that we have discussed machine learning paradigms in general terms, we turn to a discussion of the machine learning algorithms used to develop predictive models. This section provides brief summaries for many of the most frequently employed supervised learning algorithms, that is, methods that require labeled response data to train models for regression (numerical prediction) or classification (categorical prediction) problems. Methods described in this section include the following:

- *k*-nearest neighbors (KNN)
- Regression
- Classification and regression trees (CART)

- Time series forecasting
- Decision trees
- Support vector machines (SVM)
- Artificial neural networks
- Ensemble methods

This is not a comprehensive list of all of the supervised learning algorithms but rather a representative sample of procedures frequently used. For each of these methods, we briefly describe how the method works, describe some relevant extensions where applicable, and provide some example use cases.

7.6.1 k-Nearest Neighbors (KNN)

The KNN algorithm's simplicity belies its excellent performance for many real-world regression and classification problems.[23] In this approach, the training data provide the model. Any new observation is simply compared to the *k*-nearest observations (measured in high-dimensional feature space) in the training data set. For regression problems, the response values for the *k*-nearest training observations are usually averaged to provide the prediction for the new observation. In a classification problem, each of the *k*-nearest observations provides a vote for their own class, which provides an empirical estimate of the probability of belonging to each class. In the example provided in Figure 7.11, when the parameter *k* is set to three or less, the predicted class would be "red triangle," because two out of two of the nearest neighbors are from the red triangle class. When the parameter *k* is set to five, the predicted class would be "blue square" with a probability of 60% (i.e., 3/5), because three out of five of the nearest neighbors are blue squares.

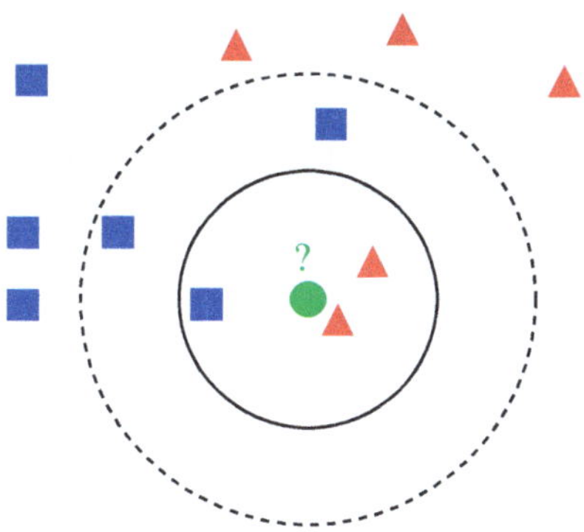

Figure 7.11 Illustration of *k*-nearest neighbors algorithm. In this illustration, when *k* is less than 3, the classification would be "red triangle" because two out of three of the nearest observations are from the red triangle class. When $k = 5$, the classification would be "blue square" with a probability of 60% (i.e., 3/5) because three out of five of the nearest neighbors are blue squares. (*Source: Ajanki*, https://commons.wikimedia.org/wiki/File:KnnClassification.svg. Used under CC BY-SA 3.0, https://creativecommons.org/licenses/by-sa/3.0/deed.en.)

23 https://en.wikipedia.org/wiki/K-nearest_neighbors_algorithm

Key modeling decisions to be made in "fitting" these models include

- selection of the parameter *k* (i.e., the number of neighbors to consider),
- the features to include in the feature space (i.e., feature selection), and
- consideration of various approaches for weighting the contribution of neighbors as a function of their distance from the target observation (i.e., near neighbor votes can factor more heavily that far neighbors).

The calculation of the distance between a new (out-of-sample) observation to each of the observations in the training data set for prediction (so that the *k*-nearest neighbors can be found) can be computationally expensive on large data sets with many features and so often unsupervised dimension reduction algorithms are used to shrink the considered feature space. One common use of the KNN algorithm is in online shopping recommendation systems that suggest products to customers that "near-neighbor" customers bought.

7.6.2 Extensions to Regression

Regression is widely used in machine learning for prediction, time series forecasting, and classification problems. In most applications, extensions or variations on basic linear regression are applied to improve predictive performance. Commonly applied extensions to linear regression include *ridge regression,*[24] *least absolute shrinkage and selection operator* (LASSO) regression,[25] and *logistic regression.*[26] Ridge regression and LASSO provide feature selection and regularization procedures used to improve the performance of regression models. Logistic regression (and other *generalized linear models*) provides the opportunity to fit models for classification problems and other problems for which the assumptions of linear regression are not met.[27]

Figure 7.12 provides one example of a relationship described using logistic regression (note that the fitted relationship is nonlinear). This figure shows the modeled change in the relative likelihood (i.e., probability) of observing a crime committed by a criminal gang as distance from a claimed gang territory increases [7]. As can be seen in the figure, the probability of observing a gang crime ½ mile from a gang's territory is only about 25% of the probability of observing a gang crime if you are actually in a gang territory. The circles at the top of the figure plot actual crime occurrence (i.e., the response variable = 1), while the plotted circles at the bottom plot the distance from a gang territory for the locations where crimes did not occur (i.e., the response variable = 0).

24 https://en.wikipedia.org/wiki/Tikhonov_regularization
25 https://en.wikipedia.org/wiki/Lasso_(statistics)
26 https://en.wikipedia.org/wiki/Logistic_regression
27 https://en.wikipedia.org/wiki/Generalized_linear_model

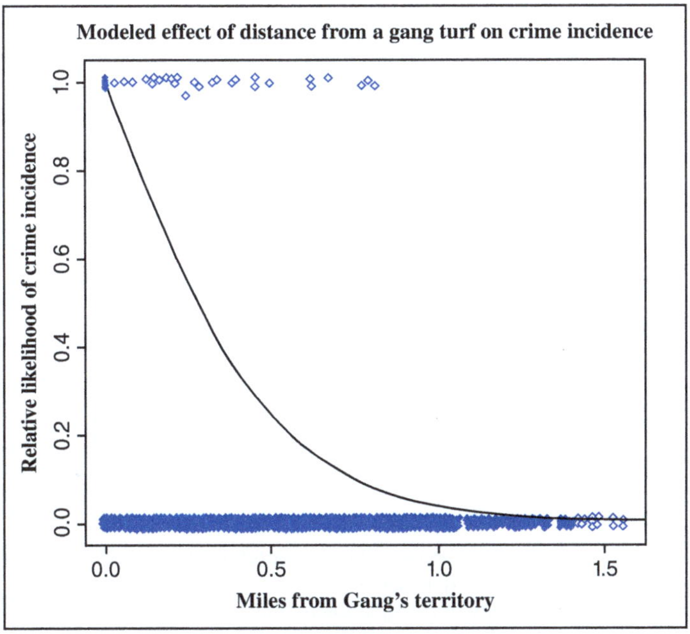

Figure 7.12 An example of a relationship model using logistic regression. This figure shows the fitted relationship asserting that the probability of observing a gang crime ½ mile from a gang's territory is only about 25% of the probability of observing a gang crime if you are actually in a gang territory. The circles at the top of the figure plot actual crime occurrence in the training data set (i.e., the response variable = 1), while the plotted circles at the bottom plot the distance from a gang territory for the locations where crimes did not occur in the training data set (i.e., the response variable = 0).

7.6.3 Classification and Regression Trees

Classification and regression tree (CART) models provide a highly visual and human-interpretable method for regression and classification problems.[28] They are closely related to the decision trees used for probabilistic decision-making. CART models are built by splitting the data from the top down into subgroups that are less "impure," meaning that each subsequent feature-based split as you move down the tree produces a grouping of observations that is more homogenous in class (for classification) or has less variance (for regression) than the parent node above it.

A completely naïve application of decision trees partitions the training data based on its features until each of the terminal leaves (i.e., the nodes at the end of the branches) of the tree is absolutely pure (i.e., all observations in the leaf have the same class or value). Such a tree is usually overfitted, so the focus during

28 https://en.wikipedia.org/wiki/Decision_tree_learning

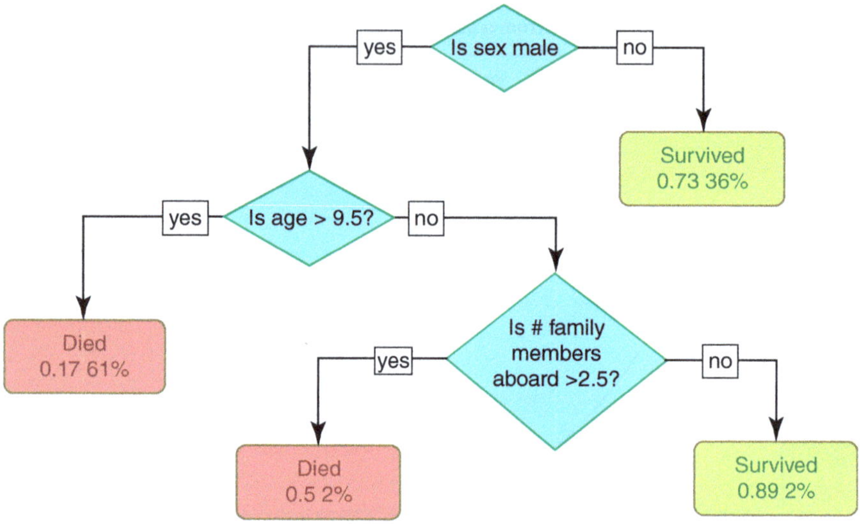

Figure 7.13 Illustration of a classification tree describing the probability of survival on the titanic based on the features of the passengers. Each terminal node provides the probability of survival (i.e., the probability of being in Class 1) as a number between 0 and 1 and the percentage of observations in the data set described by that node. A male passenger over the age of 9.5 is predicted to have a 17% probability of survival under similar circumstances. (*Source*: Dlary, https://commons.wikimedia.org/wiki/File:Titanic_Survival_Decison_Tree_SVG.png. Used under CC BY_SA 4.0, https://creativecommons.org/licenses/by-sa/4.0/deed.en/)

model fitting is on selecting an appropriate algorithm for partitioning the tree and choosing the method used to prune the tree back to an appropriate level such that the tree describes what is generally true rather than simply reflecting the data (and the noise) in the training data set.

Figure 7.13 provides an illustration of a classification tree fitted for predicting survival in a shipwreck using data from the Titanic. As this figure illustrates, in a pruned tree, the terminal leaves will usually include observations from multiple classes. Classification trees predict by identifying the correct terminal node for new observations and return the class probability derived via "class voting" by the observations in the training data set in that node. Figure 7.13 asserts that if you are a male passenger over the age of 9.5, you are estimated to have a 17% probability of survival in a similar shipwreck. Regression procedures predict in a similar manner by returning the average of the response variable for the training observations in each terminal node.

CART models are highly interpretable and therefore are an excellent method for applications in public policy decision-making where it is necessary to explain how employed models make their recommendations (i.e., CART models are not unexplainable "black boxes"). However, other machine learning methods often outperform them in practice. CART models are often used in ensemble methods such as *random forest modeling* that combine the predictions of many (perhaps thousands) of fitted CART models together. Ensemble procedures such as *boosting* and *bagging* can greatly improve upon CART modeling's predictive performance at the cost of reducing interpretability.

INTERVIEW WITH WEI-YIN LOH

Wei-Yin Loh, Professor of Statistics with the University of Wisconsin, discusses the original purpose of classification and regression trees and how classification and regression trees have evolved over time.

The original purpose of classification and regression trees was to fit models to data for which linear regression was inadequate, due to presence of collinearity, nonlinearity, and interaction effects as well as large numbers of predictor variables. AID, THAID, and CHAID were the first generation of algorithms. One of their main weaknesses was lack of an effective way to determine the size of a tree. CART, ID5, C4.5, and M5 addressed this problem by employing various methods of tree pruning. But they, as well as AID, THAID, and CHAID, have another weakness, namely, a bias toward selecting variables that generate more splits. This was addressed by QUEST and CRUISE, which use significance tests to rank the variables for splitting each node. This approach was refined in GUIDE, which has additional improvements for handling missing values in data and for fitting piecewise linear regression tree models. CTREE and MOB use permutation tests instead of the chi-squared tests in GUIDE to control selection bias. Random forest and BART are ensembles of trees. They sacrifice the interpretability of single trees for potentially higher prediction accuracy.

This is an excerpt from one of a series of interviews with analytics professionals and educators commissioned by the *INFORMS Analytics Body of Knowledge* Committee.

7.6.4 Time Series Forecasting

Although many machine learning and statistical learning texts do not provide a discussion of time series forecasting methods, these problems occur frequently in business, military, and public policy domains and so are briefly discussed here because analytics teams often spend a considerable amount of time working on these problems. A *time series* is a series of numerical values indexed by time.[29] This could include many different values of interest such as product sales volume, call volumes for a call center, or crime counts in a city.

29 https://en.wikipedia.org/wiki/Time_series

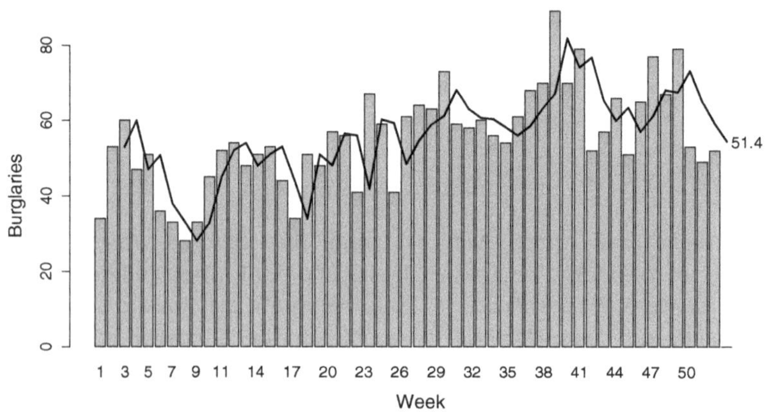

Figure 7.14 Illustration of a one-step ahead forecast for burglaries in Pittsburgh. Actual observations are depicted with gray bars and the forecast fit with an exponential smoothing model is depicted with the black line. A rolling-horizon design is used to make predictions for weeks 3 through 51.

In the context of time series forecasting, we make a distinction between a prediction and a forecast. A prediction is defined as an assertion (often probabilistic) that a specific event will take place, whereas a forecast is an assertion of how much of something will occur over a specified geographic area and period of time. In this context, the nightly weather "forecast" might include a prediction about the high temperature for the following day and a forecast for the amount of rain. We provide a brief discussion of some frequently used approaches here and recommend *Forecasting: Principles and Practice* [8], which is freely available online, as a first reference for the techniques presented here.[30] There are many more in-depth treatments of time series forecasting in the statistical literature.

Figure 7.14 provides an example use case for time series forecasting methods. This figure illustrates the results obtained by applying a rolling-horizon design to forecast the number of weekly burglaries in Pittsburgh in the year 2008. The first forecast is plotted for week 3 (with weeks 1 and 2 used as a training data set), and then each week the model is refit using the new data and a new forecast is made for the following week. This figure illustrates the intended use of the model–forecasting the burglary count for future periods, which in this case would be week 51, where the model forecasts an expected burglary count of 51.4.

Time series model fitting is similar in many ways to predictive regression modeling but is unique in that previous observations of the response variable are often the most important predictive feature for the forecast. The statistical term for this use of previous observations of the response variable is *autoregression.*[31]

30 Online text available at https://www.otexts.org/fpp.

31 https://en.wikipedia.org/wiki/Autoregressive_model

Fitting time series models often involves modeling the four components of a time series: the *trend*, *seasonality effects* (rises and falls in the data on a fixed pattern, often due to the effects of weather), *cycles* (rises and falls in the data that fall outside of a fixed pattern), and *noise*.[32] Fitting and evaluating time series models also requires the use of rolling-horizon design due to the time dependency inherent in these problems.

The three most common methods used for time series forecasting are *time series regression*, *exponential smoothing*, and *autoregressive integrated moving average (* ARIMA*)* models.[33] Time series regression extends basic regression to include auto-regression against previous observations of the response variable (i.e., burglary counts in previous weeks as illustrated in Figure 7.14). Time series regression also facilitates the use of other variables (i.e., other time series) as predictive features. For example, various studies have related temperature and weather effects to crime occurrence and so predicted temperature over the next week could be incorporated into a forecasting model using time series regression. ARIMA models extend basic autoregressive modeling to account for trends and other effects.

Exponential smoothing is a nonparametric technique that develops a forecast for the next period by using the immediately previous forecast and the immediately previous observation as the predictor variables according to the following formula:

$$\text{New forecast} = \alpha(\text{previous forecast}) + (1 - \alpha)(\text{observed value}).$$

The parameter α is a tuning parameter that places more weight either on the previous forecast or the previous observation, taking on values between 0 and 1. This model form results in a recursive relationship with previous forecasts, with the effects of previous forecasts decaying exponentially backward in time. Hence, the name for the algorithm. Fitting an exponential smoothing model is relatively simple and straightforward as it requires only the optimization of the weighting parameter α. Basic exponential smoothing has been extended to account for trend and seasonal effects in an algorithm now known as Holt–Winters exponential smoothing [9,10].

7.6.5 Support Vector Machines

Support vector machines are a machine learning technique originally developed for classification problems, although extensions have now been developed to facilitate their use for regression problems. The motivating principle for support vector machines is to find a maximum separating hyperplane in feature space that separates classes. This hyperplane is illustrated for a two-dimensional feature space in the right panel of Figure 7.15.

---

32 https://en.wikipedia.org/wiki/Decomposition_of_time_series
33 https://en.wikipedia.org/wiki/Time_series

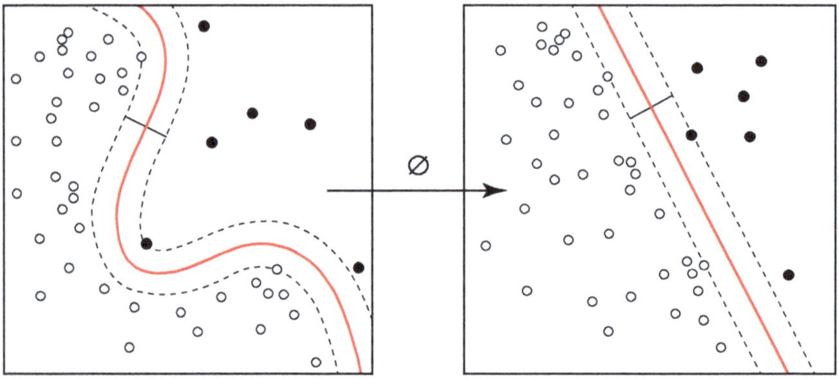

Figure 7.15 Illustration of support vector machine classification. Observations are remapped into a higher dimensional space where a linear separator can be drawn between classes. The dashed lines depict the margins, which define the edge of each class. The red line depicts the maximum separating hyperplane, which is used to define the decision rule for classification. (*Source*: Alisneaky, https://commons.wikimedia.org/wiki/File:Kernel_Machine.svg, Used under CC BY-SA 4.0, https://creativecommons.org/licenses/by-sa/4.0/deed.en.)

The idea for finding the maximum separating plane is that if we can find the plane (i.e., line in a two-dimensional case) with the greatest space between the two classes, then the decision rule for classification that corresponds to that line will be most generalizable to new cases because we have separated the two classes by as much distance as possible. The linear vectors in high dimensional space that define the "edges" of each class are known as the margins. The margins in the right panel of Figure 7.15 are depicted as dashed lines, and the central red line depicts the classification threshold (the separating hyperplane) that defines the decision rule. The points from each class that lie on and define the margins are known as the support points.

Support vector machines can also be applied to nonlinear classification by mapping training observations into a new, usually higher dimensional feature space where a linear separating hyperplane can be drawn between the classes in the training data. An example of this remapping is illustrated in Figure 7.15, where the nonlinear relationship in the left panel is transformed and projected into a new space (the right panel), where a linear hyperplane exists that separates the two classes. New out-of-sample observations are projected into this same higher dimensional space and classified based on which side of the linear separating hyperplane they fall.

7.6.6 Artificial Neural Networks

Artificial neural networks are motivated by trying to mimic the way neurons in the brain fire to influence human learning and decision-making. This modeling

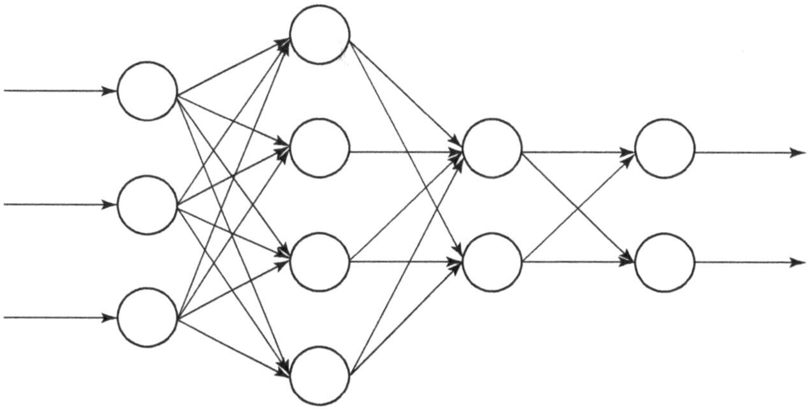

Figure 7.16 Illustration of a multilayer neural network that translates input values (predictors) into output values (predictions). (*Source*: John Salatas, https://commons.wikimedia.org/wiki/File:Multilayer_Neural_Network.png. Used under CC BY-SA 3.0, https://creativecommons.org/licenses/by-sa/3.0/deed.en.)

framework is very flexible and can be adapted for classification problems, regression problems, and unsupervised learning applications such as clustering. They are also applied extensively in reinforcement learning, and they underpin artificial intelligence systems used for image recognition and self-driving cars.

In the basic operation of artificial neural networks, input data (such as predictive features) are fed into hidden layers of nodes (neurons), which perform transformations of input values and pass them on to other hidden layers of nodes, until eventually the network of connected nodes emits the output values (predictions or recommendations). In this way, at each hidden layer, a neural network automatically generates a new set of features that are functions of the features in the preceding layer. The features in these hidden layers are better able to predict the response variable than the original set of input features. The automated process of building these hidden layers is an example of the feature learning previously discussed. Figure 7.16 provides an illustration of how hidden layers of neurons transform input values (i.e., predictive features) into output values (predictions).

Model fitting for neural networks involves making decisions about the number of hidden layers to include in the network, the number of nodes in each hidden layer, which nodes should be linked together (i.e., defining all of the network paths), and the transformation functions for every node in the network. Most of these decisions are made iteratively during the model-fitting stage based on minimization of some penalty function–which is sometimes referred to as a loss function, cost function, or error function. The most common method for iterative improvement is *backpropagation*, or feeding the errors made by the neural network back into the network and allowing it to iteratively improve its performance (or learn from its errors) by revising the weights placed on the connections between nodes.

The strength of artificial neural networks lies in their incredible adaptability to a vast array of problems. However, the decision space to be explored for fitting an artificial neural network is very large, and so fitting artificial neural networks often requires considerably more training data and computational resources than other machine learning algorithms. Artificial neural networks also suffer from being *almost completely uninterpretable.* The use of artificial neural networks and the many extensions to the basic framework, such as *recurrent neural networks* and *deep learning,* has exploded in recent years due to the availability of large cloud-based computing clusters that can be leveraged to fit these highly adaptable but complex models.[34,35]

Interview With Katya Scheinberg

Katya Scheinberg, the Harvey E. Wagner Endowed Chair Professor with the Industrial and Systems Engineering Department at Lehigh University, explains backpropagation and alternatives for facilitating learning by neural networks.

A neural network (or a deep learning network, popular these days) is a graph composed of consecutively connected layers. Each layer consists of *neurons,* and each neuron is associated with a weight and an *activation* function. The number of layers and neurons and the choice of the activation functions define a neural network.

The process of selecting specific values of the weights, let's call them *w*, is called *training* a neural network. The weights are chosen to optimize the accuracy of the network on a particular data set. This accuracy is thus a function, let's call it *f*, that depends on the data set and the weights *w*. This function is nonconvex and very high-dimensional, and it typically has complex structure.

The traditional way to optimize this function is to apply basic gradient descent steps where the gradient of *f* (*w*) is computed by essentially applying automatic differentiation to *f*. This method is called backpropagation. Computing the exact gradient of *f* requires applying this automatic differentiation process to each data point in the input data set. Since the input sets and the deep neural networks are very large in modern applications, each backpropagation step is very expensive.

An alternative, which is the current industry standard, is to randomly select a small subset of input data and apply backpropagation only based on this subset. This is known as stochastic gradient descent or SDG. Unlike the traditional gradient

34 https://en.wikipedia.org/wiki/Recurrent_neural_network
35 https://en.wikipedia.org/wiki/Deep_learning

descent, this method is very efficient in some cases but is not very robust, since its performance depends substantially on the choice of parameters, such as the step size and the sizes of data subsets. The most efficient and commonly used variants, known as momentum SGD and Adam, are heuristics that try to reduce the dependence on parameters and improve stability of the SGD method.

Recently, a lot of research has been dedicated to applying second-order methods that are variants of Newton's method to training deep neural networks. Other traditional methods from deterministic optimization are also being explored. Applying these methods in a straightforward way, using evaluations of the entire data set at each iteration, is not only very costly but also seems to lead to undesirable local solutions in some cases. The goal is to adapt these methods in such a way that, similarly to SGD, only a handful of data are used at each iteration. The results of applying these methods to neural networks are so far mixed and not clearly understood. However, this area of research is expected to see a very significant progress in the next few years.

This is an excerpt from one of a series of interviews with analytics professionals and educators commissioned by the *INFORMS Analytics Body of Knowledge* Committee.

7.6.7 Ensemble Methods

One of the most powerful techniques in machine learning is the *ensemble*, or a model federation that combines predictions from several (or many) machine learning algorithms.[36] As a general rule, ensembles improve predictive performance at the cost of making the process for producing predictions far less interpretable. There are many techniques used in machine learning for combining the predictions of machine learning models. We will briefly discuss several of the most used including the following:

- Simple averaging
- Stacking
- Bootstrap aggregating (i.e., bagging)
- Boosting

The simplest approach to ensembles involves averaging the outputs of several different models to form a consensus prediction or classification. Figure 7.17 provides an illustration of the results of combining the forecasts from three different time series forecasting algorithms into an ensemble forecast. As can be seen in Figure 7.17, this method is a variance-reduction technique because the resulting ensemble forecast, as an average of the other models, reigns in the most extreme forecasts of each if the different algorithms.

36 https://en.wikipedia.org/wiki/Ensemble_learning

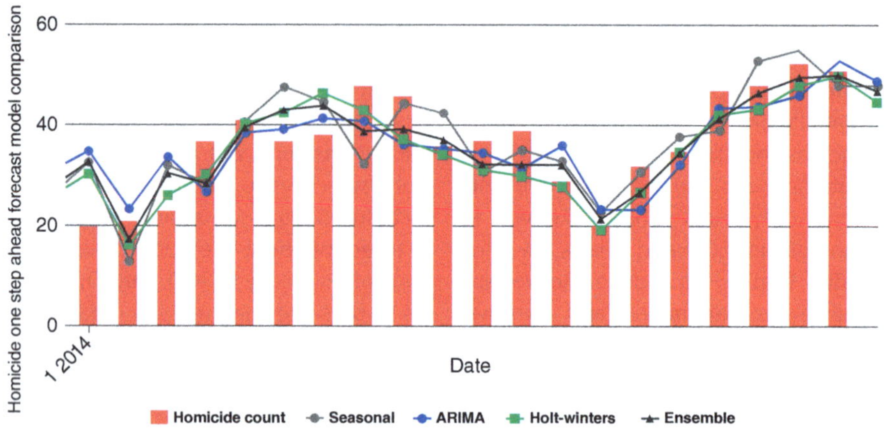

Figure 7.17 Illustration of the performance of an ensemble model developed from three time series forecasting models for forecasting the weekly homicide count in Chicago during the year 2014. The ensemble formed by the average of the three other models reigns in the extreme predictions of each method, reducing the variance. Although it can't be seen by visual inspection, this improves the predictive performance.

One can often improve on simple averaged ensembles by developing performance-based weights for the contributions of different models. This is known as *stacking* because you are stacking one machine learning algorithm on top of others. The default stacking approach uses linear regression to assign coefficient weights to the predictions of other machine learning algorithms, but there are many other schemes for implementing this; any supervised machine learning algorithm can serve as the model at the top of the stack.

Bootstrap aggregating, more frequently referred to as *bagging*, involves iteratively sampling from the training data set to create many training samples of the same size. A model is built from each sample, resulting in an ensemble of many prediction or classification models that have been fit with the same machine learning algorithm using slightly different training data sets. The bagged ensemble's predictions are averaged (for regression problems) or serve as class votes (for classification problems). The *random forest algorithm*, which is one of the most frequently employed machine learning techniques, is a bagging ensemble formed by iteratively fitting many (i.e., a forest of) classification or regression trees.[37] The random forest performs the basic bagging procedure for creating training data sets but extends it by also sampling for which features will be included in the training data set. Bagging procedures tend to reduce model error variance and improve model generalization.

Boosting is designed to reduce model bias; it was originally intended for classification problems and then extended to regression problems. Boosting

37 https://en.wikipedia.org/wiki/Random_forest

builds a sequence of models where each new model in the sequence is designed to improve performance on observations in the training data set that were misclassified in the previous model. This is done by applying weights to misclassified instances from the previous step and then refitting the model (in essence items that were misclassified in the previous step are now prioritized more heavily in the next model). The predictions of all of these models are aggregated together to form a final prediction. A common approach to boosting is an algorithm known as *Adaboost,* which has been referred to as the "best off-the-shelf classifier in the world" when applied to decision trees ([3], pp. 302) [11].[38]

7.7 Unsupervised Learning Algorithms

Unsupervised methods are designed to identify the latent (i.e., underlying or hidden) structures in data. These structures may be groups, underlying probability distributions, or variance structures. This section provides a brief description of several of the most frequently employed unsupervised learning techniques, although there are many more than are discussed here. Models discussed here include the following:

- Kernel density estimation
- Association rule mining
- Principal components analysis (PCA)
- Clustering methods
- Bag-of-words or vector space models.

7.7.1 Kernel Density Estimation

Kernel density estimation is a nonparametric statistical method for estimating an underlying probability distribution.[39] Kernel density estimates are in essence a smoothing technique that uses observations that are assumed to have been generated by some underlying parent distribution to estimate that parent distribution. Figure 7.18 illustrates how kernel density is used to provide a density estimate in a two-dimensional space given some sample observations. A kernel smoothing function (in this case, a two-dimensional Gaussian density function) is applied to each of the observations. All of the functions are then combined, providing a smooth density estimate for the underlying probability density function, which is often plotted as a "heat map" or hot-spot map.

---

38 This quote is attributed to Leo Breiman in [3].
39 https://en.wikipedia.org/wiki/Kernel_density_estimation

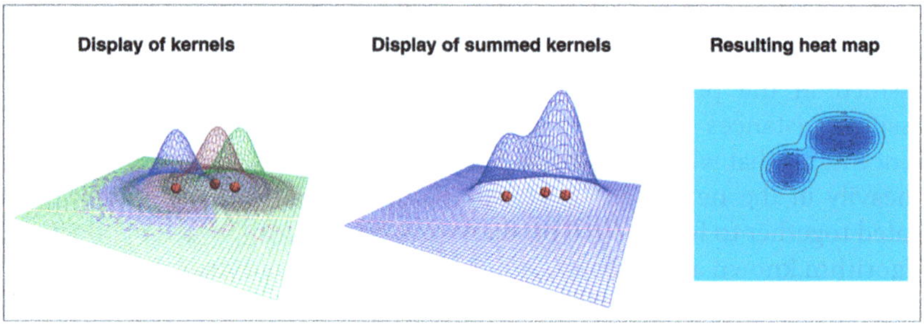

Figure 7.18 Fitting a kernel density estimate in two dimensions. Kernel functions are placed on top of each observation and then summed. The resulting density estimate is often plotted as a heat map or hot-spot map to depict the estimate for the underlying density function. This is often employed for crime prediction.

Fitting kernel density estimates requires the selection of the kernel function, which is a nonnegative function that integrates to 1.0 (so probability density functions are natural kernel functions). Using a kernel function requires selection of parameters such as a *bandwidth* and the parameters that define the probability density function used as the kernel function.[40] The optimal setting of these parameters remains an open research question, but there are some widely employed "rules of thumb" for automatically selecting these parameters based on the observed data.

Although kernel density estimation is an unsupervised learning technique designed to reveal underlying probability distributions, the technique is widely employed in predictive policing to predict future "crime hot-spots" and make resourcing decisions about the deployment of security forces (i.e., police and military patrols) [12,13]. However, other machine learning algorithms, such as logistic regression, have been shown to substantially outperform kernel density estimation when an appropriate feature set can be built, and at the cost of a much more involved model fitting process [14,15]. Kernel density remains widely used in predictive military and police applications because the relatively simple model fitting procedure can be performed using mapping software that virtually every police or military organization fields.

7.7.2 Association Rule Mining

Association rule mining is often colloquially referred to as "market basket analysis" because one of its primary applications is in analyzing the shopping cart activity of retail customers using point-of-sale data. An association rule is a statement such as, "On Friday nights, customers who buy beer also frequently

40 Ibid.

buy diapers." More formally, another association rule might defined as

$$\{\text{rum, pineapple}\} \Rightarrow \{\text{coconut}\}.$$

The goal of association rule mining is to discover relationships that are useful or important in the desired business context but for which there is no target or response variable against which to model. Rather, the available data are investigated for rules that meet criteria for "interestingness" such as *support* (frequency of appearance), *confidence* (how often the rule is true), *lift* (a measure of how much more frequently the combination of items occurs than it would under conditions of statistical independence), and others. Usually, due to the vast number of rules that can be developed on a large data set of point-of-sale records, analysts define minimum thresholds on multiple criteria that must be met simultaneously for rules to be considered for further investigation.

7.7.3 Clustering Methods

Clustering methods are designed to reveal hidden groupings of the observations in a data set and is sometimes referred to as data segmentation.[41] A cluster is a set of observations in a data set that contains observations that are more similar to each other than they are to other observations in the data set. Clustering methods typically require a means to measure similarity or dissimilarity (and there are a variety of approaches for this) and an algorithm for grouping similar items. Two frequently employed techniques for clustering are *k-means clustering* and *hierarchical clustering.*

k-means clustering is an iterative algorithm for identifying clusters that requires that all features in the data set are numerical and that the analyst specify the number of clusters. The k-means algorithm starts with an initial guess for the center point of the clusters and then iteratively minimizes the sum of the squared distances between the cluster center points and the nearest data points to that temporary cluster center point. After several (perhaps many) iterations, the feature space will be partitioned based upon the locations of the k center points, with clusters defined by grouping every observation with the nearest center point, usually measured in *Euclidean distance.*[42]

Hierarchical clustering partitions the data into a format similar to that shown for a decision tree. The top-level cluster includes all of the data. The deepest leaves of the hierarchy contain individual observations. Hierarchical clustering algorithms require the analyst to define a measure of dissimilarity to be used for partitioning the clusters. There are different strategies for partitioning the data, some of which use a top-down splitting approach and others that form clusters

---

41 https://en.wikipedia.org/wiki/Cluster_analysis
42 https://en.wikipedia.org/wiki/Euclidean_distance

by building from the bottom up. The result is a complete mapping of the "similarity" for all items presented in a hierarchical format.

7.7.4 Principal Components Analysis (PCA)

PCA is a dimension-reduction technique used to represent data in a more compact yet more descriptive form. The key idea behind principal components analysis is to remap observations in a high-dimensional space (i.e., a feature space) into a new space where the first principal component is a vector that describes the direction of maximum variance in the data and subsequent components are linearly uncorrelated with each other. In the lexicon of *linear algebra,* the first principal component is the *eigenvector* with the highest *eigenvalue.*[43] Each subsequent component is orthogonal to all preceding components and describes the direction of the highest remaining variance under that condition. Figure 7.19 provides an illustration of the first (largest) principal component (the long vector pointing toward the top right) and the second

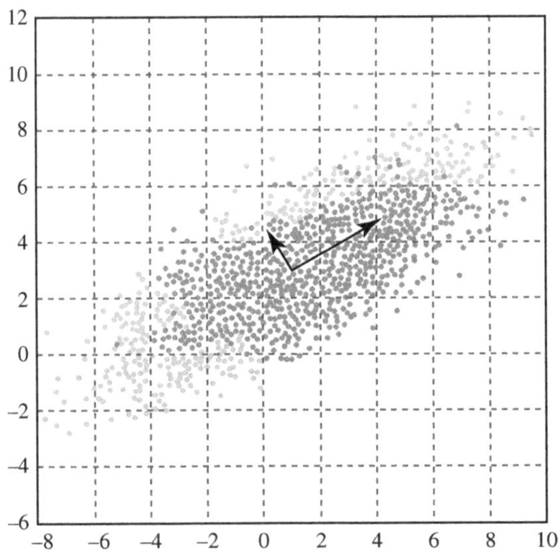

Figure 7.19 An illustration of principal component mapping in two dimensions. The first principal component maps the direction of highest variance and in this figure points to the top right. The second principal component (by definition the smallest in this case) lies orthogonal to the first. (*Source*: Nicoguaro, https://commons.wikimedia.org/wiki/File:GaussianScatterPCA.svg. Used under CC BY 4.0, https://creativecommons.org/licenses/by/4.0/deed.en.)

43 https://en.wikipedia.org/wiki/Linear_algebra

(smallest by definition in two dimensions) principal component that lies orthogonal to first principal component. As can be seen in the figure, the first principal component points in the direction of maximum variance.

PCA produces a new feature space with the same number of components as the original dimensions of the feature space, and the PCA dimensions will explain all of the variation in the original data. PCA works best when most of the variance in the data set can be described with relatively few components and so a data set can be summarized using these most descriptive principal components. Dimensions with high variance often provide distinguishing features for regression and classification problems, and thus these dimensions provide a concise and useful summary of the data set. These few highly descriptive principal components are often used as the predictive features for subsequent applications of supervised machine learning algorithms such as *principal components regression*.[44] Ridge regression employs principal components analysis as a standard part of the model-fitting algorithm.

7.7.5 Bag-of-Words and Vector Space Models

Bag-of-words is employed for a frequent analytics problem known as "find similar items."[45] This applies when the items to be found and matched contain blocks of text or other characters that make up "words" such as phone numbers or emojis. This technique is frequently applied to Web sites, tweets, e-mails, or text documents like books and articles. Bag-of-words is a dimension-reduction technique for representing text in a more compact form that also facilitates a matching procedure with other items of interest.

The basic idea is to convert the item of interest into a set list of all of the unique "words" in the document. This set of words is known as its "bag-of-words." If every document in a large corpus (such as a library's holdings) is converted into a bag-of-words, then similarity measures such as the *Jaccard Index* can be used to match documents in the corpus to new documents (such as student thesis proposals) based on word similarity in the bag-of-words list. In this way, you can provide a list of items that are similar to the item offered as the "query" (i.e., help students find relevant references for their research in the library).

The Jaccard Index for this application is calculated as follows:

$$J(A,B) = \frac{A \cap B}{A \cup B} = \frac{\#\text{ of unique words appearing in both documents}}{\#\text{ of unique words that appear in either document}}.$$

A *vector space model* extends the basic set representation of bag-of-words to include dimensions that represent how frequently a word appears.[46] A

44 https://en.wikipedia.org/wiki/Principal_component_regression
45 https://en.wikipedia.org/wiki/Bag-of-words_model
46 https://en.wikipedia.org/wiki/Bag-of-words_model

document can be represented as a vector in a large "global" feature space in which dimensions (i.e., features) of the space are words and the location of a document in a word's dimension is represented by the number of appearances of that word. When a 0 appears in a document's vector, it indicates that the word representing that dimension does not appear anywhere in the document. Vector-based similarity measures such *cosine similarity* can be used to find similar items based on the angle between any two documents' vector representations in this high-dimensional space according to the following formula (which uses vector notation):

$$\text{cosine similarity} = \frac{\mathbf{A} \cdot \mathbf{B}}{\|\mathbf{A}\|_2 \|\mathbf{B}\|_2} = \frac{\sum_{i=1}^{n} A_i B_i}{\sqrt{\sum_{i=1}^{n} A_i^2}\sqrt{\sum_{i=1}^{n} B_i^2}}.$$

There are many extensions to this basic construct that vary the way documents are represented in this feature space.

7.8 Conclusion

As has been shown in the many examples in this chapter, training machines (i.e., computers) to "learn" via the application of algorithmic modeling has a wide variety of very diverse applications. This chapter has also demonstrated that even though the algorithmic procedures employed to "fit" these models can be automated to a certain extent, the machines still require considerable input from analysts for their "training." While individual machine learning algorithms provide a framework for approaching a particular class of problem, choosing the right machine learning algorithm for any particular problem is a highly complex and iterative process that requires considerable expertise, judgment, and often the active participation of domain experts and users of your results. Often, for best results, multiple machine learning algorithms, as well as best practices for data storage, data engineering, and computing will be needed. Practitioners are well advised to algorithmically model in teams that incorporate statisticians, operations research analysts, computer scientists, data engineers, data scientists, and domain experts to form a comprehensive unit dedicated to training the machines to "learn" to solve the right problems in best way.

7.9 Acknowledgments

The authors are grateful for helpful comments and suggested examples from colleagues Lyn Whitaker (modeling process, support vector machines, neural networks, and the bias-variance trade-off) and Ruriko Yoshida (principal components analysis).

References

1 Breiman L (2001) Statistical modeling: the two cultures. *Stat. Sci.* 16(3): 199–231.
2 James G, Witten D, Hastie T, Tibshirani R (2013) *An Introduction to Statistical Learning with Applications in R* (Springer New York, NY).
3 Hastie T, Tibshirani R, Friedman J (2001) *The Elements of Statistical Learning: Data Mining, Inference, and Prediction* (Springer).
4 Huddleston SH, Brown DE (2009) A statistical threat assessment. *IEEE Trans. Syst. Man Cybern. A Syst. Hum.* 39(6): 1307–1315.
5 Huddleston SH, Brown DE (2012) Mapping gang spheres of influence. *Crime Mapp. J. Res. Pract.* 4(2): 39–67.
6 Brighton H, Gigerenzer G (2015) The bias bias. *J. Bus. Res.* 68(8): 1772–1784.
7 Huddleston SH, Porter JH, Brown DE (2015) Improving forecasts for noisy geographic time series. *J. Bus. Res.* 68(8): 1810–1818.
8 Hyndman RJ, Athanasopoulos G (2014) *Forecasting: Principles and Practice.* OTEXTS.com.
9 Holt CC (1957) Forecasting seasonals and trends by exponentially weighted moving averages. *Int. J. Forecast.* 20, 5–10.
10 Winters PR (1960) Forecasting sales by exponentially weighted moving averages. *Manage. Sci.* doi: https://doi.org/10.1287/mnsc.6.3.324.
11 Breiman L (1998) Arcing classifiers. *Ann. Stat.* 26(3): 801–849.
12 Eck JE, Chainey S, Cameron JG, Leitner M, Wilson RE (2005) *Mapping Crime: Understanding Hot Spots* (National Institute of Justice).
13 Boba R (2005) *Crime Analysis and Crime Mapping,* (Sage Publications, London).
14 Liu H, Brown DE (2004) A new point process transition density model for space-time event prediction. *IEEE Trans. Syst. Man Cybern. C Appl. Rev.* 34 (3): 310–324.
15 Smith MA, Brown DE (2007) Discrete choice analysis of spatial attack sites. *Inform. Syst. e-Bus. Manage.* 5 (3): 255–274.

8

Deployment and Life Cycle Management

Arnie Greenland

Robert H. Smith School of Business, University of Maryland, College Park, MD, USA

8.1 Introduction

This chapter is appropriately placed near the end of this book because it pulls together all of the various ideas presented in earlier chapters. Ultimately, the goal of analytics professionals is to create and implement meaningful, successful, and sustainable analytics solutions using the tools of analytics, which have been discussed in Chapters 5 and 6. The successful implementation of an analytics project relies on strong project management skills and on leveraging the analytic and data insights that are the unique contributions of analytics professionals. These have been covered in Chapters 1–4. The life cycle described in this chapter focuses on joining those pieces together in a structured and ordered way, while detailing the special nature, complexities, and challenges of delivering analytics projects.

A life cycle is defined as a sequence of phases in the process of developing an analytics model or system. It is very similar to the term used in the more general context of a software or systems development life cycle common in the discipline of information technology (IT) management. The intent of this chapter is to focus very specifically on those phases that have been identified by the developers of the CAP certification process and that are also closely related to standard methodologies accepted broadly by practicing professionals. The analytics project life cycle may actually be a component of, or integrated with, a larger information technology life cycle.

The analytics system/model life cycle is composed of several phases: initial design, development, testing, implementation, deployment, and postdeployment monitoring. The actual calendar time that transpires in each of these various phases can vary widely but depends on the specific characteristics of the model or system being developed. Total time for this entire process can range from months

INFORMS Analytics Body of Knowledge, First Edition. Edited by James J. Cochran.

to years, depending on complexity of the data relationships being modeled. The postdeployment monitoring may span multiple years.

As with any complex project, the analytics professional (AP) leading a project shoulders the responsibility of understanding the specific steps or phases and the order of those steps he/she will traverse to accomplish the stated goals. Typically, professionals refer to such a set of steps, and any complex interactions, as a *methodology*. There are a multitude of methodologies in use for project management in general, and for analytics projects in particular.

8.2 The Analytics Methodology: Understanding the Critical Steps in Deployment and Life Cycle Management

A popular and accepted methodology in the analytics community is CRoss Industry Standard Process for Data Mining (CRISP-DM). This methodology was created initially as a cooperative effort of a number of companies that were interested in data mining, including SPSS (IBM), Teradata, Daimler AG, NCR Corporation, and OHRA, an insurance company. While it originally focused on data mining, analytics professionals in many fields have found it useful for projects of all types and is the closest thing at this point to a "standard." For example, a recent article in the online site KDNuggets, written by George Piatetsky,[1] states "CRISP-DM remains the top methodology for data mining projects, with essentially the same percentage as in 2007 (43% vs 42%)." The only other data mining standard named in this article was SEMMA,[2] though its use reported in the article fell from 13% in 2007 to 8.5% in the 2014 survey. While not used universally, it is about five times more often used than the next methodology mentioned, so it is the closest thing we have at the moment to a "standard." The CRISP-DM methodology is captured in the following commonly available diagram.[3]

There are six major components of the methodology:

- Business understanding
- Data understanding
- Data preparation
- Modeling
- Evaluation
- Deployment

1 Piatestsky, G, KDDNuggets (2014) CRISP-DM, still the top methodology for analytics, data mining, or data science projects. Available at https://www.kdnuggets.com/2014/10/crisp-dm-top-methodology-analytics-data-mining-data-science-projects.html (accessed October 2014).

2 A data mining methodology developed by SAS Institute, Inc.

3 This specific version of diagram on CRISP-DM is obtained from https://en.wikipedia.org/wiki/Cross-industry_standard_process_for_data_mining

Figure 8.1 CRISP-DM diagram.

A very important feature of the CRISP-DM, shown visually in Figure 8.1, is that there are many feedback loops in the process. In simple terms, what you learn or encounter in one step of the process often impacts an earlier phase of the process, so the diagram shows either an arrow in both directions or an arrow from a later phase coming back to an earlier phase. For example, the dual arrows between business understanding and data understanding emphasize the important notion that you cannot understand the data unless you also have a very deep understanding of the business issues, and as you have questions about the data, it is best to reach back into the sponsoring organization to ask those questions and obtain understanding.

A very important point here is that these six components map nicely to the CAP Job Task Analysis (JTA),[4] so throughout this description of the phases of CRISP-DM, we will freely integrate detail from the JTA as a way to explain the phases but also as a way to link the methodology directly to the types of information that analytics professionals, working toward the CAP Certification, need to know from the JTA.

4 See Appendix A of this document.

The CRISP-DM methodology has been repeatedly praised because of its clear recognition of the importance of a strong connection to the sponsoring business or other type of organization that is seeking to create an analytics model or solution, so it is only natural that we begin this discussion with the phase of the CRISP-DM focused on understanding the business situation.

8.2.1 CRISP-DM Phase 1: Business Understanding

The first phase of the methodology in CRISP-DM is business understanding, and this phase corresponds to the first two domains of the JTA: Business Problem Framing and Analytics Problem Framing. Both the CRISP-DM and the JTA recognize the critical importance of understanding the business issues and focusing specifically on how analytics could have a real (and measurable) positive impact on that business situation.

Consider Domain I of the JTA, Business Problem (Question) Framing. The tasks enumerated therein are as follows:

- Task 1: Obtain or receive problem statement and usability
- Task 2: Identify stakeholders
- Task 3: Determine if the problem is amenable to an analytics solution
- Task 4: Refine the problem statement and delineate constraints
- Task 5: Define an initial set of business benefits
- Task 6: Obtain stakeholder agreement on the problem statement

This set of tasks lays out an excellent set to follow. Upon completion of these steps is a clearly defined and documented business problem statement.

8.2.2 JTA Domain I, Task 1: Obtain or Receive Problem Statement and Usability

A business problem statement is a clear and concise description, typically written in business terms, of what the business or organizational objectives the sponsor wants to reach. The business problem statement defines the key outcomes or accomplishments that are desired, how the organization will measure whether these outcomes have been reached (for example by specifying business metrics that would be measured and the levels or targets for those metrics that represent success), and all other relevant business issues, such as time frame, cost constraints, and other business requirements.

Clearly, the best case scenario is that the organization or business that is sponsoring the analytics project can simply deliver a complete and fully thought out Business Problem Statement. In practice, this rarely happens. It is much more common that the analytics professional (or AP) and team must create the Business Problem Statement working together with the sponsor. The work involved in doing this ranges over a wide set of possibilities. It may turn out to require a small amount of work to take an initial version of the business problem statement to a good initial place, but it is actually more common that it requires creating the document totally from scratch.

We complete this discussion focusing on the common situation of not having a starting version of the document at all. Even if the sponsor can provide an initial version, the same set of activities would be done, though maybe at lower levels of intensity. An important component of creating such a document is conducting detailed interviews to discover and document the business situation. The skills required in such an endeavor are many. As a baseline, the analytics professional needs to know what constitutes a clear and usable problem statement. While project management experience in general is needed for this activity, it is also important to be attuned to issues that impact the use of analytics as a possible solution such as time requirements for completion, availability of data, the structure and form of the data, and availability of business experts at the level required to assist in the process. Next, the AP needs to have the skills to interview knowledgeable individuals, ask appropriate questions, and ultimately document the findings. Successfully obtaining the information needed requires the ability to be persistent in making sure that there is clarity in the understanding of the business issues and that follow-up be done to fill in holes in understanding. The AP will also need to possess basic business knowledge such as how businesses are typically organized, how this sponsoring organization is organized in particular, how business processes are defined, and how they work together to fulfill business objectives.

The deliverable from this step in the process is better described as a starting point for the development of the Business Problem Statement. It is a document that describes the business situation, lays out the problem that is intended to be solved, and the basic business metrics that will be used to measure success. This document, as will be discussed in continually increasing detail in the following sections, will be enhanced by focusing on different components and aspects of successfully documenting the problem to be solved.

8.2.3 JTA Domain I, Task 2: Identify Stakeholders

This task reinforces the notion that success of any project requires involvement of key players in the organization, both at senior levels to provide the organizational backing, funding, and other resources, and to assure involvement of junior stakeholders. It is a commonly accepted understanding that the higher level involvement in, knowledge of, or enthusiastic support for a project, particularly an analytics project because of its complexity, is closely related to project success. It is also critical to have involvement at lower levels in the organization because at those levels you can find individuals with time and access to information (e.g., the data you are hoping to get) that will be essential to proper functioning of the project.

It is recommended to reach out to the full range of stakeholders early in a project, to meet with and interview as many as possible, to acquire and document their views and expectations. In some cases, it is beneficial to bring stakeholders together to share ideas and work toward an organizational consensus. However, one should be very careful in planning such a meeting, especially if there are highly divergent views in the organization that could result in a more difficult situation than if you did not bring stakeholders together.

Interview With Alan Taber

When asked about what constrains the analytics professional's choice of problem-solving approaches/methods/models, Lockheed Martin Missiles and Fire Control's System Engineer Alan Taber offered the following thoughts.

Time and money are two key constraints–that goes without saying. The third constraint is how data-driven the stakeholders are–or aren't. Obviously the stakeholders are willing to listen or they wouldn't be willing to talk to you in the first place, but some stakeholders are more intuitive, while others are more skeptical. In almost all cases, when you are working with executives, you should bear in mind that they got to where they are because they have been successful in the milieu their companies or organizations have been operating in for the last several years.

This leads us to a discussion of change management. The solution to a problem typically requires change. So change management is inherently bound into analytics. The analytics professional is ultimately going to suggest that someone do something differently, invest differently, organize differently, go after a different market segment, or go after the same market segment been in a different way. When you are managing change, you have to know how aggressively and how quickly you can guide someone along the path of change. This is a nebulous concept, and understanding and appreciating it comes with experience.

The analytics professional needs to support his or her clients in their decision-making, help them better understand the decisions they are trying to make, help them better understand the vision of where the organization could be if these decisions are made and plans carried out effectively. The analytics professional has to do so without burning out clients and making them dread the sight of him or her walking through their door. If the client feels the analytics professional always wants more time, more money, more this, more that–or is pushing them faster than they want to go–communication starts to shut down. And if the client and the analytics professional are not communicating, the analytics professional cannot solve the client's problems because the client has been open about his or her business problems.

This is an excerpt from one of a series of interviews with analytics professionals and educators commissioned by the *INFORMS Analytics Body of Knowledge* Committee.

8.2.4 JTA Domain I, Task 3: Determine if the Problem Is Amenable to an Analytics Solution

This is one of those project tasks that are unique to an analytics project. As the AP begins to understand the problem, it is important to begin thinking about, but not finalizing, whether analytics makes sense for this problem or business situation at all. Since nearly every analytics solution (discussed in more detail in Chapters 5 and 6) requires data, this is the place where the AP begins to determine whether there are data available, whether they are easily accessible, and whether there are issues relating to obtaining the data (classification of data as confidential or higher levels of classification, or whether there are privacy issues). Assuming, the AP believes that there are data and that they could be obtained, it is also important to begin thinking at this point about whether the possible size and scope of the problem is tractable. For example, is it possibly too big, too complex, or requiring more time and resources than are available? Finally, the AP should begin considering quality and timeliness of a possible solution. It might be that to perform the analysis that the key stakeholders expect will take more time and consume more resources than that organization is willing to commit. If that is the case, the AP should be very careful in committing to taking on the project.

8.2.5 JTA Domain I, Task 4: Refine the Problem Statement and Delineate Constraints

This task is essentially a recognition that as one learns more about the business problem; the problem statement may change or become more clearly understood. With additional information obtained from Tasks 2 and 3, the document created in Task 1 should, therefore, be modified. Many practitioners see a Business Problem Statement as an organic document, meaning it is a document that develops or grows over time as the authors learn more about the business situation, the expectations of key stakeholders, and other critical factors. We will see that the problem statement document may be revisited many times, further justifying considering it, from the beginning, as an organic document.

8.2.6 JTA Domain I, Task 5: Define an Initial Set of Business Benefits

One of the most critical initial steps in any project, and a key success factor, is to make crystal clear, right up front, the business goals and accompanying business benefits that are expected. It is possible that the organization seeking analytics expertise has created a written document to define the expected benefit; but it is also common that no such document exists, and even if one does exist, it may require work to clearly understand the proposed benefits. The analytics professional will very likely need to rely heavily on communication skills, such as interviewing and information extraction skills, to obtain and document the expected business benefits.

Typically, the business goals and benefits will be discussed, and more importantly measured, with reference to the business metrics that the organization as a whole uses to manage the business. It is not uncommon for businesses to have dozens, sometimes many hundreds, of metrics that are obtained, studied, and communicated, as they see fit, in their organization. So, it is also likely that the goals of the project you are in the process of designing will measure the benefits in terms of those same standard business metrics. For example, if you are working on a project to improve the efficiency of a fulfillment process within a retail business that is linked to a carefully managed inventory system, you would expect goals such as the following:

1) Lower the rate at which orders requiring a backorder to be required by a stated percent
2) Lower the error in demand forecasts by some stated percent or absolute error bound
3) Lower personnel costs in the warehouse (through better demand forecasting and more efficient staff scheduling) by a stated percent

The specific goals and associated metrics will be tightly linked to the goals of the organization or business and will use the same set of metrics that are fundamental to the efficient operation of that organization or business. The stakeholder interviews are the right place for the AP to delve into how the business metrics are determined or computed, where the data to do so come from, and how the resulting metrics are used to manage the business.

The outcome of this task is a clearly written section of the business statement that describes the expected business benefits. This section should be communicated in terms of the core business metrics documented in this task but would go further to lay out how, after the analytics project being considered is complete, those business metrics would be altered and, hopefully, improved. This is also a place for the analytics professional to proceed very carefully. The planned or expected business benefits need to be reasonable to achieve, considering all of the factors already discussed. These include the availability and quality of data required for the project as well as the time and resources that the organization has to dedicate to the project.

8.2.7 JTA Domain I, Task 6: Obtain Stakeholder Agreement on the Business Statement

The final task within Domain I closes the loop with the sponsoring organization or client by communicating all that was learned during prior five tasks. A critical success criterion here is that the language and terminology of this Business Problem Statement be consistent with that typically used by the stakeholders in managing their business. Using analytics jargon, complex mathematical or statistical words, or concepts that are foreign to the sponsoring organization may create a feeling among the stakeholders that the analytics team does not understand their business, or worse that the stakeholder community may lose faith in the ability of the analytics team to solve their business problem. It should be a document that is readily understandable to a mid-level manager in the sponsoring organization. The AP relies primarily on their information harvesting, learning, and communication skills, either written or verbal, to communicate the essential aspects of the business problem and the business needs to be satisfied in the project to the key stakeholders identified in this phase. A good strategy to avoid communication problems is the kind of regular communication throughout this phase, which has already been discussed. The goal in this task, of course, is to receive the written approval from the primary contact or key stakeholder on the project to proceed to the next step.

Creating the Business Problem Statement is a very important milestone for every project, whether it be an analytics project or not; and we have included this domain of the JTA as part of Business Understanding, the initial phase of the CRISP-DM methodology. However, as we will emphasize throughout this chapter, analytics projects have a number of unique characteristics that fall squarely in the hands of the analytics professional. One of those unique characteristics is the second Domain of the JTA: Analytics Problem Framing. We are including this domain along with Business Problem Framing (Domain I) as part of the Business Understanding phase of the CRISP-DM methodology because we see it as a critical aspect of the initial activities of the project and is clearly included in the full understanding the business problem, and more particularly whether there is a credible analytics problem and reasonable solution that is underlying this business situation. As indicated in the JTA, this domain includes the following five tasks:

- Task 1: Reformulate the problem statement as an analytics problem
- Task 2: Develop a proposed set of drivers and relationships to outputs
- Task 3: State the set of assumptions related to the problem
- Task 4: Define key metrics of success
- Task 5: Obtain stakeholder agreement

8.2.8 JTA Domain II, Task 1: Reformulate the Problem Statement as an Analytics Problem

After completing Domain I, we have an initial business problem statement document. To the analytics professional, this begs the question: What is the analytics problem? Indeed, in some cases there may not actually be an analytics problem. It may become apparent after evaluating the information in the problem statement that a different solution, other than an analytics approach, is the right course of action. For example, it may be clear from the problem statement that the solution will emerge by viewing this business situation from a purely management, information technology, business process, organizational, or personnel perspective. In which case, a different team may be best to carry the project forward.

However, fortunately for the analytics profession, many of the types of business problems we encounter are best solvable by bringing business analytics tools and techniques to their solution, and Domain II of the JTA is then relevant. This step is important to be done primarily by the analytics professional and his/her team of other analytics professionals, but it is also important to make sure that the level of communication started in Domain I continues into this part of the project as well. So, as the tasks in this domain evolve, they should be presented to and discussed with the key stakeholders identified earlier in the Business Understanding phase.

Consider now the specific Task 1 of reformulating the problem statement. The skills and knowledge on the part of the AP of creating the Analytics Problem Statement are similar to those for the phase of the CRISP-DM called Modeling, and there are two domains of the JTA (Methodology Selection and Model Building) that are relevant. There are also two other chapters in this book (Chapter 5 on Solution Methodology and Chapter 6 on Model Building) that cover a great deal of the same territory. To resolve this apparent overlap, we focus in this section only on the life cycle issues in the creation of the Analytics Problem Statement document, as this section is intended only to lay out the modeling approach at a high level. We will leave the activity of selecting the specific model and building that model to later parts of the project life cycle that happen *after* the data have been obtained and carefully analyzed.

The focus of this task is to review the modeling concepts (that have been discussed in Chapters 5 and 6) and, based primarily on knowledge of the capabilities of those modeling approaches and the experience of the analytics team, create a plan. Since this is happening before one has data, or the chance to analyze that data, by necessity, this plan should include a range of possible modeling approaches that will be considered. For example, suppose you are analyzing a problem that appears to require classification of a newly arrived transaction or "case" in some business situation into one of several treatments by the business. From this simple description, it appears that one of the classification algorithms common in the data mining or machine learning literature would be appropriate. But it would be premature to specify which specific model from a broad range of possible models (e.g., logistic regression, classification trees, neural networks, or ensemble models) is best to use. Such decisions are best left to the time in the life cycle where the data are available, properly prepared, and analysis of that data has begun. Therefore, the Analytics Problem Framing should include this range of possible approaches and avoid getting too specific on the solution approach.

Before one can successfully frame the analytics problem, there are a number of important activities to complete, and we suggest that these activities are exactly what are shown in the JTA as Tasks 2, 3, and 4, shown in the list above. We suggest, here, that one cannot actually complete the drafting of Analytics Problem Framing document without completing those tasks. These tasks include the following:

- Thinking through (and enumerating if possible) the key drivers or sets of relationships in the data that will allow the model to reach an acceptable solution
- Enumerating the assumptions that are needed in the modeling activity
- Defining metrics of success

Rather than changing the order of presentation, we postpone the description of those activities to later in this subsection but suffice it to say that we will need them to be completed before attempting to create the Analytics Problem Framing document that is discussed in the next paragraph.

The key contribution of this task is to take the business problem statement created in Domain I, along with the results of Tasks 2, 3, and 4 below, and expand (or append) a description of how analytical methods can be brought to bear to reach a solution. Many of the same skills required for the selection and building of the models, at later phases of the project, are also needed at this stage. Also, the resulting expanded or separate document should be created and then communicated in writing or through a presentation to those in the sponsoring organization for the project, most likely a subset of the key stakeholders mentioned earlier. Because this is a task that needs to be accomplished without having the benefit of analyzing data, it is potentially more difficult (maybe we should say, more risky) than what comes later. It is important that those performing this task have experience in building models of the type they will recommend, to avoid any pitfalls or issues in implementation that might surface at a later phase of the process when data are on hand to work with. Also, it is a good policy to provide a wide range of possible solution approaches if it is not crystal clear which analytical method will work the best prior to getting into the data.

8.2.9 JTA Domain II, Task 2: Develop a Proposed Set of Drivers and Relationships to Outputs

This task focuses on the structure of the model that we propose to build. A fundamental aspect of all modeling is the notion of the logical organization and presentation of the things we know (the inputs) being used by an analytics model to obtain the things we want to know at the end (the outputs). A big part of the modeling process is to sort through all of the data and discover the key drivers and important relationships that will be exploited so that the model we build will produce the required output. Such outputs include, for example, a correct estimate or prediction, an optimal allocation of resources, a more efficient process, or a critical business decision that needs to be made. This task is intended to lay out what we know about the drivers and relationships based on information that we can obtain from prior work, that can be obtained by interviews or discussions with key stakeholders, or can be suggested based on prior experience with modeling similar business situations. The output of this task is a technical write-up that can be included in the Analytics Problem Framing document being compiled for this task. The content of this write-up should be a clear description of these drivers and relationships, and how, when they are put together in a model, the goals of the modeling activity can be achieved.

Analytics skills required to perform this task are knowledge of the modeling tools, some initial knowledge and understanding of the data that are available, and experience in building models of the type being recommended in the past. In addition, as with nearly every task in the JTA, the analytics professional needs softer skills of written and verbal communication and persuasion to be effective in creating and effectively communicating this new write-up that will appear as a section of the Analytics Problem Statement.

8.2.10 JTA Domain II, Task 3: State the Set of Assumptions Related to the Problem

Similar to thinking through the structure of the proposed modeling approach, it is important to make clear to the sponsoring organization what assumptions are being made. The assumptions we are thinking of in this context are primarily technical or analytic assumptions that impact the modeling. For example, if we are applying a predictive tool, such models often have underlying assumptions such as normality of errors. While it is typically not useful to try to explain complex statistical or other modeling assumptions to stakeholders in the sponsoring organization who do not have training in those areas, it is important to communicate that models come with assumptions and focus on assumptions that are generally accessible to a larger audience. An example of such an assumption is that the future (e.g., demand for a product) will behave similar to the past. This is clearly an assumption analytics professionals accept in many models, such as forecasting models and other predictive models; but it is one that business professionals can understand as well. They will also understand the risk associated with that assumption, that is, there is a chance this assumption may not be fulfilled in practice.

As with Task 2, the written output of this task would be a component or section of the larger Analytics Problem Framing document. The skills and knowledge required to perform this task are similar to those required for the model structure task that involves enumerating of drivers and relationships for the proposed model (Task 2).

8.2.11 JTA Domain II, Task 4: Define the Key Metrics of Success

This activity is closely related to Domain I, Task 5, in which we define business benefits. In this task, we focus on how the modeling activity can improve the business situation. The key metrics of success will employ the same set of metrics developed in that prior task but will focus on how the model or set of models being considered operate. This includes how the specific improvements that are contemplated will be measured, reported, and interpreted. Again, the output from this task is a written section to be included in the business problem statement as modified to incorporate the analytics problem framing.

We cannot emphasize enough the importance of this task. Many projects, whether analytics projects or not, run into issues at future stages because the key success metrics were not written down clearly and goals of achievement for each of the key metrics not crisply defined. Therefore, great care and attention to details needs to be focused on this task. Along with the definition of these success metrics, it is also important at this stage to think how difficult it will be to access the data needed to compute or estimate these metrics and whether those data will be available at the time the success metrics are to be presented for evaluation.

8.2.12 JTA Domain II, Task 5: Obtain Stakeholder Agreement

The output from Tasks 1–4 is a revised business problem statement. It includes the original business problem statement as well as the analytics problem statement. These documents may be separate or a combined document based on the preferences and discussions of the stakeholders and the analytics professionals involved. The analytics problem framing is intended to be at an appropriate level of detail so that the organizational stakeholders can confirm that the analytics team understands both the business problem and the analytical solution statement, and that it is consistent and supportive of the type of solution and business outcomes that they are seeking.

The goal of this final task is to present the stakeholders with a written document, possibly along with a presentation, to explain the plan to the stakeholder group and answer questions. This is also an appropriate time to focus very clearly and decidedly on the notion of stakeholder expectations. The set of expectations includes both expectations about the outcomes, as far as business benefits, improved metrics, and so on; and expectations about the process of creating the model and the postdeployment requirements for maintenance.

As far as the process of creating the model is concerned, all analytics professionals are aware that modeling is complex, takes time, requires good data, and cannot promise success (in advance). Also, the modeling process may be foreign to many business people. A good example of this is building an optimization model. Practitioners who work in this area all know that despite spending a great deal of time working to extract the full set of constraints, the first time the analytics team "runs" the newly created optimization model, it is very common that the stakeholders working closely with the analytics team will say something like: "that cannot happen in our business." Of course, what really occurred was that the constraints that would preclude that model outcome were not included in the model run. So, the AP and team go back to the model, add these new constraints, and the process continues, possibly with additional such iterations until the right model is created and running.

Experienced professionals know that the process just described for optimization models and, indeed, for nearly every type of model that analytics professionals build, is common and in fact expected. However, it may not be what the stakeholder team monitoring the model building process expects. Therefore, this is the time to begin to communicate those sorts of expectations. The Analytics Problem Statement should have enough understanding of the types of modeling under consideration to set expectations; and it is critical for part of the communication process to the stakeholders at this phase to include examples of the type described above, so that when the need for iteration between analytics and business experts happens, it will not be a surprise or a cause for concern on the part of the sponsoring organization.

One other expectation to set: the postdeployment needs of the model. When the team is thinking about creating a new model, all of the attention is on the present: designing the solution, getting data, building the model, getting the improved business outcomes; but, as we will describe in the last phase of the life cycle, models also need regular maintenance, and this may not be what key stakeholders at the sponsoring organization expect. The Business Problem Statement document is a good place to plant that seed, so that when the focus turns to deployment and postdeployment phases, the stakeholders will have heard this before and are ready to plan for it at that time.

The desired outcome of this task is to receive agreement from the stakeholders in the sponsoring organization that both problem statements are acceptable and will be supported by this stakeholder group moving forward.

8.2.13 CRISP-DM Phases 2 and 3: Data Understanding and Data Preparation

Data have been called the "oil" that both powers and lubricates the analytics engine, so being successful at these phases of the project is very critical to success. This is also part of the analytics project life cycle where, rather than discussing and researching the data with secondary information (e.g., using a data dictionary that describes the data), we actually get our hands on the data, begin to look at it, clean it up and start to discover, and document the relationships that exist between different data items.

The CRISP-DM methodology includes two separate phases dedicated to data. The first phase is called Data Understanding, and the second phase is Data

Preparation. By contrast, the JTA merges all of the activities associated with data into a single Domain called simply, Data. What both of these slightly different methodological structures agree on is that data are a core component of all analytics projects. The CRISP-DM shows this visually (in Figure 8.1) by placing the image of the data in the very center of the diagram, invoking the notion that all of the phases of an analytics project revolve around the data used to create the analytics model and is critical to the success and effectiveness of what is created.

As was mentioned at the introduction of the CRISP-DM, the methodology recognizes the nonlinear characteristic of this type of work, where information obtained at later stages of the process, for example, about how the structure of the data impacts the business situation, may require revisiting some earlier stages of the business understanding activities (and vice versa). The idea that one obtains the data and retreats to his or her office to analyze it has been shown repeatedly, in practice, to be doomed to failure. The understanding of the data requires regular, and often intensive, interaction with business experts to reach the level of data understanding that is required to create a successful and sustainable business analytics solution. In addition, the link to the business needs to continue literally throughout the entire process of planning and implementation of the analytics solution.

We find again that the detail in the JTA for the data domain provides a deeper understanding of both of the CRISP-DM phases of data understanding and preparation. The JTA Data domain does not order the tasks into those that involve data understanding and those that focus on data preparation. To make this discussion simpler, we will discuss the tasks of the JTA in the order they appear in the CAP JTA document, but we will make it clear which aspects of those tasks apply to either understanding or preparation. Frankly, any activity in which we work with data, either directly or indirectly, brings better data understanding. Certainly, some of the tasks focus more on one than the other, for example, Task 3 is called "Harmonize, rescale, clean and share data" is focusing mostly on preparing the data for use.

The tasks included in the JTA Data Domain are as follows:

- Task 1: Identify and prioritize data needs and sources
- Task 2: Acquire data
- Task 3: Harmonize, rescale, clean, and share data
- Task 4: Identify relationships in the data
- Task 5: Document and report findings (e.g., insights, results, business performance)
- Task 6: Refine the business and analytics problem statements

While Chapter 4–The Data–already had provided a great deal of important information about data, in this chapter, we will focus on the process steps and their role in the full life cycle of an analytics project.

8.2.14 JTA Domain III, Task 1: Identify and Prioritize Data Needs and Sources

Before beginning the hands-on part of the data work, the AP pauses one more time to make sure that the data needs are clear and the sources are known and available. This task falls squarely into the CRISP-DM notion of Data Understanding, and the analytics team's emphasis, here, is learning as much as possible from documentation provided by stakeholders as well as interviews and workshops with stakeholders to make sure the data needs are clear and the level of understanding of the data is high. This is yet another task that focuses on the softer skills of communication, interviewing, information acquisition, and business understanding.

The expectation is always that the process of identifying the data that is needed and prioritizing the acquisition process will proceed smoothly and that stakeholders are totally forthcoming with sharing information needed at this point. However, the team should be ready for issues that might arise. One common issue relates to security and privacy, but it is also possible to encounter issues related to who is the keeper of the data, who controls its use (and distribution), and who needs to be involved in approving the process of moving forward. The key stakeholders will be critical in deciding where to put the priorities for data acquisition, which comes next, who to contact (the sources), and when more senior stakeholder involvement may be required.

The output of this task should be a plan, preferably a written document, or possibly a slide presentation or less formal document. The document should lay out exactly which data items are required, where they will come from, what form they will be transferred in, and, if possible, a time frame for accomplishing the data transfer process.

8.2.15 JTA Domain III, Task 2: Acquire Data

This task continues Task 1 toward the process of actually acquiring the data. With a plan obtained from the prior task, this task is the implementation of that plan. The range of experiences in acquiring data is very broad. In some cases, the organization is fully prepared, the appropriate senior management involvement took place, and, when asked to provide the data, it is simply provided–the ideal situation. Another possibility is that the data may be "public data" and the access to data has been set up to be seamless and can be initiated by the analytics team through a known public process–another ideal situation. However, the other end of the spectrum is also possible. It is not uncommon to encounter organizations in which control of data is closely linked to the organization's power structure. So, even if there are senior management approvals, the data owner may not simply jump to provide what is requested. This is one of the places in a project where "the rubber meets the road" in terms of whether the stakeholder involvement is at the appropriately high or influential enough level. If the stakeholder clout, when carried as far into the organization as is possible, turns out to be insufficient to break through a log jam, it is possible that the project may find its end right at this point. If the stakeholder clout is sufficient, the worst case situation is that a reluctant "data owner" can delay but not stop the sharing of the data that is required. However, experience indicates that the process may be much longer than the newcomer to these situations would expect. Data owner tactics that the analytics team might encounter include the following:

1) Seeking additional approvals and invoking a much longer process.
2) Seeking to redo the process of justification of the entire project for the data owning group within the organization, again bringing delays.
3) Micro managing the specific data items that were requested, hoping to exclude as much as possible from being released.
4) Limiting, or extending into the future, the times they are available to meet to actually exchange information about the data request and providing the actual data.
5) Setting up complex data usage requirements, for example, that the data can only be accessed on the organization's IT systems, and finding that there are limited physical resources for your team to sit and do this work.

While the prior discussion is clearly unfortunate when it happens, the AP should be heartened in that this "worst-case scenario" is not the norm; but it is always prudent to be prepared for the data acquisition task to take longer than one might imagine and require multiple steps, many meetings, repeated involvement of the stakeholders within the organization that you are working with, and could have possible impacts to schedule and cost of a project. Again, success at this stage relies primarily on soft skills such as communication, persuasion, and negotiation to get the result that is desired.

8.2.16 JTA Domain III, Task 3: Harmonize, Rescale, Clean, and Share Data

Clearing Tasks 1 and 2 of this domain means that you actually have the data you requested and it is time to begin working with it. Nearly every analytics professional has spent endless hours in the process of "cleaning" data, and, as with the acquisition task, the amount of time this can take is both difficult to estimate and almost always longer than you might think. This task falls clearly under the CRISP-DM notion of data preparation, in that we are working to get the data ready to use. It is not reasonable to include all of the types of issues that one might encounter in the process of harmonizing, rescaling, and cleaning data, but the following are some of the most commonly encountered ones:

- *Structure of the Data:* It is typical to get data extracted from a relational database, and therefore in a structure that is efficient for storage and, even, extraction but may not be structured optimally for analysis. A first step in many data harmonization tasks is to create a "normalized" data structure, typically laid out in a rectangular data structure with rows and columns where rows are the records (cases) and columns are the fields (variables, features). This structure is generally accepted to be the one best suited for analysis, though there are exceptions, such as spare matrices in an optimization where alternative data structures are actually better.

- *Missing Data:* Real data sets routinely have missing values; sometimes they are marked as such, and sometimes the field is simply empty. In other cases, the lack of a value is "coded," maybe as 0, when in fact it is missing. The analyst needs to make sure they fully understand how the data were prepared (and how missing values are "coded") and also must formulate a clear plan to handle missing data (e.g., will a record with missing values be dropped? will missing fields be imputed?). There are many issues that may need to be resolved in this regard.
- *Merging Data Sets:* It is typical that data come from a number of different data sources, and it is also common that the values for a particular data item differ between those sources. The reasons for this are many: the data could be outdated, there could be an error, the data may have been recorded in different units, and so on. In any case, the AP must resolve all of these issues.
- There are many other issues that can arise–many based on very specific knowledge of the data and their use in the business, requiring the depth of business or organizational knowledge that we have discussed often in this chapter. Such issues need to be resolved in close collaboration with the organizational stakeholders.

Successful completion of the cleaning process requires skill in manipulating the data, knowledge of the business, frequent and substantial interaction with key stakeholders (especially those with deep knowledge of the data), and enough time to get it all done. The outcome of this task will be a data set that is much better prepared for the analytical work that is ahead. Though, as with nearly every task in this methodology, it is likely that the analytics team will ALSO obtain greater data and business understanding from the process of cleaning the data, and further, the findings obtained are likely to impact both the business and analytics problem statements. Such data findings should be documented and will be collected (in Task 5) into a full set of data findings.

8.2.17 JTA Domain III, Task 4: Identify Relationships in the Data

With initially clean data in hand, the analyst can turn attention to the modeling and analysis questions that are critical to the success of the project. As discussed earlier (but in the situation when the data was not actually in hand), discovering key drivers and identifying relationships in the data is one of the fundamental activities that is performed by the AP; and it is squarely in the category of data understanding. The specifics of how to discover the set of relationships depend on the structure of the data and on the proposed model or range of models that are under consideration. Earlier chapters in this book cover a multitude of models, and how they are analyzed or approached, so we will not try to repeat those discussions here. Suffice it to say that the task of finding the key data drivers and understanding data relationships in the data is an important milestone along the analytics project life cycle.

8.2.18 JTA Domain III, Task 5: Document and Report Finding

The data work that has been described, acquisition, cleaning, and data exploration (where we are seeking key internal relationships) represent a great deal of work, and it is a good practice to write down all that transpired in performing those tasks and, then, to communicate those findings to the stakeholder community. The type of communication tool, report, presentation, or meeting depends on the preferences of both the stakeholder community and the analytic professionals involved; but it is critical that the passing of this information happen in one form or another.

8.2.19 JTA Domain III, Task 6: Refine the Business and Analytics Problem Statements

At this phase of the life cycle, the data are acquired, cleaned up, and much better understood. Throughout these tasks, a great deal of data understanding was obtained, and it is likely and expected that this understanding will have an impact on the plan for the project. Therefore, the final task in the domain is to modify both the business and analytics problem statement documents with the updated information, plans, and ideas; and then, present the revised documents to the stakeholder group. The objective of that activity is to seek an updated approval and/or agreement to move forward from the appropriate organizational structure monitoring the analytics project.

8.2.20 CRISP-DM Phase 4: Modeling

Modeling is the phase within the CRISP-DM process that analytics professionals are typically most energized about. It is the place in a project that we are most looking forward to, because this is where we bring to the business situation our "secret sauce." We are hopeful that with the business and analytics problem statements we have developed and have come to deeply understand, and the data we have acquired and prepared for use, we will create a solution that meets the needs of the sponsoring organization, improves the overall business performance (following the business metrics set forward in the plan), and turns out to be sustainable for the expected lifetime of the application.

The CAP JTA has two domains that cover the core activities of the modeling phase of the project. Those domains and the tasks encompassed by them are as follows:

- Domain IV: Methodology (Approach) Selection
  - Task 1: Identify available problem solving approaches
  - Task 2: Select software tools
  - Task 3: Test approaches (methods)
  - Task 4: Approaches (methods)
- Domain 5: Model Building
  - Task 1: Identify model structures
  - Task 2: Run and evaluate the models
  - Task 3: Calibrate models and data
  - Task 4: Integrate the models
  - Task 5: Document and communicate findings (including assumptions, limitations, and constraints)

This book contains two very detailed chapters that cover this part of the life cycle in excellent detail; therefore, we refer the reader to those chapters at this time. Chapter 5–Solution Methodology–is focused on JTA Domain IV (Methodology Selection), and Chapter 6 is focused on JTA Domain 5 (Model Building).

8.2.21 CRISP-DM Phase 5: Evaluation

While it is important in any business project to evaluate performance at each stage, it is especially important to focus on evaluation for an analytics project. Analytics projects are driven by data in that we require data to build such models. But, in addition, the output or results of those models are also typically numeric. Of course, each specific model or modeling approach (as discussed in Chapters 5 and 6) is different as to what types of results are produced. A predictive model typically has a resulting error estimate using one of the many standard methods. Classification models are evaluated looking at the percentage of classification errors and other popular measures such as precision, recall, and F-score. Prescriptive models produce "optimal solutions" such as resource allocations or product production plans that can be analyzed by business models that compare the "optimal" solution with a current or other possible solutions. The key is that each model will have a standard set of measures that can be employed for evaluation. As with the prior phase of the CRISP-DM, we will refer the reader to the specifics of Chapters 5 and 6 that cover each of the models and that include clearly defined methods for evaluation for each.

We focus primarily in this life cycle discussion on what the analytics professional does with the evaluation results that are obtained at the end of the modeling process. With respect to alignment with the JTA, it is a little less clean in this situation. Looking at the tasks in the Model Building Domain of the JTA, Task 2 mentions evaluation and Task 4 mentions reporting, both of which are clearly important to evaluation as a phase in the life cycle of a project. However, Domain 5 of the JTA, Deployment, starts off with two similar tasks: These first two tasks are as follows:

- Task 1: Perform business validation of the model
- Task 2: Deliver report with findings

We will discuss both of these important aspects of evaluation in the remainder of this Evaluation phase of the project.

In each of the earlier phases of this life cycle, the documenting of business goals, measured by relevant business metrics, were singled out as critical activities and are important activities in the creation of such documents as the Business and Analytics Problem Statements. Common examples of those metrics are as follows:

- An expected/targeted percent or actual dollar reduction in cost
- An expected/targeted improvement in efficiency
- An expected/targeted increase in revenues (sales) or profits
- And for public sector organizations, an expected increase in coverage, speed of performance, accuracy, or quality of operations

One important point to bring forward at this time is that the evaluation process often produces an ESTIMATE for the performance rather than actual performance. By this we mean, an organization cannot know for sure that a proposed model or "decision," in the case of an optimization solution, will produce the business benefits promised (e.g., sales, profit, improved quality) until that solution was put into use, data collected over the period of operation of the new approach, and analyzed to see what the actual business improvement was. The evaluation process at the immediate conclusion of the modeling process must, therefore, rely on business models, statistical methods, and other standard methods for assessing performance PRIOR to the actual implementation and use of the model. This is exactly the activity that Task 1 of the Deployment JTA describes as Perform Business Validation of the model. However, when the sponsoring organization decides to move forward with implementation, it is also important to put in place the business infrastructure, methods, tools, and reporting procedures, so the data will be available in a reasonable amount of time, to measure ACTUAL business performance obtained when using the business model, system, or decision that was the result of the analytics modeling activity we are discussing.

We will come back to the notion of evaluation of the models in their operating environments at this critically important later phase in the analytics project life cycle.

After the validation task is completed and the results obtained, it is time to communicate the results. In Task 2 of the Deployment Domain, the description is: Deliver Report with Findings. The results of the evaluation process at the model development phase can have many outcomes. Of course, we all hope that the model developed turns out to be exactly what the sponsor expected, that the results exceed all of the stakeholder expectations, and that the only sensible next step is to move forward with implementation or operation of the model/system. We will pick up the thread of the life cycle when "go forward to implementation" is the decided upon, in the CRISP-DM phase called Deployment.

It is also important now, while we are still discussing the Evaluation phase of the project, to consider what happens if the evaluation produces issues. By "issues" we mean something that requires further consideration, discussion, or analysis on the part of the entire project team, both the analytics team and the sponsoring organization. Such issues, in particular, mean that the project is not ready to move to the next phase. Among the issues that may occur are the following:

- The expected improvement in cost, revenue, profit, or time (efficiency) is below what was expected or required.
- The model takes a much longer time to reach its solutions than expected, and may need continued work to remedy this problem.
- The accuracy, variability, or sensitivity (to model inputs) for the model is such that the answers engender less confidence in the results than the sponsor and the modelers had hoped for.
- There are many other possible issues.

Before moving forward (whether there are serious issues or not), the generally accepted approach is to stop at this point and document the results of the evaluation process to the sponsoring organization. The information to communicate should include findings and recommendations of what should be done next. It is important to make sure that the sponsor continues to see the analytics team as a good partner by being forthcoming, and in particular not withholding or delaying communication, of these intermediate findings at the end of the initial modeling stage. The analytics professional, as the leader of the analytics team, should find the best medium for communication: a report, a presentation, maybe just an agenda for discussion at a meeting, and come together with the sponsor stakeholders to present the findings. If the findings are sufficiently positive, the expected recommendation is to go to the next project phase. If the findings show issues, as already mentioned, we revert back to the structure of the CRISP-DM methodology.

The CRISP-DM methodology includes an important feedback loop that is, for that very reason, located at the Evaluation phase. As seen in Figure 8.1, there is an arrow from the Evaluation phase to the Business Understanding phase. The message here is take the findings of the Evaluation phase, especially if those findings include any issues, and link back with the sponsoring organization, the business, to reconsider some or all of the assumptions, expectations, business issues, and goals of the project. Projects do not, typically, require scrapping all of the work to that point and starting over. More often the evaluation will point out specific issues, for example, clarifying expectations, getting deeper understanding of some of the data, and formulating alternative business strategies or goals. Here are a few possible outcomes or recommendations typically uncovered:

- Gaps in the data were found, so the team will take on the task of obtaining more or different data (going back to the data phases of the methodology).
- The model did not perform as well as expected, so the team may want to look at alternative modeling approaches (going back to the modeling phase).
- The variability of results has led the key decision-makers in the sponsoring organization to rethink the risk issues underlying the use of the model (going back to the business and data understanding phases).

Or course, there are many other possible outcomes, but the key is that the analytics modeling team and the organizational stakeholders will come together and decide the next stage in the process collaboratively, and then the analytics team will implement that jointly arrived at plan.

Hopefully, after the analytics project team has completed revisiting prior phases of the process, and after completing another round of evaluation, the issues that surfaced in earlier evaluation processes are resolved, that the business performance metrics related to the project are found to be in acceptable ranges for all involved, and the final decision is to move forward to the next project phase of deployment.

8.2.22 CRISP-DM Phase 6: Deployment

This phase of the CRISP-DM process is described simply as deployment, but the JTA has two remaining domains: deployment and life cycle management. Frankly, the linkage between the CRISP-DM methodology and the JTA falls apart somewhat at this point. One reason is that the final domain, model life cycle management, includes essentially everything that is in this chapter, and some of the tasks included there involve topics we have already discussed, such as documentation of the initial model, and other activities include tasks that intended to take place during and after deployment. To simplify the discussion, we will break this section into two parts:

- Activities up to and including delivery of the model (deployment)
- Activities that take place from the time of delivery forward (postdeployment)

In Domains VI and VII, the JTA contains activities that fall into both of those two categories and even some activities that happened earlier in the process. The following bullets contain these two JTA domains:

- Domain VI. Deployment (the ability to deploy the selected model to help solve the business problem)
  - Task 1: Perform business validation of the model
  - Task 2: Deliver report with findings
  - Task 3: Create model, usability, and system requirements for production
  - Task 4: Deliver production model/system*
  - Task 5: Support deployment
- Domain VII. Model life cycle management (the ability to manage the model life cycle to evaluate business benefit of the model over time)
  - Task 1: Document initial structure
  - Task 2: Track model quality
  - Task 3: Recalibrate and maintain the* model
  - Task 4: Support training activities
  - Task 5: Evaluate the business benefit of the model over time

The following discussion will refer back to specific tasks in both of these domains, but the discussion will focus simply on before delivery of the model and after.

8.2.23 Deployment of the Analytics Model (Up to Delivery)

Whether the analytics team (led by the analytics professional) is the group responsible for implementation of the model or system they have designed often depends on the size and complexity of the model created, the complexity of the data involved, and whether the analytics model fits into larger or existing IT systems within the sponsoring organization. In cases where the model size is smaller than medium size and when the system would be described as "standalone," the management within the sponsoring organization may decide to have the analytics team assume the responsibility for implementation (building, coding, and delivery) of the model they developed. However, when size, complexity, and organizational interactions are larger (and often no matter what the size), the sponsoring organization may alternatively place the responsibility for implementation to the division or group of the sponsoring organization that is responsible for IT. In that situation, the analytics team would typically move to a support role, rather than the leadership role they had in the design phases.

*Note * here indicates a task in the JTA that is not included in the CAP certification exam questions.

The first activity of deployment is to document production requirements. Since at this phase of the project the model or system is built, fitted, or "learned" from the data, and recently tested, the analytics team should fully understand the production requirements. The goal then is to write them down clearly and completely so that the implementers can move forward with clarity to a successful implementation. Even if implementation is the responsibility of the analytics team, it is important to take the time to create written production requirements as they are also useful for testing and other needs later in the life cycle.

The activity of creating a production requirements document is a standard systems design or system architecture activity that is generally the responsibility of IT professionals to do, but certainly to manage and oversee. That team will make clear what sorts of documents are required, provide a form or template in which to create them, and, often, aid the analytics team in creating the documents so that they are acceptable to the implementation team. One important role in this process is, often, to create a test plan. IT professionals usually set up carefully planned and staged procedures to test whether the software or system developed is working as expected. However, when the system is an analytical model, it often requires someone with deep understanding of that model to create worthwhile tests.

Another typical area of collaboration between analytics teams and IT implementation teams relates to the technical code related to the model. For example, if the model implements an optimization model, the analytics team likely used one of the standard commercial or open-source solvers; and it is likely that the IT professionals are not expert in that software. In those cases, the analytics team may be responsible for coding and testing a module or executable component of the larger system following guidelines and procedures specified by the IT team.

The next activity in the JTA (chronologically) is the Deliver Project Model/System, (Task 4 of Domain VI). You will notice that this task is marked in the JTA as a component that is not included in the specific certification process and, therefore, the analytics professional seeking the CAP certification will not find questions on the certification exam on this topic. This task falls more specifically in the area of IT system implementation, but it is typical that detailed knowledge of the analytics models being deployed will continue to be required by the implementation team doing the coding, testing, and final delivery of the model or system. The JTA does include Task 5 titled "Support deployment," and we will discuss this task now.

This task is a recognition that the analytics professionals need to be engaged and available as required by the implementation team to solve problems or deal with issues that arise in that activity relating to the analytics model being deployed.

Example of the types of model support that come up in many implementations are as follows:

- The data used in driving the analytics model have errors or other issues that are shutting down or crashing the system.
- The model is taking too long to run or in some other way impeding the operation of the system environment where it is located.
- Interactions with other parts of the larger IT architecture are encountering issues that the IT team cannot diagnose.

Those examples and many others that might occur require knowledge of the data and of the model, and that knowledge resides with the analytics professionals who created the model. There may be a process in place to engage the analytics team when an issue comes up, such as the creation of a "ticket" that the team is required to respond to in a particular manner, or it may be less formal where a manager or member of the implementation team calls or e-mails someone on the analytics team to seek help. In any event, the analytics professionals need to be there to help with implementation all the way through to the final testing and handoff to an operational entity to oversee.

The entire set of skills required for design and building of the model are required for this phase as well. They include all of the modeling and analytical skills it took to create the model, but more importantly they also include all of the communication or softer skills of collaboration and persuasion that an analytics professional needs to be successful in his or her work.

The next topic to discuss is training. The only mention of training in the JTA is in Task 4 of Domain VII (as you see the activities are not mentioned necessarily in the order that they often occur). We are including this as prior to handoff of the system to the operating entity because we see training as a critical and important part of any deployment. The responsibility for training may reside with either the analytics team or the larger IT team, but what is most important here is that the analytical knowledge of using and, more importantly, interpreting and explaining the model outputs will typically reside in the analytics team. Therefore, creating training materials and possibly being the individual to deliver training modules (be the "trainers"), create tests, or other ways to measure how users are faring in using the model are good activities residing with the analytics team.

As with many other aspects of the entire process, successful completion of the training activity requires an appropriate mix of strong analytics input and excellent communication skills. The individuals preparing training materials need to totally understand how the model works and must be able to explain these workings, both in writing and verbally, to the individuals from the sponsoring organization who are responsible for using the model. Further, they should be able to break down the separate aspects of the model into manageable chunks of information so they can be presented in the training activity, intersperse practice scenarios, and provide a method by which a person (if tested) can demonstrate sufficient mastery of the model.

8.2.24 Post-deployment Activities (Domain VI: Model Life Cycle Management)

We come now to the second aspect of deployment: postdelivery monitoring and reporting. Several of the important tasks included here show up in the JTA in domain VII, model life cycle management. It is reasonable for the reader to ask at this point: "Hasn't this entire chapter been about life cycle management?" And of course that answer is yes. Indeed, it would have made sense to point out the importance of the life cycle right up front in the JTA, but we see that it is mentioned for the first time as the final domain of the JTA. One good explanation is that, as you come to the end of any project and especially an analytics project, it is important to focus attention on the sustainability and continued usefulness of the model that has been created.

Therefore, we move now to the phase of the project AFTER the model is delivered and taken over by the operating organization. Three tasks in particular from Domain VII focus attention to the postdelivery time frame. They include tasks for tracking model quality (performance), recalibrating and maintaining the model, and evaluating the business benefit of the model over time. The second of those (recalibrating and maintaining the model) is marked with an asterisk in the JTA, indicating it is not intended to be included in the certification exam; however, the tasks and skills required to recalibrate and maintain are essentially those same skills that were required to build, test, and deploy the original model. To be consistent with this document being focused on the basic CAP certification process, we will focus on the other two tasks.

Consider first tracking the model performance. At an earlier stage in the life cycle modeling discussion, we mentioned the importance of preparing the stakeholders within the sponsoring organization to the need for model maintenance. Nearly all analytics models require regular maintenance of some sort or another, and that specific maintenance activity depends on what type of model was built. For example, suppose the team delivered a demand forecasting model for a manufacturing or retail entity. It is well know that demand can change over time, and not recalibrating, re-estimating, or redoing such a predictive model may lead to larger and larger errors in forecasts over time. It is important to stay on top of such loss of accuracy or quality in a model quickly as it is not uncommon for users of such a model to blame the concept or process of modeling for the problem rather than the fact that the data or underlying relationships between predictor variables and the quantities being forecasted have changed or evolved over time.

Hopefully, during the earlier communication processes and delivered documents, the sponsoring organization was prepared for the need for maintenance activities and they have planned them into the program on an ongoing basis. A key component of the maintenance process is monitoring and tracking model performance over time, regularly (weekly, monthly, quarterly, maybe even yearly) depending on the specific model and the types of decisions that it is designed to make. The findings of the monitoring process should then be reported to cognizant individuals in the management chain at the sponsoring organization, so that any degradation in performance can be highlighted, discussed, and, when it reaches the level requiring action, all of the individuals or organizations involved are ready to perform the required recalibration or maintenance activities.

Being successful at this postdeployment phase of the life cycle is often the most critical reason why some models are sustainable for long periods of time and why others are either stopped quickly or slowly fade away. The key issue here is setting expectations. Some sponsoring organizations believe they are bringing an internal or external (consulting) team of data scientists to the table to build them a model and, then, that team goes away, leaving them to monitor and run the model or system. This situation is exactly the scenario that was warned against earlier: the model becomes out of tune, the outputs created by the model begin to produce less and less useful results, and managers decide the problem is the MODEL or even the entire process of creating a model, and they abandon it. However, if management has set in place the cycle of monitoring performance, evaluating whether the model is in need of maintenance, and implementing maintenance activities when needed, the worst-case scenario described previously should not happen.

We have not yet mentioned the notion of evaluating business benefit over time. This area is quite similar to monitoring the model performance and quality over time; but the times between these evaluations may be larger than the times involved in model performance monitoring. The driver of when to do business performance measurements is how long it takes to begin to see changes or impacts in how the larger business if operating. It is possible to see these in as early as a month, but, mainly because of business variations being impacted by many factors (general business performance, economic conditions, competition, politics, etc.) it is prudent to think about quarterly and possibly yearly as the proper time frame for such major evaluations. The business or possibly the entire company has many metrics to monitor business value, including such things as market share, revenue, profit, quality, and so on. The business should set out appropriate time frames for examining these factors, relating them to the specifics of the model.

Take the example mentioned earlier of creating a forecasting model. If that forecasting model is working as expected, there are a number of related business metrics that should begin to improve. These include such things as reduction in inventory costs, a larger number of inventory turns, reductions in back orders or lost orders because of stock-outs, improved revenue, larger market share, and even greater profits. As much as possible, it is important to link the operation of the model built to these core business metrics and to put together a case (in the time frame decided upon) demonstrating how the model created is bringing business benefit. Of course, we hope that the model does indeed bring the business value expected. If, however, the results of this evaluation do not produce the results expected, it may require revisiting the model's content, data, or use. At this point, it is typically in the hands of the management organization to chart the appropriate course of action, but at least they will have the information to make an informed decision.

8.3 Overarching Issues of Life Cycle Management

While the prior section of this chapter walked through the entire life cycle of an analytics project, that discussion repeatedly included a number of terms and concepts that, we believe, need to be separated out for special emphasis. They include the following:

- Documentation
- Communication
- Testing
- Creation and use of metrics, including success criteria

Each of these concepts will be discussed in separate sections.

8.3.1 Documentation

At many places in the life cycle already described, we mentioned the importance of documentation: primarily for the Business Understanding and Data Understanding phases, carrying through to the Deployment phase. In those phases, we described the value of specific components of the Job Task Analysis, including the Problem Statement and the Analytics Problem Statement, and the need to have these documents regularly updated throughout the process as new or better information became available to the analytics professionals. Finally, as we came to the deployment process, we focused on the need for the analytics professionals to layout a clear path for implementation and use of the model through documentation to the team of implementers (who often come from a different organization), even if they participate in or retain some of the deployment tasks.

While documentation is a mainstay of business processes, industry and professional standards, and a major focus of academic and other research endeavors, we believe that the task of developing documentation is especially important in the context of a Business Analytics project for one simple reason: stakeholders involved in such projects have markedly different skills, experiences, training, and roles. So, reaching a clear understanding across all of those different backgrounds and roles is a particularly difficult goal to achieve. Good documentation is the major tool we have to meet that challenge.

Often the organization that needs the assistance is focused on the business issues, ranging from financial operations or supply chain management through sales and marketing. They are focused on the most basic business metrics of cost, revenue, profit, or other measures of organizational performance. We have made the case for why analytics professionals need to be focused on these metrics as well, but it is also clear that they need to be thinking also about the data, modeling formulation, model building, and performance of the models built. The place where all of these priorities come together is in the documentation that drives the process.

The business leaders who are requesting or using the results of the analytics project need to understand what the analytics team will be facing: the complexity of the data, the challenges of model size or other complexities, the certainty that any model developed will need to be maintained over time so that resources will be committed to insure that models created in the initial process will be sustained throughout the projected lifetime of that model. At the same time, the analytics professionals need to understand the business needs and weave those into every decision they make on the modeling and implementation path.

The documents described throughout the life cycle of a project are, therefore, critical to success. We have proposed the idea of an "organic" document development approach. By this we mean the concept that one creates the core document described above in the early stages of the project, but changes and improves, as the team obtains better understanding of data or the business needs, early results of modeling activities and testing that is done on the model. The final documentation should include the best understanding of all of these components as the model comes to deployment. The deployment document continues the development of a clear path, in this case for the team responsible for implementation of the model, to assure that the model that was created functions as designed.

While documentation is very important, it is also very difficult to do well. It is well known that IT systems fail most often because the requirements of the project are not clearly understood (documented) during the implementation of a project. They are also subject to what is typically referred to as "project creep," which is the situation in which the specific performance or functional requirements tend to increase in scope through the process of trying to design and build a system. This tendency to have project creep can be attributed most often to a deficiency in understanding of the goals and challenges of the project from one of the key partners in the project (the sponsoring organization or the analytics team building the model). Good documentation is designed to address and prevent the various misunderstandings from happening.

Unfortunately, there is no secret solution to avoiding the problems or difficulties described earlier. It comes down to old-fashioned values and work ethic. The analytics team needs to develop a passion for success and a commitment to hard work. The main task for the analytics professional is to take the process of creating documentation very seriously. The AP must fully own the process of creating, editing, and improving the documentation. They need to regularly discuss it with the stakeholders from the sponsoring organization to work tirelessly to make sure that there are no areas of misunderstanding. Frankly, the AP needs to be passionate about the documentation, so that it will fulfill its intended goal that all parties involved understand the project objectives, the process, the limitations, and the ongoing need and longer term responsibilities of taking on such a project.

8.3.2 Communication

Communication is closely aligned with the documentation section above, and the reasons for having regular and quality communication between the sponsoring organization and the analytics team are important to ensure that all parties have the same understanding of the project, it goals, how it is progressing in meeting those goals, and ultimately the business benefit that comes from the project. But there is a subtle difference between documentation and communication. Documentation is primarily a document or other tangible written or visual artifact (like a graph or sketch). A more general term for communication includes verbal communication, something that requires more care, simply because you cannot go back and make judicious edits to something already communicated verbally. Of course, it is possible to revisit something discussed in the past and adjust or change what you communicated; but doing so too often can breed a lack of trust in the business relationship which can, in and of itself, be a problem for the successful completion of a project. Therefore, we suggest that the same high level of passion for verbal communication be a core focus on the analytics professional, and the specifics of how that type of passion is implemented is discussed here.

There are number of simple strategies that will foster better communications:

- Prepare carefully for meetings, including practicing what you intend to say. It is good to go through with others the specific words you will use to make sure that others perceive what you are saying correctly.
- Agree with others on the team how best to describe a specific issue. We are not talking here about "spinning" a problem you are facing. We believe that it is always best to be transparent, clear, and unambiguous, even if the issue being communicated is of a problem encountered or even an error made by the analytics team. The goal should be to lay out an issue, describe the options for moving forward, and work to get consensus between both the sponsoring organization and the analytics team.

- Follow up verbal communications with written communication, such as an e-mail or formal minutes of a meeting where the topics discussed and conclusions reached are documented. Alternatively, the follow-up could be a revision of a document that was discussed.
- Err of the side of overcommunication. You will seldom experience a client who says, "You told me about this requirement too many times." However, it is very common to hear words like: "you never told me that before" or "I did not understand that could be a problem."

Lastly, for this topic there is a word about the staff with whom you may be working. The types of individuals drawn to business analytics are generally individuals comfortable with mathematics, statistics, engineering, and, in general, analytics-oriented projects. It will not be a surprise to many who studied these areas in college or graduate school, or who worked with professionals in these fields, that sometimes this population is less comfortable with communication than those who are, for example, in general business, the social sciences, not to mention fields such as literature, languages, or related disciplines.

We are not suggesting that analytics professionals do not have the capability to be outstanding communicators. Quite the opposite, we believe the training in the analytic fields focus on clarifying of ideas, sensible organization of information, and getting quickly to the heart of the matter. It is critically important for people in this field to focus on being successful communicators. Some in this field find themselves working in a business environment where the preferred language for business communication is not the one they grew up speaking. In such cases, extra effort needs to be put on clarity in communication. The understanding of how some in this field are challenged in the communications areas applies both to the AP and to those who the AP is leading or managing in such projects. As a leader, the AP also has a role as mentor, to encourage their staff to develop better communications skills, and to become more effective professionals themselves.

In summary, the analytics professional needs to be passionate about communicating, both in written and verbal forms, with the sponsoring organization that is the beneficiary of the analytics work they are doing. It is hard work, but the payoff will be worth it.

Interview With Russell Walker

Russell Walker, Clinical Associate Professor at Kellogg School of Management, responded in the following manner when asked about the importance of the analytics professional's communications with her or his stakeholders.

It is enormous. I say this to many people that I've worked with and to my students–If you have a great model but the people above you who are responsible for making decisions and allocating resources don't understand your model or don't have confidence in how your model operates, your work was largely useless. So the process of being a successful analytics professional does not end once the model has been developed. The analytics professional must communicate it, maybe even sell it, to the executives who will be responsible for taking action, and must gain these executives' confidence.

I have seen many people in my career, students and other professionals, say, "Well, that's not my job, the users of my model should take a class in statistics." Although I can relate to that attitude to some degree, unfortunately that is not how a corporation may work. Suppose you are the CEO and you've deployed a team to investigate some issue. If the team develops an analytical model that you don't feel comfortable using, guess what, you won't use it, and you make your decision based on some other paradigm. And that's unfortunate because you might make an inferior decision.

We make decisions based on what we feel most comfortable with. So a very important component of being a successful analytics professional and producing results and models that will be used, and so having an impact on an organization, especially a for-profit organization, is indeed that communication. Sometimes you have to simplify, take a complex model, and build a presentation that says here is the input, here is the output, and here are the most important factors. Building a graphic, some sort of schematic shows what the model does, can be much more effective than presenting the model as a black box or providing a collection of *p*-values and coefficients. The analytics professionals who provide their stakeholders with insights will be much more successful in having their models and ideas accepted.

This is an excerpt from one of a series of interviews with analytics professionals and educators commissioned by the *INFORMS Analytics Body of Knowledge* Committee.

8.3.3 Testing

The area of testing is another one of the often discussed areas within project management. And we have all heard of "horror stories" about how lack of carefully planned and implemented testing resulted in failed projects. It is common, especially in situations when implementation resides with an enterprise or centralized IT organization, that testing is the responsibility of that organization. In those situations, the analytics professionals, and the larger analytics team, have a support role rather than a leadership one. However, the analytics team role is critical because it is likely that this analytics team is the only group working on the project deployment team that actually understands how the model works, where its vulnerabilities might be, and how to effectively test them.

We do not intend to provide a complete explanation of the testing process. Suffice it to say that testing methodologies typically proceed from testing individual components (*unit testing*) to looking at how a system (including one focused around an analytics model) functions from end-to-end. Concepts such as *string testing* or *stress testing* are words that are commonly used in these environments, and there are other possible specialized testing approaches that go beyond the basic ones mentioned here.

In each of the components of testing mentioned, the analytics professional has an important role. For unit testing, the AP must lay out how a specific piece of the model works. They may be asked to define inputs and the corresponding outputs. The understanding of the model is critical to planning for a successful test. For example, the test should cover the full range of inputs that the model will see, so there are no surprises as the testing moves forward.

The idea of string testing is to come up with how all of the individual components of the model work in tandem as they typically go from beginning to end. Again, the role of the analytics professional is to define a complete set of test scenarios that track how the model proceeds step-by-step through its normal or planned set of operations.

Finally, the notion of stress testing is the idea of pushing the model to its limits. Can the model handle a particularly large or complex situation? Can the model operate as quickly as is required? Does the model performance degrade in stressful or complex situations? Clearly, success in defining what this means requires a very deep understanding of the model, and this knowledge and understanding will certainly come from the lead analytics professional as well as specialized analytics professionals on that person's team.

We have included this as a separate section to emphasize its importance. We suggest, as we have with the other topics in this final section, that special care and attention is needed from analytics professionals in this area, because good testing is, yet again, another critical factor in analytics project success. It may be more challenging, as well, because the ownership of the deployment process may reside in a different part of the sponsoring organization. This introduces the need for an additional set of skills: working as a support group to a larger and more complex team. This requires all of the same communication and documentation skills described in the sections above on those topics, but also it requires skills of persuasion, compromise, and passion for success.

8.3.4 Metrics

Metrics are numerical measures that represent important summary information about the operations of a business or organization. There are many different words to describe the same concept of measuring results. For example, it is common to hear the term *key performance indications* (KPIs) to mean essentially the same thing as business metrics. Furthermore, organizations use these metrics to monitor the performance of components of that organization or business, to measure and compensate employees, and track the course of their future plans. These concepts have become so standard that the U.S. Congress in

1993 passed the Government Performance and Results Act, which requires government agencies to develop, publish, plan, and report on performance metrics in their operations since they are publicly funded agencies.[5] It is also a standard practice of private sector organizations to create and use such systems of business metrics in the running of their businesses.

Successful models must support and provide information that is consistent with the sponsoring organizations business metrics and produce outputs that directly support their calculation and credibility. In most cases, the sponsoring organization will be driving the identification of the key metrics. For example, that organization might say right up front, "we expect this model to increase profits by 10%" or similar requirement. Also, it is common for organizations to have a very complex set of metrics, and might expect a particular project to impact many of them.

What is critical for the analytics professional is to be laser focused on these metrics, asking questions such as follows: How are they defined? What data are used to compute them? How often are they updated?

In the rare case, where the sponsoring organization does not lay out an array of their set of metrics, the analytics professional step should be pro-active, working with that organization to identify and define how they will be computed and presented throughout the running of the analytics project and, often, afterward. Having no credible, reasonable metrics at all would be a formula for disaster for any project, because, at the end, it is not possible to assess the success.

We end this discussion with a simple, but strong recommendation: Be focused on metrics. Use them every step of the way through the description, implementation, deployment, and operation of the analytics project. More often than not, they will be a critical factor in success of a project.

5 See https://www.gpo.gov/fdsys/pkg/STATUTE-107/pdf/STATUTE-107-Pg285.pdf

9

The Blossoming Analytics Talent Pool: An Overview of the Analytics Ecosystem

Ramesh Sharda[1] *and Pankush Kalgotra*[2]

[1]*Spears School of Business, Oklahoma State University, Stillwater, OK, USA*
[2]*Graduate School of Management, Clark University, Worcester, MA, USA*

9.1 Introduction

With the rising need for analytics in businesses, the demand for analytics professionals is surpassing the supply. A survey published in *MIT Sloan Management Review* recorded that 43% companies lack the appropriate analytical skills [1]. According to Deloitte.com, International Data Corporation (IDC) predicted that U.S. companies will require 181,000 people with deep analytical skills by 2018 and nearly one million employees with data management and interpretation abilities [2]. In response to this industry demand, academic programs in analytics are being developed all over the world. A catalog of programs on INFORMS Web site (https://www.informs.org/Resource-Center/Search-Education-Database) lists more than hundred programs in place already, with more being added regularly. But the need for skilled analytics professionals is still noted in the industry media.

Many thought leaders, including professionals and academicians, have suggested innovative ways to address the analytics talent shortage. For instance, Hiltbrand and Hart [3] suggest filling the analytics skill gap with crowdsourcing. Another suggestion by the industry leaders is to leverage nonanalytics employees in the organization to perform analytics [4]. The term used to label these analysts is "citizen data scientist." Of course, it is easy for a company to hire talent from competing companies in the same industry to build any new initiative. However, Young [5] suggested that the talent gap can be filled by defining the talent ecosystem more broadly because one may not find an appropriate candidate by limiting the search to a specific pool of aspirants. Therefore, expanding the talent ecosystem to find the required skillset may be an answer to the talent shortage. An ecosystem may include a company's vendors, consultants, customers, regulators, and so on. Defining this group broadly can help

INFORMS Analytics Body of Knowledge, First Edition. Edited by James J. Cochran.

identify a much bigger and better pool. In this chapter, we take this challenge up for the analytics industry. We identify and explain various components of the analytics ecosystem where hiring managers may find the required talent, especially experienced talent.

The skills needed in the field of data analytics are varied. Hiltbrand and Hart [3] discussed the competitions conducted by Netflix, Kaggle, and Idaho National Laboratory in which they witnessed that the contestants participated from different areas such as computer science, information technology, data analytics, engineering, and so on. Since the data analytics skillset is diverse, analysts can possibly emerge from multiple disciplines. Thus, the talent shortage can be addressed by casting a wide net encompassing the allied areas of analytics. Our aim in this chapter is to identify specific clusters in the analytics ecosystem in which hiring managers can find the required workforce. A secondary purpose of understanding the analytics ecosystem for the analytics professionals is to become aware of organizations, new offerings, and opportunities in sectors allied with analytics.

Interview With Robert Clark

Robert Clark, Senior Research Biologist with RTI International, offered these thoughts on the skills one needs to work effectively on an analytics team.

Obviously, you need to understand the problem. You need to understand what the client wants, what data are currently available and how they can be accessed and used, what additional data need to be collected, and what tools could be applied to the problem,. We've had issues with that in the past, trying to use a specific machine learning algorithms and things like that, say with genomics, and it has been a learning process trying to use some of those tools in that arena.

Non-tool-specific skills are also hugely important. Communication is the key in many ways. Being able to discuss what you know, and then of course be able to listen and understand, is paramount. And it is important to able and willing to ask for clarification when you do not understand! That is critical to success across all fields, not just data science.

This is an excerpt from one of a series of interviews with analytics professionals and educators commissioned by the *INFORMS Analytics Body of Knowledge* Committee.

9.2 Analytics Industry Ecosystem

Although some researchers have distinguished business analytics professionals from data scientists [6], for the purpose of understanding the overall analytics ecosystem, we treat them as one broad profession. Although skill needs can vary from mathematicians to a programmer to a modeler to a communicator, we

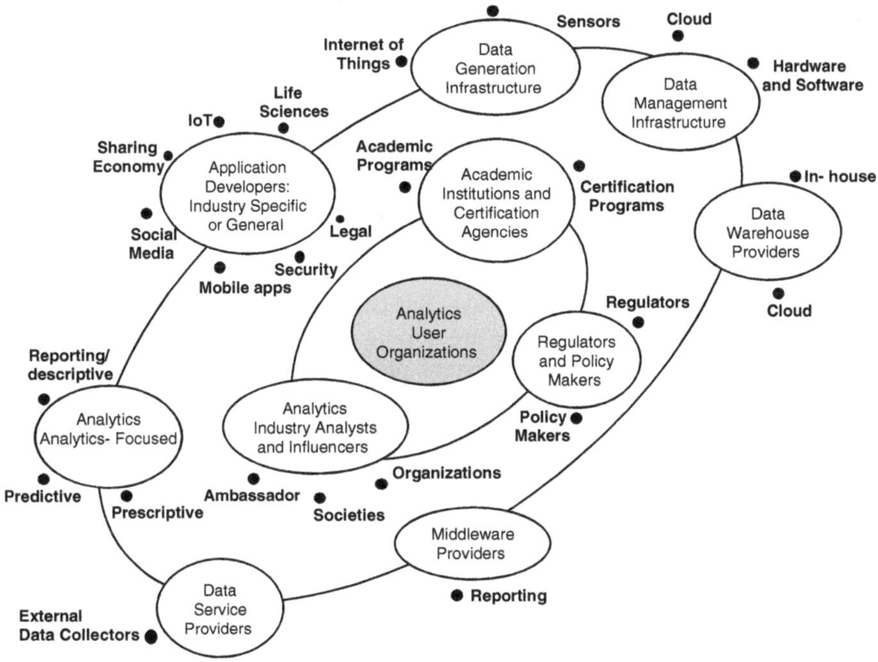

Figure 9.1 Analytics ecosystem.

believe this issue is resolved at a more micro/individual level rather than at a macro level of expanding the talent pool. We also take the widest definition of analytics to include all three types as defined by INFORMS–descriptive/reporting/visualization, predictive, and prescriptive [7].

Figure 9.1 illustrates one view of the analytics ecosystem. The components of ecosystem are represented by the planets of a planetary system. The subcomponents of each planet/component are depicted as the satellites of the planet. Eleven key sectors or clusters or planets in the analytics space are identified. The components of the analytics ecosystem are grouped into three categories represented by the outer orbit, inner orbit, and the core (middle part) of the system.

The outer orbit contains seven planets broadly termed as technology providers. Their primary revenue comes from providing technology, solutions, and training to analytics user organizations so that they can employ these technologies in the most effective and efficient manner. The inner orbit can be generally defined as the analytics accelerators. The accelerators work with both technology providers and users. Finally, the core of ecosystem is comprised of analytics user organizations. This is the most important component as every analytics industry facet is driven by the user organizations. Being at the center of system, this component is the driving force of the ecosystem. User organizations create demand for analytics applications and their success flourish rest of the ecosystem. Of course, analytics talent is needed not by just this group of users but also all the other players in the ecosystem.

In the past, we have used the metaphor of a flower to describe the analytics ecosystem [7]. Other authors (e.g., Ref. [8]) identify players in the analytics industry through groups in tables. The metaphor of a planetary system is well suited to the analytics ecosystem as multiple components at the same level can be placed on the same orbit. Similar to a planetary system in which planets are held together by gravity, all the planets of our analytics ecosystem interact with each other and move together. Different components of ecosystem exchange information and overlap in many ways. But these are also different because of their focus on a specific value they are adding to the analytics value chain.

We use the terms components, clusters, planets, and sectors interchangeably to describe the various players in the analytics space. We introduce each of the industry sectors below, and give some examples of players in each sector. The list of company names included in any planet is not exhaustive. The representative list of companies in each cluster is meant to illustrate that cluster's unique offering so as to describe where analytics talent may be used or hired away from. Also, mention of a company's name or its capability in one specific group does not imply that it is the only activity/offering of that organization. The main goal is to focus on the different analytic capabilities within each component of the analytics space. Many companies play in multiple sectors within the analytics industry, and thus offer opportunities for movement within the field both horizontally and vertically.

9.2.1 Data Generation Infrastructure Providers

Perhaps the first place to begin identifying the clusters is by noting a new group of companies that enable generation and collection of data that may be used for developing analytical insights. Although this group could include all the traditional points of sale systems, inventory management systems, and technology providers for every step in a company's supply/value chain and operations, we mainly consider new players where the primary focus has been on enabling an organization to develop new insights into its operations as opposed to running its core operations. Thus, this group includes companies creating the infrastructure for collecting data from different sources.

One of the emerging components of such infrastructure is the "sensor." Sensors collect massive amount of data at a faster rate and have been adopted by various sectors such as health care, sports, and energy. For instance, some of the major players manufacturing sensors to collect health information are AliveCor, Apple, Google, Garmin, Shimmer, Jawbone, Kinsa, Netatmo, and Fitbit. Likewise, the sports industry is using sensors to collect data from the players and field to develop strategies to improve performance. Examples of the companies producing sports-related sensors include Sports Sensors, Zepp, Shockbox, and others.

Similarly, sensors used for traffic management are produced by Advantech B+B SmartWorx, Garmin, Sensys Network, and many others.

Sensors play a major role in Internet of Things (IoT), and are an essential part of smart objects. Sensors comprising machine-to-machine communication are driving growth of IoT. The leading players in the infrastructure of IoT are Intel, Microsoft, Google, IBM, Cisco, General Electric, Smartbin, SIKO Products, Omega Engineering, Apple, and SAP. There are many industrial Internet of Things providers that develop industry-specific sensors, but those are too numerous to mention. This cluster is probably the most technical group in the ecosystem.

9.2.2 Data Management Infrastructure Providers

This group includes all of the major organizations that provide hardware and software targeting the basic foundation for all data management solutions. Obvious examples of these include all major hardware players that provide the infrastructure for database computing–IBM, Dell, HP, Oracle, and so on; storage solution providers such as EMC (recently bought by Dell) and NetApp; companies providing indigenous hardware and software platforms such as IBM, Oracle, and Teradata; data solution providers offering hardware and platform-independent database management systems such as SQL Server family of Microsoft; and specialized integrated software providers such as SAP fall under this group. This group also includes other organizations such as database appliance providers, service providers, integrators, developers, and so on that support each of these companies' ecosystems.

Several other companies are emerging as major players in a related space, thanks to the network infrastructure enabling cloud computing. Companies such as Amazon (Amazon Web Services), IBM (Bluemix), Microsoft (Azure), General Electric (Predix), and Salesforce.com pioneered to offer full data storage and analytics solutions through the cloud that now have been adopted by several companies listed above.

A recent crop of companies in the Big Data space are also part of this group. Companies such as Cloudera, Hortonworks, and many others do not necessarily offer their own hardware but provide infrastructure services and training to create the Big Data platform. This would include Hadoop clusters, MapReduce, NoSQL, Spark, Kafka, Tez, Flume, and other related technologies for analytics. Thus, they could also be grouped under industry consultants or trainers enabling the basic infrastructure. Full ecosystems of consultants, software integrators, training providers, and other value-added providers have evolved around many of the large players in data management infrastructure cluster. Some of the clusters below will identify these players because many of them are moving to analytics as the industry shifts its focus from efficient transaction processing to deriving analytical value from the data.

9.2.3 Data Warehouse Providers

Companies with a data warehousing focus provide technologies and services aimed toward integrating data from multiple sources, thus enabling organizations to derive and deliver value from their data assets. Many companies in this space include their own hardware to provide efficient data storage, retrieval, and processing. Companies such as IBM, Oracle, and Teradata are major players in this arena. Recent developments in this space include performing analytics on the data directly in memory.

Another major growth sector has been data warehousing in the cloud. Examples of such companies include Snowflake and Redshift. Major companies from other related sectors are also moving into this space–SAS and Tableau are good examples. Companies in this cluster clearly work with all the other sector players in providing data warehouse solutions and services within their ecosystem, and hence act as the backbone of analytics industry. It has been a major industry in its own right and, thus, a supplier and consumer of analytics talent.

9.2.4 Middleware Providers

Data warehousing began with a focus on bringing all the data stores into an enterprise-wide platform. Making sense of these data has become an industry in itself. The general goal of middleware industry is to provide easy-to-use tools for reporting or descriptive analytics, which forms a core part of BI or analytics employed at organizations. Examples of companies in this space include Microstrategy, Plum, and many others. A few of the major players that were independent middleware players have been acquired by companies in the first two groups. For example, Hyperion became a part of Oracle, SAP acquired Business Objects, and IBM acquired Cognos. This sector has been largely synonymous with the Business Intelligence providers offering dashboarding, reporting, visualization services to industry, building on top of the transaction processing data, and the database and data warehouse providers. Thus, many companies have moved into this space over the years, including general analytics software vendors such as SAS, or new visualization providers such as Tableau, or many niche application providers. A product directory at TDWI.org lists 201 vendors just in this category (http://www.tdwidirectory.com/category/business-intelligence-services) as of June 2016, so the sector has been robust. This sector is attempting to move toward the data science segment of the industry. On the other hand, software companies that have focused on visualization are incorporating capabilities that were once the domain of middleware in terms of customized reports and aggregate-to-detail analyses.

9.2.5 Data Service Providers

Much of the data an organization uses for analytics is generated internally through its operations, but there are many external data sources that play a major role in any organization's decision-making. Examples of such data sources include demographic data, weather, data collected by third parties that could inform an organization's decision-making, and so on. Several companies realized the opportunity to develop specialized data collection, aggregation, and distribution mechanisms. These companies typically focus on a specific industry sector and build upon their existing relationships in that industry through their niche platforms and services for data collection. For example, Nielsen provides data sources to their clients on customer retail purchase behavior. Another example is Experian, which includes data on each household in the United States. Omniture has developed technology to collect web clicks and share such data with their clients. Comscore is another major company in this space. Google compiles data for individual Web sites and makes a summary available through Google Analytics services. Other examples are Equifax, TransUnion, Acxiom, Merkle, Epsilon, and Avention. This can also include organizations such as ESRI.com, which provides location-oriented data to their customers. There are hundreds of other companies that are developing niche platforms and services to collect, aggregate, and share such data with their clients. As noted earlier, many industry-specific data aggregators and distributors exist, which are moving to offer their own analytics services. Thus, this sector also is a growing user and potential supplier of analytics talent, especially with specific niche expertise.

9.2.6 Analytics-Focused Software Developers

Companies in this category have developed analytics software to analyze data that have been collected in a data warehouse or are available through one of the platforms identified earlier (including Big Data). It can also include inventors and researchers in universities and other organizations that have developed machine learning algorithms for specific types of analytics applications. We identify major industry players in this space along the three types of analytics: descriptive, predictive, and prescriptive analytics.

Reporting/Descriptive Analytics

Reporting or descriptive analytics is enabled by the tools available from the Middleware industry players identified earlier or unique capabilities offered by focused providers. For example, Microsoft's SQL Server BI tool kit includes reporting as well as predictive analytics capabilities. On the other hand, specialized software is available from companies such as Tableau for visualization. SAS also offers a Visual Analytics tool with similar capacity. There are many open-source visualization tools as well. Literally, hundreds of data visualization tools have been developed around the world, and many such tools focus on visualization of data from a specific industry or domain. Because visualization is the primary way thus far for exploring analytics in industry, this sector has witnessed the most growth. Many new companies are being formed. For example, Gephi, a free and open-source software, focuses on visualizing networks.

Predictive Analytics

Perhaps the biggest recent growth in analytics has been in predictive analytics and machine learning. Many statistical software companies such as SAS and SPSS embraced predictive analytics early on and developed software capabilities as well as industry practices to employ data mining and classical statistical techniques for analytics. IBM-SPSS Modeler from IBM and Enterprise Miner from SAS are some of the examples of tools used for predictive analytics. Other players in this space include KXEN, Statsoft (recently acquired by Dell), Salford Systems, MATLAB, and scores of other companies that may sell their software broadly or use it for their own consulting practices (next group of companies).

Four open-source platforms (R, RapidMiner, Weka, and KNIME) have also emerged as popular industrial strength software tools for predictive analytics and have companies that support training and implementation of these open-source tools. Revolution Analytics (now a part of Microsoft) is an example of a company focused on R development and training. R integration is now possible with most analytics software. A company called Alteryx uses R extensions for reporting and predictive analytics, but its strength is in shared delivery of analytics solutions processes to customers and other users. Similarly, RapidMiner, Weka, and KNIME are also examples of open-source providers. In addition, companies such as Rulequest (sells proprietary variants of Decision Tree software) and NeuroDimensions (a Neural Network software company) are examples of houses that have developed specialized software around a specific technique of data mining.

Prescriptive Analytics

Software providers in this category offer modeling tools and algorithms for optimization of operations usually called management science/operations research (MS/OR) software. This field has had its own set of major software providers. IBM, for example, has classic linear and mixed-integer programming software. Several years ago, IBM also acquired a company called ILOG, which provides prescriptive analysis software and services to complement their other offerings. Analytics providers such as SAS have their own OR/MS tools–SAS/OR. FICO acquired another company called XPRESS that offers optimization software. Other major players in this domain include companies such as AIIMS, AMPL, Frontline, GAMS, Gurobi, Lindo Systems, Maximal, NGData, Ayata, and many others. A detailed delineation and description of these companies' offerings is beyond the scope of our goals here. Suffice it to note that this industry sector has also seen much growth recently.

Of course, there are many techniques that fall under the category of prescriptive analytics and each has its own set of providers. For example, simulation software is provided by major companies such as Rockwell (ARENA) and Simio. Palisade provides tools that include many software categories. Similarly, Frontline offers tools for optimization with Excel spreadsheet as well as predictive analytics. Decision analysis in multiobjective settings can be performed using tools such as Expert Choice. There are also tools from companies such as Exsys, XpertRule, and others for generating rules directly from data or expert inputs. *ORMS Today* publishes surveys of software in a specific category periodically, and is thus a good source of information about companies specializing in prescriptive analytics.

Some new companies are evolving to combine multiple analytics models in the Big Data space, including social network analysis and stream mining. For example, Teradata Aster includes its own predictive and prescriptive analytics capabilities in processing Big Data streams. Several companies have developed complex event processing (CEP) engines that make decisions using streaming data, such as IBM's Infosphere Streams, Microsoft's StreamInsight, and Oracle's Event Processor. Other major companies that have CEP products include Apache, Tibco, Informatica, SAP, Druid, and Hitachi. It is worthwhile to note again that the provider groups for all three categories of analytics are not mutually exclusive. In most cases, a provider can play in multiple components of analytics.

9.2.7 Application Developers: Industry-Specific or General

The organizations in this group use their industry knowledge, analytical expertise, solutions available from the data infrastructure, data warehouse, middleware, data aggregators, and analytics software providers to develop custom solutions for a specific industry. Thus, this industry group makes it possible for the analytics technology to be used in a specific industry. Of course, such groups may also exist in specific user organizations. Most major analytics technology providers such as IBM, SAS, and Teradata clearly recognize the opportunity to connect to a specific industry or client and offer analytic consulting services. Companies that have traditionally provided application/data solutions to specific sectors are now developing industry-specific analytics offerings. For example, Cerner provides electronic medical records (EMR) solutions to medical providers and their offerings now include many analytics reports and visualizations. Similarly, IBM offers a fraud detection engine for the health insurance industry, and it is working with an insurance company to employ their Watson analytics platform in assisting medical providers and insurance companies with diagnosis and disease management. Another example of a vertical application provider is Sabre Technologies, which provides analytical solutions to the travel industry, including fare pricing for revenue optimization and dispatch planning.

This cluster also includes companies that have developed their own domain-specific analytics solutions and market them broadly to a client base. For example, Nike, IBM, and Sportvision develop applications in Sports Analytics to improve the play and increase the viewership. Acxiom has developed clusters for virtually all households in the United States based upon the data they collect about households from many different sources. Credit score and classification reporting companies (FICO, Experian, etc.) also belong in this group. This field represents an entrepreneurial opportunity to develop industry-specific applications. Many companies emerging in web/social media/location analytics are trying to profile users for better targeting of promotional campaigns in real time. Examples of such companies and their activities are as follows: YP.com employs location data for developing user/group profiles and targeting mobile advertisements, Towerdata profiles users on the basis of e-mail usage, Qualia aims to identify users through all device usage, and Simulmedia targets advertisements on TV on the basis of analysis of a user's TV-watching habits.

Growth of smartphones has spawned a complete industry focused on specific analytics applications for consumers as well as organizations. For example, smartphone apps such as Shazam, Soundhound, or Musixmatch are able to identify a song on the basis of first few notes and then let the user select it from their song base to play/download/purchase. Waze uses real-time traffic information shared by users, in addition to the location data, for improving navigation. Voice recognition tools such as Siri on iPhone, Google Now, and Amazon Alexa are leading to specialized applications for very specific purposes in analytics applied to images, videos, audio, and other data that can be captured through smartphones and/or connected sensors. Smartphones have also elevated the sharing economy providers. Recently, the sharing economy has gained an immense popularity for the transportation services and has given rise to companies such as Uber, Lyft, Curb, and Ola. The sharing economy concept has also been used by companies such as Airbnb, VRBO, and Couchsurfing for hospitality services. Many of these companies are exemplars of analytics leading to new business opportunities.

Online social media is another hot area in this cluster. Undoubtedly, Facebook is the leading player in the space of online social networking followed by Twitter and LinkedIn. Moreover, the public access to their data has given rise to multiple other companies that analyze their data. For example, Unmetric analyzes the Twitter data and provides solutions to their clients. Similarly, there are several other companies that focus on social network analysis.

Other satellites around this planet are the group of industries in specific industries. To illustrate a few, consider finance, legal, life sciences, and security sector. Companies such as Affirm, Lending Club, Payoff, OnDeck, ZestFinance, Cignify, Wonga, and Dataminr provide financial services to their clients. Several companies are providing legal services to their clients using data analytics. Some of these companies are Brightleaf, Counselytics, Everlaw, Judicata, Premonition.ai, DiligenceEngine, eBrevia, Lex Machina (now acquired by LexisNexis), and Ravel. The group of industries producing life sciences applications include 3scan, 23andMe, Deep Genomics, HumanDX, Kyruus, HealthTap, Metabiota, uBiome, Vital Labs, Ovuline, Tute Genomics, Zephyr Health, Zymergen, and many others. Due to the increasing cases related to cyber security, several companies have emerged in this area. Companies creating security applications for their clients are Area 1 Security, CounterTack, Cybereason, Cylance, Feedzai, Fortscale, Guardian Analytics, Keybase, Recorded Future, Sift Science, Signifyd, ThreatMetrix, and so on. New companies keep coming up, which focus on applications targeted at a specific industry. Turck (2017) provides names of many additional industry-focused analytics providers.

A trending area in the application development industry is the Internet of Things. Several companies are building applications to make smart objects. For example, SmartBin has made Intelligent Remote Monitoring Systems for the waste and recycling sectors. Several other organizations are working on building smart meters, smart grids, smart cities, connected cars, smart home, smart supply chain, connected health, smart retail, and other smart objects.

One of the emerging trends in the analytics industry is deep learning. It involves use of hierarchical algorithms to model higher level abstractions in the data. The industry players in this group include Google, with their products named Tensorflow, Apache Singa, Microsoft Cognitive Toolkit, and several other open-source packages such as MXNet, Theano, and OpenNN. Another evolving label in the analytics industry is virtual reality (VR) analytics. Companies such as Google, Facebook, and GE are investing and showing great interest in this emerging area.

The start-up activity in this sector is growing and is in major transition due to technology/venture funding and security/privacy issues. Nevertheless, the application developer sector is perhaps the biggest growth industry within analytics at this point. This cluster provides an innovative pool of talent to the hiring managers.

We next introduce the "inner orbit" of the analytics planetary system. These clusters can be called analytics accelerators. Although they may not be involved in developing the technology directly, these organizations have played a key role in shaping the industry.

9.2.8 Analytics Industry Analysts and Influencers

This cluster includes three types of organizations or professionals. The first group is the set of professional organizations that provide advice to the analytics industry providers and users. Their services include marketing analyses, coverage of new developments, evaluation of specific technologies and development of training/white papers and so on. Examples of such players include organizations such as the Gartner Group, The Data Warehousing Institute, Forrester, McKinsey, and many of the general and technical publications and Web sites that cover the analytics industry. Gartner Group's Magic Quadrants are highly influential and are based on industry surveys. Similarly, TDWI.org professionals provide excellent industry overview and are very aware of current and future trends of this industry.

The second group includes professional societies or organizations that also provide some of the same services but are membership based and organized. For example, INFORMS is focused on promoting analytics. Special Interest Group on Decision Support and Analytics (SIGDSA), a subgroup of the Association for Information Systems, also focuses on analytics. Most of the major vendors (e.g., Teradata and SAS) also have their own membership-based user groups. These entities promote the use of analytics and enable sharing of the lessons learned through their publications and conferences. They may also provide recruiting services and are thus good sources for locating talent.

A third group of analytics industry analysts is what we call analytics ambassadors, influencers, or evangelists. These folks have presented their enthusiasm for analytics through their seminars, books, and other publications. Illustrative examples include Steve Baker, Tom Davenport, Charles Duhigg, Wayne Eckerson, Bill Franks, Malcolm Gladwell, Claudia Imhoff, Bill Inman, and many others. Again, the list is not inclusive. All of these ambassadors have written books (some of them bestsellers!) and/or given many presentations to promote the analytics applications. Perhaps another group of evangelists to include here is the authors of textbooks on business intelligence/analytics who aim to assist the next cluster to produce professionals for the analytics industry.

9.2.9 Academic Institutions and Certification Agencies

In any knowledge-intensive industry such as analytics, the fundamental strength comes from having students who are interested in the technology and choose that industry as their profession. Universities play a key role in making this possible. This cluster, then, represents the academic programs that prepare professionals for the industry. It includes various components of business schools such as information systems, marketing, management sciences, and so on. It also extends far beyond business schools to include computer science, statistics, mathematics, and industrial engineering departments. Universities are offering undergraduate and graduate programs in analytics in all of these disciplines, though they may be labeled differently. A major growth frontier has been certificate programs in analytics to enable current professionals to retrain and retool themselves for analytics careers. Certificate programs enable practicing analysts to gain basic proficiency in specific software by taking a few critical courses of schools offering. INFORMS Web site includes a list of many such programs, with new ones being added daily.

Another group of players assists with developing competency in analytics. These are certification programs to award a certificate of expertise in specific software. Virtually every major technology provider (IBM, Microsoft, Microstrategy, Oracle, SAS, Tableau, and Teradata) has its own certification program.

These certificates ensure that potential new hires have a certain level of tool skills. On the other hand, INFORMS offers a Certified Analytics Professional (CAP) certificate program that is aimed at testing an individual's general analytics competency. Any of these certifications give a college student additional marketable skills.

The growth of academic programs in analytics is staggering. Only time will tell if this cluster is overbuilding the capacity that can be consumed by the other clusters, but at this point the demand appears to outstrip the supply of qualified analytics graduates and this is the most obvious place to find at least entry level analytics hires.

9.2.10 Regulators and Policy Makers

The players in this component are responsible for defining rules and regulations for protecting employees, customers, and shareholders of the analytics organizations. The collection and sharing of the users' data require strict laws for securing privacy. Several organizations in this space regulate the data transfer and protect users' rights. For example, the Federal Communications Commission (FCC) regulates interstate and international communications. Similarly, the Federal Trade Commission (FTC) is responsible for preventing data-related unfair business practices. The International Telecommunication Union (ITU) regulates the access to Information and Communication Technologies (ICT) to underserved communities worldwide. On the other hand, a nonregulatory federal agency named the National Institute of Standards and Technology (NIST) helps advance the technology infrastructure. There are several other organizations across the globe that regulate the data security. This is a very important component in the ecosystem so that no one can misuse the consumers' information.

For anyone developing or using analytics application, it is crucial to have someone on the team who is aware of the regulatory framework. These agencies and professionals who work with them clearly offer unique analytics talents and skills.

9.2.11 Analytics User Organizations

Clearly, this is the economic engine of the whole analytics industry, and therefore, we represent this cluster as the core of the analytics planetary system. If there were no users, there would be no analytics industry. Organizations in every industry, regardless of size, shape, and location, are using analytics or exploring the use of analytics in their operations. These include private sector, government, education, military, and so on around the world. Companies are exploring opportunities in analytics space to try to gain/retain a competitive advantage. Specific companies are not identified in this section. The goal here is to see what type of roles analytics professionals can play within a user organization.

Of course, the top leadership of an organization, especially in the information technology group (Chief Information Officer, etc.), is critically important in applying analytics to its operations. Reportedly, Forrest Mars of the Mars Chocolate Empire said that all management boiled down to applying mathematics to a company's operations and economics. Although not enough senior managers subscribe to this view, the awareness of applying analytics within an organization is growing everywhere. A health insurance company executive once told us that his boss (the CEO) viewed the company as an IT-enabled organization that collected money from insured members and distributed it to the providers. Thus, efficiency in this process was the premium they could earn over a competitor. This led the company to develop several analytics applications to reduce fraud and overpayment to providers, promote wellness among those insured so they would use the providers less often, generate more efficiency in processing, and thus be more profitable.

Virtually all major organizations in every industry that we are aware of are hiring analytical professionals under various titles. Figure 9.2 is a word cloud of the selected titles of our program graduates at Oklahoma State University from 2013 to 2016. It clearly shows that analytics is a popular title in the organizations hiring graduates of such programs. Other key words appear to include terms such as Risk, Health, Security, Revenue, Marketing, and so on.

Of course, user organizations include career paths for analytics professionals moving into management positions. These titles include project managers, senior managers, and directors, all the way up to Chief Information Officer or Chief Executive Officer. This suggests that user organizations exist as a key cluster in the analytics ecosystem, and thus can be a good source of talent. It is perhaps the first place to find analytics professionals within the vertical industry segment.

Figure 9.2 Word cloud of titles of Analytics Program Graduates.

Other than the talent sources discussed in this chapter, the analytics talent can be found within the organization itself. It is possible that the current employees of the organization have the common analytical skills, which are not visible. Such employees can be identified by hosting analytics competitions in the organization. In addition, the individuals with Lean Six Sigma certifications can be of interest to the analytics industry.

9.3 Conclusions

The purpose of this chapter has been to present a map of the landscape of the analytics industry. Eleven different groups that play a key role in building and fostering this industry are identified. More planets/components and orbits can be added over time in the analytics ecosystem. Because data analytics requires a diverse skillset, understanding of this ecosystem provides a richer pool of analytics talent to the hiring managers. Moreover, it is possible for professionals to move from one industry cluster to another to take advantage of their skills. For example, expert professionals from providers can sometimes move to consulting positions or directly to user organizations. Overall, there is much to be excited about the analytics industry at this point.

Interview With Eric Stephens

Eric Stephens, Manager of Population Health Analytics at the Vanderbilt University Medical Center, offered these thoughts on executive support for analytics.

Every organization is different and has the ability and prerogative to decide how and where analytics will be placed. Most organizations tend to isolate analytics within a single department; IT is the typical location but other common functional areas include finance, marketing, or sales. In other words, analytics is essentially established to serve a single department, and thus usually does not get applied outside of those bounds (or, if there are multiple analytics teams isolated in various functional areas, they typically don't collaborate with each other, leading to multiple versions of the truth). Contrast that with a centrally located team with strong executive support: In this case, analytics is able to have a much broader, cross-organizational impact and can be deployed on those areas of the business in which it will have the greatest return on investment. As a practitioner who has experienced both situations, I much prefer the latter, as I know that I will have the ability to be a part of many different and interesting projects and will also be provided with the resources necessary to do my job as effectively as possible.

I firmly believe that strong executive support is critical for analytics to have maximum organizational impact. When compared to more traditional business functions such as sales, marketing, finance, and IT, analytics is still in its infancy, and thus—believe it or not—to some executives, it is still an unproven expense. Instead of recognizing the true potential value of analytics, these business leaders instead see it as a "nice to have." They're the ones who might say something like, "I read in Harvard Business Review or the Wall Street Journal that all these companies are doing this big data/analytics thing, so I guess we also need to do this big data/analytics thing. I'm going to get an analytics person and stick him or her in IT . . . analytics uses data and computers, so that's where we're going to place it." Obviously, I'm being a little facetious, but I've actually seen something like this happen. Of course, in most of these situations, analytics doesn't get the amount of organizational support necessary, leading to a less-than-optimum application of the capability. The sad fact of this, of course, is that the executive's perception that analytics is not a critical business function becomes a self-fulfilling prophecy.

This is an excerpt from one of a series of interviews with analytics professionals and educators commissioned by the *INFORMS Analytics Body of Knowledge* Committee.

References

1 Ransbotham S, Kiron D, Prentice PK (2015) The talent dividend. *MIT Sloan Manage. Rev.* 56(4): 1.
2 Deloitte (2016) Analytics trends 2016–the next evolution. Available at www2.deloitte.com/na/en/pages/risk/articles/analytics-trends-2016.html.
3 Hiltbrand T, Hart R (2014) Bridging the analytics skill gap with crowdsourcing. *Bus. Intell. J.* 19(2).
4 Burgelman L (2015) The rise of the citizen data scientist. Retrieved from http://www.ngdata.com/the-rise-of-the-citizen-data-scientist/.
5 Young MB (2015) Buy, build, borrow, or none of the above? new options for closing global talent gaps. Retrieved from https://www.conference-board.org/publications/publicationdetail.cfm?publicationid=2904.
6 Davenport TH, Patil DJ (2012) Data scientist. *Harvard Bus. Rev.* 90, 70–76.
7 Sharda R, Delen D, Turban E (2017) *Business Intelligence, Analytics, and Data Science: A Managerial Perspective*, 4th ed. (Pearson, Boston).
8 Turck M (2016) Is Big Data still a thing? (The 2016 Big Data landscape). Retrieved from http://mattturck.com/2016/02/01/big-data-landscape/.

Appendix: Writing and Teaching Analytics with Cases

James J. Cochran

Culverhouse College of Business, The University of Alabama, Tuscaloosa, AL, USA

The case teaching methodology is commonly used in professional degree programs as a way to afford students opportunities to assume responsibility for and develop appreciation of typical situations faced by practitioners. Instructors over a wide range of clinical disciplines such as medicine, business, pharmacy, and law have found that a well-conceived and well-written case can help their students develop profound insight into the practice of a discipline without exposing students and organizations to the risks associated with the case scenario. Use of the case method of teaching has spread from these disciplines to other less-clinically oriented disciplines, such as political science [1], anthropology [2], sociology [3], chemistry [4], and astronomy [5].

But what is a *case*, or more precisely, a *teaching case*? In this appendix, we will consider this question as well as discuss a classification scheme for cases. We will discuss approaches to finding material for and subsequently writing a teaching case, factors in selecting a published teaching case for classroom use, considerations in assessing student performances on cases, and development of a case discussion facilitation style. We will also briefly discuss sources of published teaching cases and outlets for teaching case authors. Finally, we will provide a relatively simple analytics case and discuss how it has been used in classrooms.

A.1 What Is a Teaching Case?

A teaching case *is* an account of a situation that provides background information, generally comprising a comprehensive history of a problem that includes

INFORMS Analytics Body of Knowledge, First Edition. Edited by James J. Cochran.

introductions to multiple players and stakeholders who have a wide range of interests and motivations.

The teaching case can require the students, working individually or in teams, to use this background information (perhaps to be augmented with additional information collected by the students) to

i) describe, interpret, and/or evaluate an action that has occurred and/or
ii) address a problem, give a recommendation, make a decision, and/or develop a strategy in order to provide guidance for the future in a realitstic context. Thus, teaching cases may be *retrospective* and/or *prospective.*

Retrospective teaching cases often share the actual outcome of the player and stakeholder decisions and strategies they describe. These types of teaching cases require students to (i) develop an understanding of the issue(s) faced by the players and stakeholders, (ii) review how the decisions made and strategies employed by these players and stakeholders lead to the outcome, and iii) compare the outcome to the outcomes that likely would have resulted from other feasible decisions and strategies.

Prospective teaching cases generally do not share the actual outcome of the decisions and strategies they describe. These types of cases require students to (i) identify the issue(s) faced by the players and stakeholders, (ii) generate and compare potential decisions that could be made and strategies that could be employed to address the issue(s), and (iii) make recommendations on the course (s) of action to be taken.

What *is not* a teaching case? A teaching case is not an *extended homework exercise* or *story problem;* a teaching case is more complex and requires far more of the student. It is not a *case study,* that is, it is not an academic examination of a specific problem or circumstance intended to generalize across populations. A teaching case provides a learning experience for students. Although the situation, players, and stakeholders described in a teaching case may not exist, a teaching case is not *contrived*; the context, scenario, and problem(s) faced by the players and stakeholders in a teaching case must be relevant to a real problem.

There is no minimum or maximum length for a teaching case, and teaching cases often contain information that is either irrelevant or only tangentially pertinent in order to provide students with experience in assessing the relevancy of information. Cases may also omit or fail to include important information in order to provide students with experience identifying and finding additional information that will be instrumental to effectively addressing a problem, making a decision, or formulating a strategy. Some cases provide data, some require the student to find or collect data, and some cases are data-free (i.e., do not require the student to use actual data in analyzing the case).

A.2 My Motivation for Using Teaching Cases[1]

I began teaching at Wright State University's business college immediately after completing my Master's degree in economics. I routinely taught three or four different courses across four sections each academic term. I often taught one or two sections each of the college's second required course in introductory statistics (inference and modelling) at the sophomore level and the college's first and second required courses in introductory operations research (deterministic modelling and stochastic modelling) at the junior level during an academic term. Class enrolments were approximately 40 students, and the math backgrounds of the students were varied and often underdeveloped or weak.

In my first year as an instructor, I quickly realized that my students did not share my enthusiasm for the courses I was teaching. Other faculty members who taught the same courses as I confided that this was common and expected. I became increasingly frustrated, and many of my students and I shared a common source of frustration; after completing these courses, students often still had a weak understanding of the concepts that had been covered.

I thought about this a great deal. The frustration I saw in many of my students was similar to what I had seen in my classmates when I took the same courses just a few years prior. I ultimately decided I could summarize the students' frustrations with three short questions:

- *When (i.e., under what circumstances) will I use these concepts and methods?*
- *Where (i.e., for what problems) will I use these concepts and methods?*
- *How will I use these concepts and methods?*

It occurred to me that I might be able to address all three of these questions effectively through teaching cases. My students had never used the case methodology as a basis for learning, so I was likely limited in what I could do–creating versions of undergraduate introductory statistics and operations research courses for business students that were entirely case-based was not feasible–especially this early in my career as an instructor. Therefore, I decided to develop and utilize a hybrid approach. In my later interactions with other instructors of operations research, statistics, and analytics who have gravitated to the case method, I found that my experience and my motivation are far from unique!

In the hybrid approach I developed, I would continue to devote most class meetings to discussion of the concepts and methodologies to be covered in the course. I would also assign two relatively short (one to two page) cases before each of my three examinations. Analysis of the two cases assigned before each examination would require students to use the concepts and methods that were to be tested over on the ensuing examination, and the students would be given no indication what concepts or methodologies were pertinent to each case. Each

1 More detail on some aspects of this discussion are provided in Cochran [6–8].

student would independently analyze the cases and submit a two-page report on her or his findings for each case. I would devote the class meeting immediately prior to each exam to class discussion of the two assigned cases. The cases would be primarily prospective and would provide students with data. In some cases, relevant facts and/or data would be omitted and students would have to make and assess the potential impact of assumptions. In some cases, irrelevant facts and/or data would be included and students would have to identify and eschew these facts and/or data. I would grade my students on the quality of their analyses, their exposition, and their participation in the class discussions.

In recognition of my students' (and my) lack of experience with the case methodology, I would proceed slowly with the first few cases and temper my expectations. As the term progressed and students gained experience and confidence, I would increase my expectations and adjust my grading accordingly (and I would explain this to my students in advance).

I expected my students' critical thinking, modelling, technical, and analytic skills to improve somewhat after implementation of the hybrid case methodology I had developed, and they did. However, the improvements I saw–in the quality of work, effort put forth, and attitude toward the courses–stunned me. I was particularly gratified to see the weakest responses to questions on exams (which were often embarrassingly poor prior to implementation of the hybrid case method) to be thoughtful, intelligent, and well developed after implementation of the hybrid case method. Students enjoyed the challenge of working on the cases, and they welcomed the opportunity to work on real problems that did not have single well-defined solutions. They also appreciated the opportunity to share their thoughts, opinions, insights, and ideas with each other and their instructor, and even the wariest of students ultimately participated in case discussions.

Of course, implementation of this hybrid case method did not proceed without difficulties. I had to develop case facilitation skills in real time; none of my colleagues used cases, so I had to develop these skills without guidance. Fortunately, my students understood that I was striving to improve their education and were extremely patient. I also had to find sources of cases. There were very few published teaching cases available in statistics or operations research when I began teaching, so I had to write my own cases–six new cases each academic term for each different course I taught–based on my experiences in private industry and consulting as well as those of colleagues, friends, and family. I had a few instances in which one or more students interpreted portions of cases in ways I had not intended. This required me to develop and further refine some interesting case facilitation skills (which I soon concluded was an important part of the case facilitator's repertoire).

I also had to grade the students' written case analyses and participation in case discussions. This was time-consuming, but I did recover some of this time when grading examinations; because my students' performances on the examinations were now much improved, the examinations were far easier to grade. I was concerned about how to cover all of the required course material after giving up approximately 10% of my class meeting time for case discussions. However, I soon found that a student who had the case assignments in hand came to class with a sharper focus, and I was able to get through the same course material in less time.

Finally, I was concerned about how students would react to this increase in their workload–6 case analyses and approximately 12 written pages per student in each course. Again, the outcome was somewhat surprising and extremely gratifying. Not one student complained–to me or in comments on anonymous teaching evaluations–about the additional work. Several thanked me for helping them understand the material in a meaningful way. In later academic terms, several students returned to take other courses from me so that they could further develop their quantitative skills through the hybrid case methodology. My colleagues on the faculty also began reporting that my former students were now more effectively applying concepts from my courses to problems in their courses.

In summary,

- my students' modelling, technical, and analytic skills improved dramatically,
- my workload increased somewhat,
- my students were happier and more frequently reported enjoying my classes,
- my former students applied concepts from my courses to problems in other courses, and
- I enjoyed teaching more than I had before I implemented a hybrid case teaching approach.

All in all, not a bad deal–for me or my students!

I have continued to experiment with the case method over the past 30+ years–using different facilitation styles, implementing teaching cases into large sections, trying different schemes for facilitating class participation–and my results have convinced me that cases are an extremely robust and effective tool for teaching students about the practice of a discipline. Since analytics is a practical discipline, it is logical to conclude that teaching cases are ideally suited to analytics courses.

A.3 Writing a Teaching Case

Authors of teaching cases find topics for teaching cases in two ways: opportunistic and intentional. Each offers its own challenges and difficulties. Once the author has found a topic, she or he must consider several factors when developing and writing the case. This section describes the important considerations that must be made in each of these phases of writing a teaching case.

Interview With Susan Martonosi

Susan Martonosi. Associate Professor and Global Clinic Director in the Department of Mathematics at Harvey Mudd College, offered these thoughts on using pedagogical goals to guide case writing:

There are several decisions that a teaching case author must make about the scope of the case. Ideally, these decisions should be motivated and guided by the pedagogical goals for the case. For instance, if one goal is for students to learn how to discern salient information from a real business situation for the purposes of decision-making, then that would dictate the inclusion of extraneous details in the written case that students must sift through. If a pedagogical goal is to help students learn how to make reasonable simplifying assumptions and parameter estimates and test them using sensitivity analysis, then this would motivate leaving out important details from the written case. In a teaching case where a pedagogical goal is to gain familiarity with several descriptive statistics and associated paradoxes, then realistic data could be fabricated to illustrate the desired phenomena simultaneously.

On the other hand, if the goal is to understand how the methods of data acquisition, cleaning, and management can affect the statistical output, then it is important to give the students access to real, messy data. Note that in the examples described here, the design considerations are motivated by pedagogical goals that are largely distinct from and independent of the specific analytics methodology used in the case. The structure of a teaching case can be designed to achieve pedagogical goals related to the problem-solving process in addition to analytics content.

This is an excerpt from one of a series of interviews with analytics professionals and educators commissioned by the *INFORMS Analytics Body of Knowledge* Committee.

A.3.1 Sources of Teaching Cases

An instructor who uses teaching cases and finds herself or himself in need of a teaching case on a specific topic has a few options. If this instructor is fortunate, he or she may have recently worked on a project or currently may be working on a project that lends itself to a teaching case that will be relevant to the desired topic. If not, she or he can look for published teaching cases that meet specific needs. The Harvard School of Business, Virginia's Darden School of Business, and the University of Western Ontario's Ivey Business School each produce and sell high-quality cases that cover a wide variety of topics across a broad range of business disciplines. *INFORMS Transactions on Education* (https://pubsonline.informs.org/journal/ited) publishes cases and articles on classroom use of the cases in an open-access environment. A brief list of topics from the dozens of teaching cases published by *INFORMS Transactions on Education* (*ITE*) includes Simpson's paradox and probability [9], integer programming [9–11], data envelopment analysis [12], revenue management [13], probability models [14], data analysis [15], logistics [16], risk management [17], and vehicle routing [18].

Faculty, students, and the public all have free access to the cases and associated articles published by *ITE*, but the journal maintains a set of teaching notes for each case it publishes on a password-protected Web site. Instructors are thoroughly vetted before being granted access to the teaching notes (which are also made available free of charge) to prevent guileful students from gaining access.

Other journals also publish teaching cases. Examples of managerial cases published by other journals include decision analysis [19] multicriteria decision-making [20], inference [21,22], social network analysis [23], structural equation modeling [24], and regression [25].

If the instructor cannot find a suitable published case or is not prepared to pay for a teaching case, then she or he must develop the case. Perhaps the instructor can recall a situation she or he or a colleague faced that would provide the basis of an effective teaching case that meets the instructor's needs. In such instances, the author may have to provide some realistic embellishments to give substance to the teaching case; if the author is proficient in the topical area of the case, she or he can generally accomplish this without jeopardizing the realism of the case. However, with embellishment comes the risk of rendering the case unrealistic, which will defeat the purpose of the teaching case. In the most risky instances, the author fabricates most or all of the critical components of the case, and what results is not a teaching case. Students are perceptive and they will see this for what it is–artificial, unrealistic, and irrelevant.

Often an instructor will confront a problem or scenario in her or his professional or private life that could be the basis of an effective teaching case. In such instances, the potential topic for the teaching case has arisen opportunistically. Although this may seem to be a fortunate occurrence, chance does favor the prepared mind. If the teaching case author does not recognize the potential for developing a teaching case from this problem or scenario, she or he may miss this opportunity. If the teaching case author does recognize the potential for a teaching case to be developed out of this problem or scenario, but is not prepared to gather all of the relevant information that would be necessary to develop the associated teaching case, she or he again may miss this opportunity.

Instructors who routinely write teaching cases have various methods for being prepared to take advantages of opportunities when they arise. They often work on applied projects. They scan newspapers, magazines, blogs, and Web sites for inspiration. They keep electronic lists of ideas to develop and topics for which they need a teaching case. And they devote time to thinking about developing cases.

If properly developed, such cases can offer rich experiences for students, but often the author will not be able to naturally manipulate the case so that it is relevant to the specific issue that the author wants to address. For example, an author/instructor who needs to develop a teaching case on quality control may be confronted with a terrific opportunity to develop a teaching case on inventory management while shopping. Rather than ignoring this opportunity or (even worse) attempting to contort the situation into the mold of a quality control problem, an opportunistic author will develop the inventory management case for her or his later use (or for use by her or his colleagues). This often happens to teaching case authors who consult with government and/or private industry; the problems to which the author is exposed through consulting opportunities may not naturally lend themselves to the case topics she or he would like to develop presently, but these problems may be interesting and could provide the basis of effective teaching cases on other topics.

Interview With Matthew J. Drake

Matthew J. Drake, Associate Professor of Supply Chain Management at Duquesne University and Editor-in-Chief of Decision Sciences Journal of Innovative Education, elaborates on identifying content for teaching cases.

The authorship of teaching cases has been a major part of my academic career. My doctoral study was even supported by a grant my advisor received to integrate the teaching of ethical decision making into the industrial engineering curriculum in part through the use of teaching cases. I wrote two teaching cases while I was a doctoral student to fulfill this grant, and I have continued to write teaching cases ever since.

One of the most challenging parts of writing a teaching case is identifying a company scenario that is rich enough to justify multiple defendable recommendations yet defined enough to provide students - especially the less-experienced - with the structure they need to avoid being overwhelmed. I can think of two suggestions for potential teaching case authors to consider for identifying effective case scenarios.

An academic mentor once gave me some particularly sage advice with respect to the intersection of traditional research and teaching cases. He said that the most fruitful kinds of applied research projects with companies are those that can generate three related intellectual contributions: (1) a top-tier research journal publication, (2) a publication in a practitioner-oriented journal such as *Sloan Management Review or California Management Review*, and (3) a teaching case. Projects that can generate high-level research articles and can be distilled into a practical managerial vernacular are often excellent sources of scenarios for teaching cases that will interest and benefit students as well. These publication-dense research projects can be extremely helpful for academics whose institutions reward or expect them to have an impact on multiple constituencies with their research.

A second source of content for teaching cases can be prior consulting projects that the academics have completed with previous sets of students. It is particularly common to incorporate consulting projects into traditional courses. If these projects prove to be successful, they are strong candidates to be turned into effective teaching cases so that future groups of students can benefit from a similar learning experience. These teaching cases can be relatively easy and quick to prepare because the instructor has all of the data and is already familiar with the decision scenario and recommended analysis.

This is an excerpt from one of a series of interviews with analytics professionals and educators commissioned by the *INFORMS Analytics Body of Knowledge* Committee.

A.3.2 The Teaching Case Writing Process

Once an author has identified a potential topic for a case, she or he must consider the intended audience for the case. Cases intended for use with sophomores who are taking service courses must be written very differently than cases that are intended for use by students in MBA programs. In conjunction with this consideration, the author must identify the pedagogical goals to be achieved through her or his use of the case. For an analytics-oriented teaching case, the author must decide specifically what analytics method(s) she or he wants the student to use to address the issue(s) of the case.

At this point, the methodical author will reflect on the finer details of the case. Considerations at this stage include the following:

- How much guidance, if any, will the case give the student with regard to
  - problem identification:
    - Will all necessary details be provided?
    - Will extraneous details be provided?
  - motivations and objectives of the players and stakeholders?
  - analytical method(s) to be employed?
- How much detail will be provided?
  - How much breadth will be provided?
  - How much depth will be provided?
- What, if any, domain-specific knowledge should the student need to address the issues of the case?
- How much data will be provided?
  - Will the data be clean (i.e., no errors)?
  - Will the data be aggregated, or will the students have to use several files (perhaps in various formats) to assemble the data needed?
  - Will all necessary data be provided?
  - Will extraneous data be included?

- What are the set of reasonable explanations, solutions, decisions, and/or strategies a student could propose?
  - What are the relative virtues of these explanations, solutions, decisions, and/or strategies?
  - What are the relative deficiencies of these explanations, solutions, decisions, and/or strategies?

Once the teaching case author has diligently considered all of these issues, she or he is ready to collect whatever additional information that is necessary, such as

- additional background information,
- additional data, and
- interviews with players and stakeholders,

and then write the first draft of the teaching case.

A.3.3 Finalizing the Teaching Case

The process of writing a teaching case should be iterative in two ways. First, the author will likely have to iterate between the draft and the issues outlined in the previous steps, revising each until she or he finds a satisfactory convergence of the draft and the objectives of the case. Second, the author may wish to revise the teaching case (perhaps several times) after using it in class to reflect what she or he has learned about the case from these experiences.

Once the author has finalized the draft, she or he should write a set of teaching notes for the case. In addition to providing the author of the case with gentle reminders of when and how to use a teaching case, the teaching notes should explain the purpose and intended audience of the case to colleagues who may wish to use the case. These notes should include the following:

- A teaching plan that suggests ways the case be used in class.
- A summary review of the case scenario:
  - The background provided, including an indication of what information is relevant, what information is irrelevant, and what (if any) additional information is necessary.
  - The players and stakeholders and their motivations.
- A review of the problem(s) that should be addressed, decision(s) that should be made, and/or strategy(ies) that should be formulated by the student in analyzing the case.
- A note on the domain-specific knowledge the student will need to address the issues of the case.
- A discussion of the data that are provided:
  - The ways the data will have to be cleaned by the student to eliminate errors.
  - The ways the data will have to be manipulated to assemble the final data set.
  - Whether all necessary data are provided and/or extraneous data are included.

- A discussion of the analytics method(s) the student should use to address the issue(s) of the case.
- Detail on the set of reasonable explanations, solutions, decisions, and/or strategies a student could propose:
  - The relative virtues of these explanations, solutions, decisions, and/or strategies.
  - The relative deficiencies of these explanations, solutions, decisions, and/or strategies.
  - Issues that could be faced in implementation of the suggested solution to the case.
- Additional questions that instructors can use to extend the case discussion in class.
- The actual outcome of the case scenario (if the case is retrospective).

This document will also need to be revised regularly to reflect the author's (and perhaps her or his colleagues') classroom experience in using the case and revisions the author makes to the case.

Although the author/instructor will attempt to anticipate all potential student reactions to all of the issues in a teaching case, it is virtually impossible to do so. It is important that authors spend a sufficient amount of time considering all issues of the case and how they are to be presented, and then be open to revising the case and/or teaching notes to reflect what she or he learns about the case through classroom use. This is also why an author/instructor should test a teaching case in several classrooms before submitting it to a journal for publication.

For examples of teaching notes for cases, the reader can request access to password-protected teaching notes that accompany cases published by *INFORMS Transactions on Education* (https://pubsonline.informs.org/journal/ited).

A.4 Using a Teaching Case

Instructors who wish to use teaching cases face many choices. In addition to selecting appropriate cases for the pedagogical objectives, the instructor must decide on how she or he will assess the students' work on the case assignment (written analyses, class discussions, and/or formal presentations) and she or he must select/develop a style for facilitating in-class case discussions.

A.4.1 Selecting a Case

An instructor who uses teaching cases must consider several factors when selecting the cases she or he will use. These factors include the following:

- How well the case
  - matches the course objectives.
  - meets instructor's pedagogical objectives.

  - offers students an opportunity to understand different perspectives/points of view.
  - facilitates application of course concepts by students.
  - fosters a better understanding of target concepts by students.
  - provides students with an opportunity to improve their
    - critical thinking skills.
    - communication skills.
    - interpersonal skills.
- What students are likely to learn from analyzing the case.
- The analytic technique(s) appropriate to the case problem.
- The technical level of the case.
- The required domain-specific knowledge.
- The ambiguity and level of student maturity/experience with the case method:
  - The existence of multiple plausible and compelling conclusions with different implications.
  - The amount of irrelevant information included.
  - The amount of relevant information omitted.
- Instructor's comfort with the
  - case scenario.
  - required analysis.
- The quality of available support material.
- The cost of the case:
  - How much time will the case take to cover?
  - How much will students have to pay to obtain access to the case?

Each instructor must decide how much weight to ascribe to each of these factors.

A.4.2 Assessing the Student

An instructor who uses teaching cases can achieve different goals through using written analyses, class discussions, and formal presentations to assess students. Written analyses allow students to develop and refine their ability to present the facts and discuss their assumptions dispassionately; make a cogent, logical, succinct, and thoughtful argument; and present the potential advantages and disadvantages of alternative strategies. This approach favors the independent and methodical student. Class discussions allow students to think in their feet and interactively debate the pros and cons of various decisions and strategies. The quick thinking and loquacious student will tend to thrive in this environment. Formal presentations achieve a combination of what can be achieved through written analyses and class discussions.

Whether the instructor assigns written analyses, class discussions, formal presentations, or some combination, students will naturally have questions about the details of the case and the nature of the assignment. With regard to students discussing the case with me, I will not respond to questions concerning the choice of an appropriate analytic approach prior to the class discussion of the cases. However, I will permit students to ask general questions about the case scenario, but they must understand that the nature of my response will depend on the nature of their question. They must also accept that they may receive any of the following responses:

- a single answer to their question;
- multiple possible answers to their question; or
- no answer to their question.

I also warn my students that I may, in order to provoke discussion, provide different students with different *appropriate and reasonable* responses to similar questions.

The instructor must also decide if students will work individually or in teams. If students are to work individually, the instructor must delineate the extent to which students are allowed to discuss the case among themselves. I generally allow students to discuss the case among themselves as much as they care to as long as their final written analyses are their own (students benefit greatly from these discussions).

If students are to work in teams, the instructor must determine the number of students that will be permitted to belong to a team, how teams are to be formed, and how conflicts within teams will be resolved. Team conflict resolution can be a particularly thorny issue, and the instructor should develop a strategy or policy to deal with this issue in advance and communicate this with students.

Written Analyses

Regardless of whether students are to work independently or in teams on their analyses and written reports, the instructor must also communicate what she or he expects with regard to

- content–what level of detail does the instructor expect?
- writing style–does the instructor expect a technical report, a business memo, or a broad overview?
- exposition–how much emphasis will the instructor place on the quality of writing?
- length–what are the minimum and maximum lengths expected by the instructor? Are these guidelines or strict limits?
- format–how does the instructor expect the report to be organized and presented?
- lead time–how much time does the student have to analyze the case?

Communicating expectations is always an important factor in classroom success, but this is particularly critical when working with students have little or no experience with the case methodology.

When I require written analyses of teaching cases in my undergraduate introductory statistics and operations research courses, I provide my students with a suggested format that consists of four sections (with appropriate appendices). The four sections are:

- Section 1–Overview
  Review the scenario and context. Identify the problem(s) to be addressed. Discuss the important players and stakeholders and their interests and motivations. Assume you are employed as an analyst for the organization in the case.
- Section 2–Methodology
  Explain and justify the approach(es) that you propose to use in addressing the problem(s)/responding to the question(s) suggested by the case. Discuss any assumptions (mathematical or otherwise) that you are making, and explain the consequences that could arise if your assumptions are invalid. If appropriate, explain why other approaches under consideration are inferior or unsuitable. Use *nontechnical terms that someone with a minimal background in operations research can understand.*
- Section 3–Results
  Present and interpret the results. Explain the potential implications of the analysis. Include graphs, displays, calculations, or printouts if appropriate, or place them in appendices and refer to them in this section. Do not include graphs, displays, calculations, or printouts if they do not provide illumination. Suggest a decision or a strategy if appropriate. If possible, discuss issues that may arise in implementation of the suggested decision or strategy. Be creative and use intuition.
- Section 4–Critical Assessment
  Examine the approaches to data collection and analysis. Discuss positive and negative aspects of this process. Suggest (i) ways *to improve the analytic process* you just completed and (ii) *directions for future analysis.*
- Appendices–Relevant Printouts, Tables, and Graphics
  Results and displays may be placed in appendices. Note that appendices should be numbered and appropriately labeled, and each appendix *should be referred to at least once in the body of the case analysis.*

I also remind students that they should

- use *nontechnical terms that someone with a minimal background in statistics or operations research can understand.*
- avoid discussions of the mechanics of the solution algorithm or software used in the analysis.
- resist the temptation to review or critique the teaching case (this is not the place for the student to explain how she or he feels about the assignment).

This is a proposed format and is not mandatory–my students have complete latitude in determining the format in which they present their case analyses. This policy recognizes that each student has an analytic style that is a culmination of his or her unique skills and experiences (both in and out of the classroom), and serves to encourage students to further develop and refine their styles.

I limit the students' final written analysis of a case to two pages of text with a 10- or 12-point font and 1 in. margins in order to provide the students with experience writing in a concise manner appropriate to business communications. However, appendices do not count against this limit, and I make allowances for students who choose to integrate tables, graphs, equations, charts, and other displays into the bodies of their written analyses.

I explain that I will base case grades on the appropriateness of the analytics technique(s) they apply to the case problem, how well they apply the analytic technique(s) they have selected, and the quality and correctness of their interpretation of their results. I also make it clear that the quality of writing is important. I reward students who use short, well-crafted sentences that flow and are easy to follow. Spelling, grammar, and usage are also factors.

Class Discussions

The instructor who uses class discussion as a basis of evaluating the students' efforts must carefully and completely communicate her or his expectations. Considerations here are similar to considerations that must be made by instructors who assign written analyses:

- Content–What level of detail does the instructor expect?
- Speaking style–Does the instructor expect technical language, business language, or conversational language?
- Exposition–How much emphasis will the instructor place on the quality of speaking?
- Contribution–How much is each student expected to contribute to each case discussion? How will students be selected to contribute to the discussion?
- Format–How will the instructor facilitate the case discussion?

The answer to each of these questions depends on the complex interaction between the instructor, the students, the course material, and the case. However, I do adhere to a basic outline with regard to the format (and again, this is similar to how I handle this issue when assigning written analyses). My class case discussions generally proceed through four broad areas in this order:

- Part 1–Overview
  Review the scenario and context. Identify the problem(s) to be addressed. Discuss the important players and stakeholders and their interests and motivations. Assume you are employed as an analyst for the organization in the case.

- Part 2–Methodology
  Explain and justify the approach(es) that you propose to use in addressing the problem(s)/responding to the question(s) suggested by the case. Discuss any assumptions (mathematical or otherwise) that you are making, and explain the consequences that could arise if your assumptions are invalid. If appropriate, explain why other approaches under consideration are inferior or unsuitable. Use *nontechnical terms that someone with a minimal background in operations research can understand.*
- Part 3–Results
  Present and interpret the results. Explain the potential implications of the analysis. Include graphs, displays, calculations, or printouts if appropriate, or place them in appendices and refer to them in this section. Do not include graphs, displays, calculations, or printouts if they do not provide illumination. Suggest a decision or a strategy if appropriate. If possible, discuss issues that may arise in implementation of the suggested decision or strategy. Be creative and use intuition (i.e., think outside of the box).
- Part 4–Critical Assessment
  Examine the approaches to data collection and analysis. Discuss positive and negative aspects of this process. Suggest (i) ways *to improve the analytic process* you just completed and (ii) *directions for future analysis.*

Depending on time and technology available, I may also allow students to present limited relevant printouts, tables, and graphics during the discussion.

Again, I remind students that they should

- use *nontechnical terms that someone with a minimal background in operations research can understand.*
- avoid discussions of the mechanics of the solution algorithm or software used in the analysis.
- resist the temptation to review or critique the teaching case (this is not the place for the student to explain how she or he feels about the assignment).

Students will occasionally attempt to deviate from the ordering of these areas of discussion–many students want to present the results of their analyses first; this is something I do not allow. However, there are instances in which deviations are not only permissible but also beneficial. For example, a discussion of the analytic results or the critical assessment may take the discussion back to further consideration of how the data were collected or assumptions that have been made.

I attempt to give each student who wishes to participate at least one opportunity during each case discussion, and I do not let a minority of the students monopolize the discussion. I also stress the importance of being direct, succinct, and considerate/polite when making a point during a case discussion.

I explain that I will base case grades on the quality of the contribution made by each student–the content and appropriateness of the contribution, the manner in which the contribution is made, and the originality of the contribution. I also explain that I not recognize (and may penalize) contributions that are empty or meaningless, rambling or incoherent, inappropriate, or rude.

Formal Presentations

Formal presentations are an interesting combination of written analyses and class discussions; they require preparation of a physical product (as do written analyses) and some oral explanation (as do class discussions). Therefore, much of the previous discussions of written analyses and class discussions in the teaching case environment apply to the use of formal presentations. The instructor who uses the formal presentation to assess student performance must also consider two other factors:

- Technology–Some instructors limit their students to the use of PowerPoint software in formal presentations of case results. This limitation has the advantage of reducing the likelihood students will produce presentations that are inappropriate for business settings. Other instructors will allow students more latitude. In these settings, students can run software or code in time and discuss the results; use audio and visual recordings, animation, and sound effects; and utilize a wide range of other visual aids. This does increase the likelihood students will produce presentations that are inappropriate for business settings, but it also allows them to be creative and learn from their mistakes. Some instructors will require students to submit/preview their presentations in advance to ensure the students are giving appropriate presentations.
- One Presentation or Multiple Presentations–Here instructors have a few interesting options. Does the entire discussion of a teaching case consist of one detailed presentation of an analysis with the remaining students asking questions of the presenter(s)? Do several (perhaps each) of the students or teams give brief presentations of their analyses with the remaining students asking questions of the presenter(s)?
- Noncompetitive or Competitive–If the students who are not presenting their results are tasked with asking questions of the presenter(s), are the students in the audience rewarded for finding flaws in the presented analysis? This approach will tend to bring important points into the discussion very quickly, albeit at the risk of some bruised feelings.

In a novel but rarely considered approach to competition, the instructor may assign every student or team the presentation of their case analysis and provide several presentation stations (perhaps projection and a screen in each corner of a classroom) and allow students or teams to competitively present their ideas and results. The author developed this approach (which he refers to as the box-and-one approach) and has found that it creates a venture-capital atmosphere that facilitates rapid (i) identification and distilment of the relevant issues and (ii) identification and evaluation of the merits of alternative solutions, decisions, and strategies. This approach requires a great deal of preparation and classroom facilitation by the instructor and works best with academically mature students.

A.4.3 Facilitating Case Discussions

There are many approaches to facilitating a case discussion, and it is critical that the instructor find the facilitation style that will work for her or him and tailor it to specific situations (cases, classes, and students). The instructor must develop a style that fits her or his personality and teaching philosophy so that she or he is comfortable facilitating class discussion of cases.

The instructor must also understand and accept that some case discussions will be superior to others; some will be more lively, some will be more thoughtful, and some will be more intense. This does not necessarily reflect on the quality of learning that is occurring during the case discussion. The instructor must therefore have reasonable expectations for each case discussion that reflect the students, the course, the teaching case under discussion, and the instructor.

In considering and developing a facilitation style (i.e., how she or he will conduct and orchestrate case discussions in her or his classrooms), there are two issues that are of primary importance:

1) How much does the instructor prompt, prod, and/or push the students during the discussion?

How much assistance will the instructor provide her or his students during the case discussion? *Early* is the key consideration when deciding how much to prompt, prod, and/or push students–early in a student's academic experience, early in the academic term, and early in the discussion of the case. When in these states, the instructor generally must prompt, prod, and/or push more frequently to initiate, provoke, and control the flow of the case discussion. As one moves out of these states, the instructor can expect more from students and can allow them more latitude in their discussions.

2) Does the instructor aim for consensus or allow for contention during the discussion?

Will the instructor attempt to help students find a single resolution to the case upon which they can all agree to a large extent, or will the instructor allow for or even encourage a variety of resolutions to develop and even flourish during the case discussion? The key consideration when deciding to aim for consensus or allow for contention is the *openness* of the case, which may be discerned through responses to the following questions:

- Does the background provided omit relevant information?
- Does the background provided include irrelevant information?

- Are the players and stakeholders and their motivations at odds with each other?
- Are there many potential problem(s), decision(s), and/or strategy(ies) to be addressed?
- Is domain-specific knowledge required?
- Do the data need to be cleaned to eliminate errors?
- Do the data have to be manipulated to assemble the final data set?
- Are necessary data provided?
- Are extraneous data included?
- Are there multiple analytics methods that could be applied?
- Are there many reasonable explanations, solutions, decisions, and/or strategies?
- Are there potential difficulties to be addressed in implementation of the suggested solution(s)?

A response of *yes* to any of these questions increases the potential need for a contentious approach. Although the contentious approach will likely intimidate many students (and perhaps some instructors), it ultimately provides the clearest path to student appreciation of the complexities and nuances of using analytics to aid in decision-making and strategy formulation.

The answers to the questions of how much to prompt, prod, and/or push the students and whether to aim for consensus or allow for contention during the discussion depends on the complex interaction between the instructor, the students, the course material, and the case.

The overarching goal in making these choices is to find a way to enable and encourage engagement and constructive participation by the students, and it is important that the instructor explain to the students that learning by everyone in the class is best facilitated by regular participation of all students in the class. The student in a case-based course must accept that she or he has the responsibility to share his or her understanding, knowledge, and judgment with the class to advance the classes' collective learning and development.

Thus, students in a course taught with cases must take complete responsibility for their learning. Because this may be a radical departure from the expectations other instructors have of their students, the instructor who is using cases must consistently stress this theme in all communications with students. Some instructors create a contract for the students that clearly explains this expectation explicitly, and some of these contracts include a section that clearly explains the expectations the students should have of the instructor. This approach, whether established through an actual contract or other dialog between the instructor and the students, establishes an important level of professionalism in the case-based class.

Another critical component of an instructor's success in facilitating the case discussion is her or his preparation. The instructor who integrates cases into her or his course must prepare exhaustively for the classroom discussion; the instructor must arrive for a case discussion with a knowledge of the case that far exceeds the understanding of the case that could be developed by any of her or his students. This means doing far more that simply reading and rereading the case and the teaching note or spending a great deal of time analyzing the case. Although these tasks are important, they are not sufficient. The instructor must take time to develop the specific teaching objectives that she or he wants to achieve; reflect on the case from the student perspective; anticipate the approaches, methodologies, and case resolutions students may suggest (and be prepared to critique these); and foresee questions students may ask (and be prepared to respond).

A.5 An Example of a Simple Case

A "Boring" Time is a relatively brief teaching case developed by the author to impart understanding and appreciation of the concept of *variation* in students taking undergraduate introductory business statistics courses. The case, which is generally used early in the academic term, also raises some basic but important issues in design of experiments. Table 1.

A "Boring" Time

Jon Weideman, second shift foreman for Cut-Rate Machining, Inc., is attempting to decide from which vendor to purchase a drilling machine. He narrows his alternatives to four vendors: The Hole-Maker, Inc. (HM), Shafts & Slips, Inc. (SS), Judge's Jigs (JJ), and Drill For Bits, Inc. (DB). Each of these vendors is offering machines of similar capabilities at similar prices, so the effectiveness of the machines is the only selection criteria that Mr. Weideman can use.

Weideman invites each vendor to ship one machine to his Richmond, Indiana, manufacturing facility for a test. He starts all four machines at 4:00 p.m. and lets them warm up for 2 hours, at which point four of his employees will each be assigned to drill 100 3 inch diameter holes in 6 inch thick stainless steel disks in one of the four machines over a 2 hour period. The diameter of each hole drilled with each machine is then measured and recorded. The results of Mr. Weideman's data collection are shown in Table B.1.

Table B.1

Hole #	Hole Maker	Shafts & Slips	Judge's Jigs	Drill for Bits
1	3.155364381	2.922524340	2.601899600	2.020303130
2	2.997547371	2.973922960	2.602222607	3.033852815
3	3.088575397	2.982287510	2.597631775	3.322239256

Table B.1 (*Continued*)

Hole #	Hole Maker	Shafts & Slips	Judge's Jigs	Drill for Bits
4	3.181965687	2.791679756	2.601851908	2.000248758
5	3.147572201	2.962144415	2.601901474	2.252476499
6	3.126991075	2.957430491	2.598021888	3.763290024
7	2.931293214	2.824897122	2.600366708	3.416787373
8	3.197907116	2.817265146	2.595383117	2.876396443
9	3.204837358	2.885608124	2.600140457	1.919411822
10	3.044548037	2.845581953	2.596201699	3.219009598
11	3.190527980	2.941790215	2.600905324	3.678588884
12	3.116415588	2.820284509	2.603322843	2.870016696
13	3.180619951	2.933137634	2.600945566	3.075640074
14	3.176394345	2.828115475	2.607790565	3.359021707
15	3.134453058	2.886864697	2.590968624	2.555330649
16	3.108766216	2.991908991	2.610265642	3.741594022
17	3.168782785	2.914440831	2.595834569	2.853519987
18	3.142101925	2.884767015	2.602127655	2.946857838
19	3.065931976	2.936737292	2.599805806	3.531605834
20	3.066546404	2.881212293	2.605580164	2.748536269
21	2.958237837	2.846133223	2.606172859	2.765770766
22	3.019540508	2.921795013	2.602436484	1.771058583
23	2.820970546	3.060279394	2.607027351	1.365669361
24	3.130088022	2.896676282	2.595333934	4.581084216
25	3.010817231	2.960039996	2.597343155	3.955788825
26	2.960007028	2.953646826	2.593119247	3.490767450
27	3.119837403	2.926863402	2.593588444	2.407792181
28	2.998937608	2.860933850	2.610195588	2.658767961
29	3.267965880	2.886545297	2.606771969	2.798071801
30	2.981308424	2.994208596	2.601520822	3.123670540
31	3.035227149	2.820618474	2.598045139	2.619224669
32	3.182983077	2.966404097	2.604896813	4.398896610
33	3.146199205	2.942281876	2.606829271	4.227886955
34	3.164555096	2.916002138	2.596627389	2.898724804
35	3.154340901	2.975004853	2.590560135	2.765779801

(*continued*)

Table B.1 *(Continued)*

Hole #	Hole Maker	Shafts & Slips	Judge's Jigs	Drill for Bits
36	3.160142986	2.920510118	2.601894604	3.268629042
37	3.056469899	2.799884500	2.599812572	3.809144229
38	3.178801971	2.833801006	2.603735613	2.347380084
39	3.211824105	2.837998294	2.606434598	3.548507952
40	3.126141696	2.811825139	2.600849208	1.768339817
41	3.214448177	2.878167692	2.597818548	3.111718220
42	3.092014213	2.964308419	2.604160620	2.786814689
43	3.199409355	2.901818545	2.601358860	3.678749257
44	3.141816141	2.916665807	2.597638550	1.483897121
45	2.946596111	2.905828001	2.601848562	1.197450108
46	3.172948477	3.026060282	2.601813392	2.525998843
47	3.044225350	2.769265229	2.601215795	2.987014879
48	3.079348417	2.887170921	2.598029960	2.836127562
49	3.001841378	2.851112061	2.598645961	2.924752579
50	3.180564841	2.911699356	2.597415826	2.322914453
51	2.996207663	2.885997590	2.598895676	3.944505498
52	3.057665310	2.956883891	2.598229519	3.247389632
53	2.995936456	2.859481109	2.605657407	2.206632522
54	3.007319483	2.958925568	2.600361524	2.959886150
55	3.135957146	2.909237098	2.591072498	2.514374436
56	3.022197909	2.776571237	2.596104043	3.592987267
57	3.251440750	2.988441771	2.596185657	2.619025196
58	3.162873684	2.859696368	2.595014302	2.851140376
59	3.251847393	2.907871491	2.595042367	3.133586364
60	3.096149722	2.920562883	2.603425598	4.565572241
61	3.118916763	2.963868387	2.593972965	4.264724864
62	2.945501606	2.898019453	2.603538526	2.396928891
63	3.079379110	2.931471621	2.605789739	3.483637791
64	2.987995641	2.952967473	2.598965562	4.294408099
65	3.110292668	2.910044124	2.604833717	4.310251392
66	3.010843266	2.919169150	2.607415524	2.419788866
67	2.901098531	3.005182353	2.607991703	2.946617039
68	3.044826268	2.942895952	2.603583303	2.781532795

Table B.1 (*Continued*)

Hole #	Hole Maker	Shafts & Slips	Judge's Jigs	Drill for Bits
69	3.083492574	2.913551530	2.598179189	2.750560455
70	3.227765141	2.922459942	2.602805028	2.651547714
71	3.140541685	2.919215122	2.595002091	2.148581557
72	2.953882834	2.903688154	2.613596999	3.015572201
73	3.024359065	2.949977940	2.603263999	3.735816835
74	2.917967065	2.962769528	2.598612555	3.497309231
75	3.024122458	2.965881997	2.601894644	2.875472463
76	3.102284437	2.839063500	2.595055976	2.76277536
77	3.096893735	2.966474455	2.599896155	2.843656554
78	3.016547594	2.904449551	2.608950253	2.762737294
79	3.015705694	2.775223566	2.604252603	2.69899297
80	3.132252113	3.004848599	2.600590155	3.72455745
81	3.285616963	2.894253371	2.604160154	2.561427079
82	3.024126035	2.998095792	2.605396497	2.215330112
83	3.146220381	2.907822434	2.596112077	3.790243151
84	3.058863382	2.957593971	2.593220693	1.773839322
85	3.037223024	2.916673787	2.606199027	2.377684665
86	3.186002638	2.828697252	2.604114466	3.011324779
87	3.027562972	2.902385335	2.593502879	4.381211209
88	3.116138318	2.795937499	2.600223968	3.560862786
89	3.222135150	2.868748135	2.594398915	2.304808833
90	3.205057904	2.884473418	2.602293104	2.902489145
91	3.179464120	2.955715257	2.600614701	1.999277817
92	3.132212300	3.034477754	2.594746012	3.349808454
93	3.175893377	2.851182778	2.592769812	3.31809983
94	3.234545447	2.916458306	2.597495742	4.227518202
95	3.138173800	2.894849985	2.598871945	2.570458293
96	3.082324587	3.000378384	2.596084369	3.457906483
97	2.913722262	2.885821324	2.59287364	2.43683616
98	2.986965817	2.916356140	2.600478599	3.307038724
99	3.164196672	2.839317401	2.603708746	2.800375463
100	3.206022871	2.884512789	2.595289133	2.605460712

Based on these results, from which vendor would you suggest Mr. Weideman purchase his new machine?

Hint: Think carefully about what qualities would make a drill desirable for Mr. Weideman.

In addition to the guidelines discussed in Section 4.2.1, students are given the following grading criteria:

1) Analyses are to be
   - typed or word-processed.
   - double-spaced.
   - two pages maximum of text (not including displays, tables, appendices, etc.).
   - one-inch margins.
   - twelve point type size.
   - Times New Roman font.
2) Each appendix must be referenced in the body of report.
3) Some statistic(s) (numerical measure such as the mean, variance, midrange; graphical display such as a line graph) are to be used.
4) No discussion of how to use software (this includes Excel).
5) The raw data are available for download in an Excel file on the classroom Web site.
6) Students receive full credit unless they egregiously violate these standards.

In analyzing the data provided in this case, students will naturally calculate some summary statistics such as those included in Table B.2.

Based on the sample means, the drill provided by Drill for Bits performed best and the drill provided by Judge's Jigs performed worst. However, students who stop at this point are missing an important characteristic of these data. A line

Table B.2

Summary Statistic	Hole Maker	Shafts & Slips	Judge's Jigs	Drill for Bits
$\overline{x}$	3.096194829	2.908347920	2.600370430	2.985636918
m_d	3.113215493	2.912625443	2.600534377	2.887560623
minimum	2.820970546	2.769265229	2.590560135	1.197450108
maximum	3.285616963	3.060279394	2.613596999	4.581084216
range	0.464646417	0.291014165	0.023036864	3.383634108
midrange	3.053293754	2.914772312	2.602078567	2.889267162
s	0.093657619	0.060980033	0.004762979	0.719022438

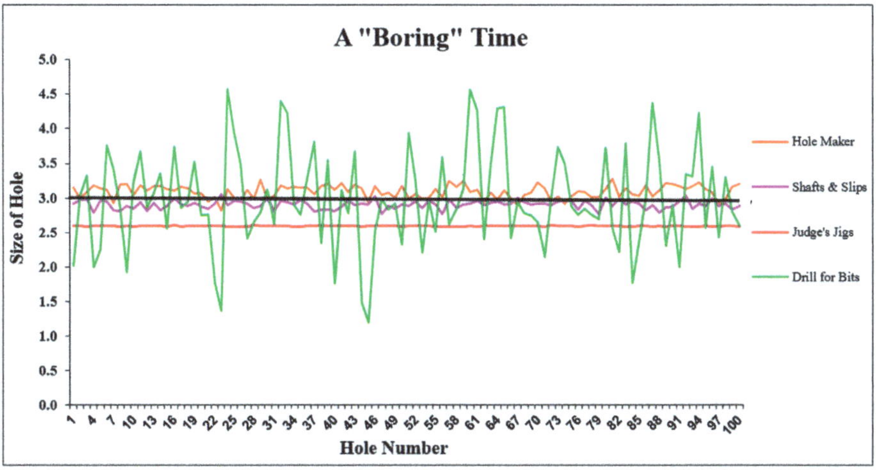

Figure B.1

graph or side-by-side boxplots also provide excellent illustrations of this point (see Figure B.1).

Through this graph, students can see both accuracy (How close are the diameters of the holes drilled by one machine to the target of 3 inches) and precision (How consistent are the diameters of the holes drilled by one machine?). They can also see that

- Hole Maker is reasonably consistent (as we can also see in the last row of the table of summary statistics) and routinely overdrills;
- Shafts & Slips is reasonably consistent (as we can also see in the last row of the table of summary statistics) and routinely underdrills;
- Judge's Jigs is extremely consistent (as we can also see in the last row of the table of summary statistics) and always underdrills; and
- Drill for Bits is wildly inconsistent (as we can also see in the last row of the table of summary statistics).

The case does not explain the physical properties of the stainless steel disks or the purpose of the hole, so students cannot discern whether (i) a hole can be rebored or (ii) a hole that is too large is acceptable. This makes assessing the drills provided by Hole Maker and Shafts & Slips difficult. However, if the drills can be recalibrated, the best drill to purchase may be the extremely consistent product produced by Judge's Jigs. This is precisely the ambiguity that I want my students to struggle with when analyzing a teaching case.

Students should also note that assigning each machine to a different employee over the entire life of the experiment could result in a serious bias; the effect of the machine and employee are perfectly confounded. Astute students will recognize this problem and suggest a rotation of the four employees assigned to the four machines. Other students may go further and suggest that Weideman use more than four employees to test the four drills. Here I can use this teaching case to introduce some basic concepts of experimental design in a meaningful way very early in the course.

A.6 Final Thoughts

Because teaching cases afford students low-risk opportunities to assume and ultimately appreciate various roles in typical situations faced by practitioners, they are ideally suited for use in analytics courses. Instructors can use teaching cases to help students understand the answers to three important questions that students frequently ask in analytics courses:

- When they will use these concepts and methods?
- Where will they use these concepts and methods?
- How will they use these concepts and methods?

Many students who have never taken a course that uses the case methodology will be somewhat apprehensive at the beginning of the academic term, but a gentle introduction and early success will quickly alleviate most of their unease. Students will also be less anxious if the instructor provides detailed guidelines and expectations.

There are many approaches to implementing cases into courses. Some instructors design entire courses around the case method, while others (including the author) integrate cases into strategic points in courses. Students can work individually or in teams, and instructors can assess student performance through written analyses, class discussions, and/or formal presentations.

Instructors (including the author) who have integrated cases into analytics courses have reported

- dramatic improvement in students' modelling, technical, and analytic skills,
- increase in instructor workload,
- happier students who enjoy class,
- more frequent application of concepts from my courses to problems in other courses by former students, and
- a more gratifying teaching experience for the instructor.

It is important to reiterate that although the author/instructor will attempt to anticipate all potential student reactions to all of the issues in a teaching case, it is virtually impossible to do so. It is critical that authors spend a sufficient amount of time considering all issues of the case and how they are to be presented, and then be open to revising the case and/or teaching notes to reflect what she or he learns about the case through classroom use. This is also why an author/

instructor should test a teaching case in several classrooms before attempting to publish the case.

Finally, an instructor who implements a case methodology into her or his course must be supremely prepared. The instructor must know the case better than any student to avoid surprises, but she or he must also be prepared to deal with surprises. This is what makes teaching with the case method exciting, challenging, fun, and effective.

Bibliography

1 Fluno, RY (1970) The Group Case Study in political science teaching: A report. *Pol. Sci. Polit.* 3(1): 22–25.
2 Spindler, G, and Spindler, L (1990) The inductive case study approach to teaching anthropology. *Anthropol. Educ. Q.* 21(2): 106–112.
3 Foran, J (2001) The case method and the interactive classroom. *Thought and Action*, 19(1): 41–50.
4 Martin, CL, Dewprashad, B, Kosky, C, and Vaz, GS (2004) Using clinical cases to teach general chemistry. *J. Chem. Educ.* 81(10): 1471.
5 Slater, TF (2015) Teaching astronomy with case studies. *Phys. Teach.* 53(8): 506–507.
6 Cochran, JJ (2000) Successful use of cases in introductory undergraduate business college operations research courses. *J. Oper. Res. Soc.* 51(12): 1378–1385.
7 Cochran, JJ (2009a) Pedagogy in operations research: Where have we been, where are we now, and where should we go? *ORiON*, 25(2): 161–184.
8 Cochran, JJ (2012) You want them to remember? Then make it memorable! *Eur. J. Oper. Res.* 219(3): 659–670.
9 Cochran, JJ (2004) Bowie Kuhn's worst nightmare: An integer programming & Simpson's paradox case. *INFORMS Trans. Educ.* 5(1): 18–36.
10 Köksalan, M, and Batun, S (2009) Assigning regions to sales representatives at Pfizer Turkey. *INFORMS Trans. Educ.* 9(2), 70–71.
11 Beliën, J, Goossens, D, Van Reeth, D, and De Boeck, L (2011) Using mixed-integer programming to win a cycling game. *INFORMS Trans. Educ.* 11(3): 93–99.
12 Gavirneni, S (2006) Teaching data envelopment analysis using Applichem–New perspective on a popular operations case. *INFORMS Trans. Educ.* 6(3): 38–45.
13 Agrawal, N, Morris, A, Cohen, MA, and Gans, N (2009) Revenue management at Harrah's Entertainment. *Inc. INFORMS Trans. Educ.* 9(3): 158–159.
14 Cochran, JJ (2009b) All of Britain must be stoned! An effective introductory probability case. *INFORMS Trans. Educ.* 10(2): 62–64.

15 Farahat, A, and Martonosi, SE (2010) Flight delays at RegionEx. *INFORMS Trans. Educ.* 11(3): 100–102.
16 Drake, MJ, Griffin, PM, and Swann, JL (2011) Keeping logistics under wraps. *INFORMS Trans. Educ.* 11(2): 57–62.
17 Keller, B, and Bayraksan, G (2012) Quantifying operational risk in financial institutions. *INFORMS Trans. Educ.* 12(2): 100–105.
18 Milburn, AB, Kirac, E, and Hadianniasar, M (2017) Growing pains: A case study for large-scale vehicle routing. *INFORMS Trans. Educ.*, 17(2): 75–80.
19 Gennaro, V (2013) Diamond dollars case competition: The Mike Trout dilemma. *Case Stud. Sports Manage.* 2(1): 60–65.
20 Lee, S, and Juravich, M (2017) Multi-criteria decision-making: Ticket sales outsourcing in an NCAA Division I athletic department. *Case Stud. Sports Manage.* 6(1): 31–38.
21 Dixon, JC, Chittle, L, and Horton, S (2015) An age-old problem in little league baseball. *Case Stud. Sports Manage.* 4(1): 1–6.
22 Sweeney, K, and Schramm-Possinger, M (2016) Do stadium upgrades influence fan attendance? The case of the Jacksonville Jaguars. *Case Stud. Sports Manage.* 5(1): 73–79.
23 Naraine, M, Kerwin, S, and Parent, MM (2016) Oh captain, my captain! Using social network analysis to help coaching staff identify the leadership of a national sports team. *Case Stud. in Sports Manage.* 5(1), 80–84.
24 Wakefield, L, and Bennett, G (2016) How do fans feel? A case analysis of measuring and modeling attitudes using partial least squares structural equation modeling. *Case Stud. Sports Manage.* 5(1): 85–88.
25 Sweeney, K, Schramm-Possinger, M, Gregg, EA, and Stranahan, H (2016) Predicting consumer commitment: A case study of the NFL and Ray Rice. *Case Stud. Sports Manage.* 5(1): 89–94.

Index

INFORMS Analytics Body of Knowledge, First Edition. Edited by James J. Cochran.

d

n

o

p